Current Developments in Mathematical Sciences

(Volume 2)

Liutex-based and Other Mathematical, Computational and Experimental Methods for Turbulence Structure

Edited by

Chaoqun Liu

Department of Mathematics,
University of Texas at Arlington,
Arlington, Texas 76019,
USA

&

`Yisheng Gao
College of Aerospace Engineering,
Nanjing University of Aeronautics and Astronautics　　　,
Nanjing 210016,
China

Current Developments in Mathematical Sciences

Volume # 2

Liutex-based and Other Mathematical, Computational and Experimental Methods for Turbulence Structure

Editor: Chaoqun Liu & Yisheng Gao

ISSN (Online): 2589-272X

ISSN (Print): 2589-2711

ISBN (Online): 978-981-14-3760-1

ISBN (Print): 978-981-14-3758-8

ISBN (Paperback): 978-981-14-3759-5

need for a court order if at any point you breach any terms of this License Agreement. In no event will any delay or failure by Bentham Science Publishers in enforcing your compliance with this License Agreement constitute a waiver of any of its rights.

3. You acknowledge that you have read this License Agreement, and agree to be bound by its terms and conditions. To the extent that any other terms and conditions presented on any website of Bentham Science Publishers conflict with, or are inconsistent with, the terms and conditions set out in this License Agreement, you acknowledge that the terms and conditions set out in this License Agreement shall prevail.

Bentham Science Publishers Pte. Ltd.
80 Robinson Road #02-00
Singapore 068898
Singapore
Email: subscriptions@benthamscience.net

BENTHAM SCIENCE

CONTENTS

PREFACE

Turbulence is a centuries-long world puzzle. Turbulence coherent structure really means vortex structure. However, there was no mathematical definition of vortex ever before and there was no mathematical definition for turbulence either. Therefore, there was no vortex science or vortex dynamics since vortex had no definition. Since turbulence is built up and driven by vortices, there was no serious scientific research on turbulence theory and turbulence structure because there was no definition of vortex. This book collected a lot of scientific efforts to give an accurate and mathematical definition of vortex, that is Liutex developed by Liu *et al* [C. Liu *et al.*, Phys. Fluids **30**, 035104 (2018)].

The core of this book is a collection of papers presented in the 13th World Congress of Computational Mechanics (WCCM2018), Symposium 704, **Mathematics and Computations for Multiscale Structures of Turbulent and Other Complex Flows**, New York, United States on July 27, 2018. This book also collects quite a number of other research papers working on the vortex definition, vortex identification and turbulence structure from different insight angles including mathematics, computations and experiments. Of course, the priority is dedicated to an accurate and mathematical definition for vortex, which was first named "RORTEX" in 2018 and was changed to "LIUTEX" approved unanimously by alliance of six universities and Alliance of Vortex Research in 2019. Besides Liutex, this book also publishes a lot of efforts to do analysis on turbulence structure by unobjectionable mathematics, incredible DNS computations, and marvelous experiments.

This book contains thirteen chapters which are briefly introduced in this preface.

The first chapter, Liutex – A New Mathematical Definition of Vortex and Vorticity Decomposition for Turbulence Research, is written by Chaoqun Liu, Yisheng Gao and Yifei Yu at University of Texas at Arlington. For long time, people recognize vortex as vorticity tube and measure the vortex rotation strength by vorticity magnitude. These misunderstandings have been carried out by thousands of research papers and almost all textbooks. Robinson (1989) has found the association between regions of strong vorticity and actual vortices can be rather weak. Many vortex identification criteria have been proposed. However, vortex still has no rigorous mathematical definition with direction and magnitude and the relationship between vortex and vorticity was unknown. Because we do not have definition for vortex, there was really no vortex science. Since vortex is the building block and the muscle of turbulence, lack of mathematical definition for vortex becomes a bottle neck for turbulence research. Really, there was no serious turbulence research without definition of vortex. In our recent work, a mathematical definition called Liutex (Previously called Rortex) is given to identify the rigid rotation of fluid motion. Since vortex core is near the rigid rotation, Liutex naturally represents the vortex cores. Liutex is a local mathematical vector definition with direction and magnitude of pure rotation without shear contamination, which is unique and Galilean invariant. The Liutex direction is defined as the local rotation axis and the Liutex magnitude is the local rotation. More important, we derive the accurate mathematical relation between vorticity and vortex, which is Vorticity = Liutex + Antisymmetric Shear (RS decomposition). This new discovery is an important breakthrough in modern fluid dynamics and is extremely important for turbulence research. In addition, the velocity gradient tensor has been decomposed to two parts, R (rigid rotation) and NR (non-rotation part) as a counterpart of the traditional Cauchy-Stokes decomposition (Helmholtz decomposition) which is improper since vorticity cannot represent flow rotation. Liutex is a new physical quantity like velocity, vorticity, temperature, pressure, which has been ignored by our founding fathers of fluid dynamics for centuries but is particularly important for vortex dynamics and turbulence research. Introduction of Liutex, RS decomposition of vorticity, and R-NR decomposition of velocity gradient would open a new era for vortex dynamics and new turbulence research, likely new fluid dynamics.

The second chapter, Liutex – a New Vortex definition, and its Calculation and Galilean Invariance, is written by Yiqian Wang, Yisheng Gao and Chaoqun Liu from University of Shanghai for Science and Technology in China and University of Texas at Arlington in USA. The Liutex (previously known Rortex) method introduces a vortex vector field to mathematically and systematically describe vortices in flow fields. In the present study, the calculation procedure of Liutex is revisited which includes two-step reference coordinate rotation. Then, for the first time, an explicit formula to calculate Liutex is derived and the physical intuition and efficiency improvement brought by this formula are discussed. In addition, the Galilean invariance, which has been widely accepted as a preliminary check for a successful vortex identification method is discussed for Liutex vector.

The third chapter, New Omega Vortex Identification Method Based on Determined Epsilon, is authored by Xiangrui Dong, Yisheng Gao, Chaoqun Liu from University of Shanghai for Science and Technology and University of Texas at Arlington. A new Omega method with ε determination is introduced to represent the ratio of vorticity square over the sum of vorticity square and deformation square, for the vortex identification. the advantages of the new Ω method can be summarized as follows: (1) Omega, as a ratio of the vorticity squared over the sum of the vorticity squared and deformation squared, is a normalized and case-independent function which satisfies $\Omega \in [0,1]$; (2) Compared with the other vortex visualization methods, which require a wide threshold change to capture the vortex structures, Ω can always be set as 0.52 to capture vortex for different cases and time steps; (3) ε is defined as a function without any adjustment on its coefficient in all cases; (4) The Ω method can capture both strong and weak vortices simultaneously. In addition, Ω method is quite robust with no obvious change in vortex visualization.

The fourth chapter, Stability Analysis on Shear Flow and Vortices in Late Boundary Layer Transition, is solely authored by Jie Tang from University of Texas at Arlington. Turbulence is still an unsolved scientific problem, which has been regarded as "the most important unsolved problem of classical physics". Liu proposed a new mechanism about turbulence generation and sustenance after decades of research on turbulence and transition. One of them is the transitional flow instability. Liu believes that inside the flow field, shear (dominant in laminar) is unstable while rotation (dominant in turbulence) is relative stable. This inherent property of flow creates the trend that non-rotational vorticity must transfer to rotational vorticity and causes the flow transition. To verify this new idea, this chapter analyzed the linear stability on two-dimensional shear flow and quasi-rotational flow. Chebyshev collocation spectral method is applied to solve Orr–Sommerfeld equation. Several typical parallel shear flows are tested as the basic-state flows in the equation. The instability of shear flow is demonstrated by the existence of positive eigenvalues associated with disturbance modes (eigenfunctions), *i.e.* the growth of these linear modes. Quasi-rotation flow is considered under cylindrical coordinates. An eigenvalue perturbation equation is derived to study the stability problem with symmetric flows. Shifted Chebyshev polynomial with Gauss collocation points is used to solve the equation. To investigate the stability of vortices in flow transition, a ring-like vortex and a leg-like vortex over time from our Direct Numerical Simulation (DNS) data are tracked. The result shows that, with the development over time, both ring-like vortex and leg-like vortex become more stable as Omega becomes close to 1.

The fifth chapter, POD and DMD Analysis in Late Flow Transition with Omega Method, is authored by Sita Charkrit and Chaoqun Liu at Department of Mathematics, University of Texas at Arlington, Arlington, Texas 76019, USA. In this chapter, the proper orthogonal decomposition (POD) and dynamic mode decomposition (DMD) are applied to analyze the 3D late transitional flow on the flat plate obtained from Direct numerical simulation (DNS). POD is used to find the most persistent spatial structures while DMD is used to find single frequency modes. The omega method is applied as a vortex identification to visualize vortices with iso-surfaces $\Omega = 0.52$. The results in POD and DMD are discussed and compared to show the same and different features such as shapes, amplitudes and time evolutions.

The sixth chapter, Comparison of Liutex and Eigenvalue-based Vortex Identification Criteria for Compressible Flows, is written by Yisheng Gao and Chaoqun Liu at Department of Mathematics, University of Texas at Arlington, Arlington, Texas 76019, USA. Most of the currently popular vortex identification methods, including the $\mathbf{Q}$ criterion, the Δ criterion and the λ_{ci} criterion, are exclusively determined by the eigenvalues or invariants of the velocity gradient tensor and thereby can be classified as eigenvalue-based criteria. However, these criteria will suffer from several shortcomings, such as inadequacy of identifying the rotational axis and contamination by shearing. Recently, a new eigenvector-based Liutex method (previously named Rortex) was proposed to overcome the issues associated with the eigenvalue-based criteria. In this paper, the comparison of Liutex and two eigenvalue-based criteria, namely the λ_{ci} criterion and a modification of the original $\mathbf{Q}$ criterion, are performed to assess these methods for compressible flows. According to the analysis of the deviatoric part of the local velocity gradient tensor, all the scalar, vector and tensor forms of Liutex are valid for compressible flows without any modification, while two eigenvalue-based criteria, though applicable to compressible flows, are prone to severe contamination by shearing as in incompressible flows. Vortex structures in the problem of shock-vortex interaction are examined to confirm the validity and superiority of Liutex in compressible flows.

The seventh chapter, Observation of Coherent Structures of Low Reynolds Number Turbulent Boundary Layer by DNS and Experiment, is written by Panpan Yan from Beijing Jiaotong University, Beijing, 100044, China, Chaoqun Liu from University of Texas at Arlington, Yanang Guo and Xiaoshu Cai from University of Shanghai for Science and Technology, Shanghai, 200093, China. In order to study the characteristics of coherent structures of the turbulent boundary layer, the motion single frame, and long exposure imaging (MSFLE) method is proposed and an elaborate direct numerical simulation experiment was also conducted. MSFLE method is a Lagrangian measurement method, the speed of the camera is kept the same as the speed of the coherent structure, and the particle trajectory was captured by long exposure. By calculating the trace of the points on a chosen plane of the DNS result, we can obtain the particle trajectory like MSFLE method. Multilayer of vortex structures was observed and the evolution of the vortex packets with time was recorded.The result of the DNS simulation agrees well with the experiment. The size of the vortex of the different layer is almost the same, and no vortex breakdown was observed. The formation of the small-scale vortex is caused by sweeps and ejections of the larger coherent structures rather than the breakdown process.

The eighth chapter, Direct Numerical Simulation of Incompressible Flow in a Channel with Rib Structures, is authored by Ting Yu, Duo Wang, Heng Li and Hongyi Xu from Aeronautics and

Astronautics Department, Fudan University Shanghai PR China. The paper applied the state-of-the-art flow simulation method, i.e. the Direct Numerical Simulation (DNS), and strongly coupled the DNS with the heat-transfer governing equation to solve the thermal-turbulence problem in both 2-dimensional(2D) and 3-dimensional(3D) channel with rib tabulator structures. An innovative approach was applied to the simulations in one case. The surface roughness effects of the cooling vane were considered by including the roughness geometry in the DNS and the immersed-boundary method were invented to handle the geometry complexities due to the roughness. Two inlet conditions, namely the uniform flow and full-developed turbulence, were applied at the inflow surface of the channel. Half height of the channel was used as the scale length. The Prandtl number was set at $Pr = 0.7$. Five Reynolds number of 1000, 2500, 5000, 7500 and 1000 were calculated in the 2D cases and the Reynolds numbers of 2500 and 5000 were applied in 3D cases where a periodical condition was applied in the span-wise direction. Additionally, Reynolds number of 10000 was set in the case with roughened surface. The stream-wise velocity, turbulence intensity, the Nusselt (Nu) number were analyzed. Results in 2D cases and 3D cases presented a great difference on flow structure. At the same time, with increasing Reynolds number, the length of recirculation zone and the enhancement of heat transfer showed a decreasing trend. A vortex identification method, the newly-defined Rortex, was applied.

The ninth chapter, Vortex and Turbulent Structure Inside Hydroturbines, is written by Yuning Zhang from Key Laboratory of Condition Monitoring and Control for Power Plant Equipment (Ministry of Education), School of Energy, Power and Mechanical Engineering, North China Electric Power University, Beijing, China and Yuning Zhang from College of Mechanical and Transportation Engineering, China University of Petroleum-Beijing, Beijing China and Beijing Key Laboratory of Process Fluid Filtration and Separation, China University of Petroleum-Beijing, Beijing China. In this chapter, various kinds of vortex in the hydroturbines are briefly introduced with a focus on the swirling vortex rope in Francis turbine and the vortex in the vaneless space of the reversible pump turbine. The vortex induced pressure fluctuation and vibrations are initially demonstrated based on the on-site measurement in the power stations. Then, detailed characteristics of the vortex in the hydroturbines are demonstrated based on the plenty of examples together with the aid of the quantitative analysis.

The tenth chapter, Comparative Study of Supersonic Turbulent Channel flows between Thermally and Calorically Perfect Gases, is written by Xiaoping Chen from National-Provincial Joint Engineering Laboratory for Fluid Transmission System Technology, Zhejiang Sci-Tech University, Hangzhou, Zhejiang, China. In this chapter, to study the effects of gas model on the turbulent statistics and flow structures, direct numerical simulations (DNSs) of supersonic turbulent channel flow for thermally perfect gas and calorically perfect gas are conducted at Mach number 3.0 and Reynolds number 4800 combined with two wall temperature of 298.15K (low temperature condition) and 596.30 K (high temperature condition). The results show that, for high temperature condition, the effects of thermally perfect gas are important because the vibrational energy excited degree exceeds 0.1. Many of turbulent statistics used to express low temperature condition for calorically perfect gas still can be generalized for high temperature condition. The gas model does not have a significant influence on the strong Reynolds analogy. Omega could capture both strong and weak vortices simultaneously for supersonic flows, even under thermally perfect gas, which is difficult to obtain by Q. Compared to the results of calorically perfect gas, the vortex structure becomes smaller, sharper and more chaotic by considering thermally perfect gas.

The eleventh chapter The Experimental Study on Vortex Structures in Low Reynolds Number Turbulent Boundary Layer, was authored by Yanang Guo, Xiaoshu Cai, Wu Zhou, Lei Zhou, Xiangrui Dong from Institute of Particle and Two-phase Flow Measurement, University of Shanghai for Science and Technology, Shanghai, China. A motion single frame and long exposure (MSFLE) imaging method, which is a Lagrangian-type measurement, is experimentally carried out to study the vortex structures in a fully developed turbulent boundary layer with a low Reynolds number on a flat plate. In order to give the process of the vortex generation and evolution, on the one hand, the measurement system moves at the substantially same velocity as the vortex structure; on the other hand, a long exposure time is selected for recording the paths of the particles. In the experiment, the vortex structure characteristics as well as the temporal-spatial development can be shown by the streamwise-normal and streamwise-spanwise images which are extracted from a fully developed turbulent boundary layer. The result shows that the interaction between high and low-speed streaks induces the generation, deformation and 'breakdown' of the vortex structures, and badly influences the vortex evolution.

The twelfth chapter, Experimental Studies on Coherent Structures in Jet Flows using Single-Frame-Long-Exposure (SFLE) Method is authored by Lei Zhou, Xiaoshu Cai, Wu Zhou and Yiqian Wang from Institute of Particle and Two-phase Flow Measurement, University of Shanghai for Science, China. An experimental investigation on the flow structures in jet entrainment boundary layer flows based on the Single-Frame-Long Exposure (SFLE) method is carefully performed. It is found that two entrainment mode of 'engulfing' and 'nibbling' alternatively appear in the region of $2d$ to $3.5d$ in the axial direction and $1d$ to $1.25d$ in the radial direction with d being the diameter of jet nozzle. The appearance probability of such a pattern and the proportion of the 'engulfing' mode increases with Reynolds number Re when $Re \geq 1981$ (the Reynolds number is based on the nozzle diameter and jet velocity). However, the influence of Reynolds number on this flow pattern becomes weaker when $Re > 2245$. The main frequency of this structure is found to be between 10-19Hz with Fourier analysis. The vortical structures are further explored with the moving SFLE (MSFLE) method, and it is found that vortices always exist near the turbulent and non-turbulent interface (TNTI).

The last chapter (thirteenth), Hybrid Compact-WENO Scheme for the Interaction of Shock Wave and Boundary Layer, is co-authored by Jianming Liu from Jiangsu Normal University of China and Chaoqun Liu from Department of Mathematics, University of Texas at Arlington, Arlington, USA. In this chapter, an introduction on hybrid Weighted Essentially non-oscillatory (WENO) method is given. The hybrid techniques including both central and compact finite difference schemes are introduced. The paper reviews the driven mechanism of the high order finite scheme required for compressible flow with shock. The detailed constructing processes of the compact and WENO schemes are given and the hybrid detector.

I hope this book will be useful to scientists and engineers who are interested in fundamental fluid dynamics, vortex science and turbulence research.

In conclusion, I want to thank the numerous authors for their incredible contributions and having patience in assisting us. Furthermore, I want to acknowledge and thank the referees for their tiresome work on making this book come to fruition. Last but not the least, I would also like to thank my family including Weilan Jin (my wife), Haiyan Liu (my daughter) and Haifeng Liu (my son) for their unconditional support. The co-editor, Yisheng Gao is also grateful to his family for the strong support.

Chaoqun Liu
Department of Mathematics,
University of Texas at Arlington,
Arlington, Texas 76019,
USA

List of Contributors

Chaoqun Liu Department of Mathematics, University of Texas at Arlington, Arlington, Texas 76019, USA

Duo Wang Department of Aeronautics and Astronautics, Fudan University, Shanghai, PR China

Heng Li Department of Aeronautics and Astronautics, Fudan University, Shanghai, PR China

Hongyi Xu Department of Aeronautics and Astronautics, Fudan University, Shanghai, PR China

Jianming Liu School of Mathematics and Statistics, Jiangsu Normal University, Xuzhou 221116, China

Jie Tang Department of Mathematics, University of Texas at Arlington, Arlington, Texas 76019, USA

Lei Zhou Institute of Particle and Two-phase Flow Measurement, University of Shanghai for Science and Technology, Shanghai 200093, China

Panpan Yan Shenyang Aircraft Design and Research Institute, Aviation Industry of China, Shenyang, 110035, China

Sita Charkrit Department of Mathematics, University of Texas at Arlington, Arlington, Texas 76019, USA

Ting Yu Department of Aeronautics and Astronautics, Fudan University, Shanghai, PR China

Xiaoshu Cai Institute of Particle and Two-phase Flow Measurement, College of Energy and Power Engineering, University of Shanghai for Science and Technology, Shanghai 200093, China

Xiaoping Chen National-Provincial Joint Engineering Laboratory for Fluid Transmission System Technology, Zhejiang Sci-Tech University, Hangzhou, Zhejiang 310018, China

Xiangrui Dong Institute of Energy and Power Engineering, University of Shanghai for Science and Technology, Shanghai, China

Yisheng Gao College of Aerospace Engineering, Nanjing University of Aeronautics and Astronautics, Nanjing 210016, China

Yanang Guo   Institute of Particle and Two-phase Flow Measurement, University of Shanghai for Science and Technology, Shanghai, China

Yiqian Wang   School of Mathematical Science, Soochow University, Suzhou 215006, China

Yifei Yu   Department of Mathematics, University of Texas at Arlington, Arlington, Texas 76019, USA

Yuning Zhang   Key Laboratory of Condition Monitoring and Control for Power Plant Equipment (Ministry of Education), School of Energy, Power and Mechanical Engineering, North China Electric Power University, Beijing 102249, China

Yuning Zhang   College of Mechanical and Transportation Engineering, China University of Petroleum-Beijing, Beijing 102249, China

Wu Zhou   Institute of Particle and Two-phase Flow Measurement, University of Shanghai for Science and Technology, Shanghai, China and School of Aerospace Engineering, Tsinghua University, Beijing, China

Liutex – A New Mathematical Definition of Vortex and Vorticity Decomposition for Turbulence Research

Chaoqun Liu[1,*], Yisheng Gao[2] and Yifei Yu[1]

[1]Department of Mathematics, University of Texas at Arlington, Arlington, Texas 76019, USA

[2] College of Aerospace Engineering, Nanjing University of Aeronautics and Astronautics, Nanjing 210016, China

Abstract: For a long time, people recognize a vortex as a vorticity tube and measure the vortex rotation strength by vorticity magnitude. These misunderstandings have been carried out by thousands of research papers and almost all textbooks. It has been found that the association between regions of strong vorticity and actual vortices can be rather weak. Accordingly, many vortex identification criteria have been proposed. However, the vortex still has no rigorous mathematical definition and the relationship between the vortex and the vorticity is still not clear. Because we do not have a definition for the vortex, there exists no vortex science. Since the vortex is the building block and the muscle of turbulence, the lack of the mathematical definition for the vortex becomes a bottleneck for turbulence research. Actually, there is no serious turbulence research without the definition of the vortex. In our recent work, a mathematical definition called Liutex (previously called Rortex) is introduced to identify the rigid rotation of fluid motion. Liutex is a local mathematical vector definition with the direction and magnitude of pure rotation without shear contamination, which is unique and Galilean invariant. The Liutex direction is defined as the local rotation axis and the Liutex magnitude is the local rotation strength. More importantly, we derive the accurate mathematical relation between the vorticity and the vortex, which is Vorticity= Liutex + Antisymmetric Shear (RS decomposition). This new discovery is an important breakthrough in modern fluid dynamics and is extremely important for turbulence research. In addition, the velocity gradient tensor has been decomposed to two parts, R (rigid rotation) and NR (non-rotation part) as a counterpart of the traditional Cauchy-Stokes decomposition which is improper since the vorticity cannot represent flow rotation. Liutex is a new physical quantity like velocity, vorticity, temperature, pressure, which has been ignored by our founding fathers of fluid dynamics for centuries but is particularly important for vortex dynamics and turbulence research. The introduction of Liutex, RS decomposition of vorticity, and R-NR decomposition of the velocity gradient tensor would open a new era for vortex dynamics and new turbulence research, likely new fluid dynamics.

*Corresponding author Chaoqun Liu: Department of Mathematics, University of Texas at Arlington, Arlington, Texas 76019, USA; Tel: +1-8172725151; Fax: +1-8172725802; E-mail: cliu@uta.edu

Keywords: Angular velocity, Coherent structures, Liutex, Turbulence, Velocity gradient tensor, Vortex identification, Vortex.

INTRODUCTION

Vortex is intuitively recognized as the rotational/swirling motion of the fluids. However, a universally accepted definition for vortex is yet to be achieved, which is probably one of the major obstacles causing considerable confusions and misunderstandings in turbulence research. Vorticity is mathematically defined as the curl of velocity. Vortices are ubiquitous in nature. As addressed by Küchemann "vortices are the sinews and muscles of turbulence" [1], some vortical structures, such as hairpin vortices, referred to as coherent turbulent structures [2], are recognized as one of the most important characteristics of turbulent flow and have been studied for more than 60 years [3]. It is generally acknowledged that intuitively, vortices represent the rotational/swirling motion of the fluids. However, a precise and rational definition of vortex is deceptively complicated and remains an open issue [4-5]. The lack of a consensus on the vortex definition has caused considerable confusions in visualizing and understanding the vortical structures, their evolution, and the interaction in complex vortical flows, especially in turbulence [4].

In classical vortex dynamics [4, 6-7], the vortex is usually associated with the vorticity which has a rigorous mathematical definition (the curl of velocity). Wu *et al*., [4] define a vortex as "a connected region with high concentration of vorticity compared with its surrounding." Lamb [8] uses vorticity tubes to define vortices. Nitsche [9] asserts, "A vortex is commonly associated with the rotational motion of fluid around a common centerline. It is defined by the vorticity in the fluid, which measures the rate of local fluid rotation." An immediate contradiction to these definitions is that the Blasius boundary layer where the vorticity is large near the wall, but no rotational/swirling motion (considered as a vortex) is observed, as the vorticity cannot distinguish a vortical region from a shear layer region. In addition, the maximum vorticity does not necessarily represent the center of the vortex. Robinson [10] pointed out that the association between regions of strong vorticity and actual vortices can be rather weak in the turbulent boundary layer, especially in the near wall region. Wang *et al*., [11] also found that vorticity magnitude will be reduced when vorticity lines enter the vortex region and vorticity magnitude inside the vortex region is much smaller than the surrounding area, especially near the solid wall in a flat plate boundary layer, for most three-dimensional vortices like Λ-shaped vortices.

Another possible candidate for vortex definition is the one based on closed or spiraling streamlines [12]. Robinson *et al*., [13] claim, "A vortex exists when instantaneous streamlines mapped onto a plane normal to the vortex core exhibit a roughly circular or spiral pattern, when viewed from a reference frame moving with the center of the vortex core". Although it seems intuitive, Lugt [12] pointed out that "the definition and identification of a vortex in unsteady motions is difficult since streamlines and pathlines are not invariant with respect to Galilean and rotational transformations. Recirculated streamline patterns at a certain instant in time do not necessarily represent vortex motions in which fluid particles are moving around a common axis. Thus, instantaneous streamline patterns do not provide enough information to be used for the definition of a vortex."

Due to the essential requirement for visualizing vortical structures and their evolution in turbulence, several vortex identification criteria have been developed, including λ_2-criterion [14], Q-criterion [15], λ_{ci}-criterion [16], and $\lambda_{cr}/\lambda_{ci}$-criterion [17], *etc*. Nevertheless, these methods still fail to provide a rigorous definition of vortices. Moreover, these methods require proper thresholds. It is difficult to determine which threshold is proper, since different thresholds will indicate different vortical structures. For example, even if the same DNS data on the late boundary layer transition is employed, "vortex breakdown" will be exposed for some large threshold in Q-criterion while no "vortex breakdown" can be found for some smaller threshold. This will directly influence one's understanding and explanation on the mechanism of turbulence generation, *i.e.* turbulence is caused by "vortex breakdown" or not caused by "vortex breakdown". Recently, a new vortex identification method called Ω-method is given by the proposer, based on the idea that a vortex is a connected area where the vorticity overtakes the deformation [18]. The vorticiy represents the fluid's intention to rotate but deformation would resist rotation. Ω-method possesses several advantages, such as normalized from 0 to 1, no need for a case-related threshold, clear physical meaning and capability to capture both strong and weak vortices simultaneously. However, it is still not the ideal answer to the question of the mathematical definition of vortex, owing to some limitations such as the introduction of an artificial parameter ε and the incapability to identify the swirl axis and its orientation. Kolář [19] formulated a triple decomposition from which the residual vorticity can be obtained after the extraction of an effective pure shearing motion and represents a direct measure of the pure rigid-body rotation of a fluid element. However, the triple decomposition is not unique, so a so-called basic reference frame must be first determined. Searching for the basic reference frame results in an expensive optimization problem for every point in the flow field, which limits the applicability of the method. Kolář *et al.*, [20-21] also introduced the concepts of the maximum corotation and the average corotation of line segments near a point and apply these methods for vortex identification. However, the so-called maximum-corotation method suffers from the unstable behavior of the maxima and the averaged corotation vector is evaluated by integration over a unit sphere, which makes it difficult to be used to study the transport property of the vortex. In addition to the mentioned Eulerian vortex identification methods, some objective Lagrangian vortex identification methods are developed to study the vortex structures in the rotating reference frame [22-23]. For more details on currently available vortex identification methods, one can refer to recent review papers by Zhang *et al.*, [24-25] and Epps [26].

For long time, people recognize a vortex as a vorticity tube and measure the vortex rotation strength by vorticity magnitude. These misunderstandings have been carried out by thousands of research papers and almost all textbooks. Robinson [13] found the association between regions of strong vorticity and actual vortices can be rather weak. Many vortex identification criteria have been proposed. However, the vortex still has no rigorous mathematical definition with direction and magnitude and the relationship between the vortex and the vorticity was unknown. Because we do not have definition for vortex, there was really no vortex science. Since the vortex is the building block and the muscle of turbulence, the lack of mathematical definition for vortex becomes a bottle neck for turbulence research. Really, there was no serious turbulence research without definition of vortex. In our recent work, a mathematical definition called Liutex (previously called Rortex) is given to identify the rigid rotation of fluid motion [29-32]. Since the vortex core is near the rigid local rotation strength, Liutex naturally represents the vortex cores. Liutex is a local mathematical vector definition with direction and magnitude of pure rotation without shear contamination, which is unique and Galilean invariant. The vortex rotation axis is the Liutex direction and the rotation angular speed is represented by the magnitude of Liutex. More importantly, we derive the accurate mathematical relation between the vorticity and the vortex, which is Vorticity= Liutex + Antisymmetric Shear (RS decomposition). This new discovery is an important breakthrough in modern fluid dynamics and is extremely important for turbulence research. In addition, the velocity gradient tensor has been decomposed to two parts, R (rigid rotation) and NR (non-rotation part) as a counterpart of the traditional Cauchy-Stokes decomposition which is improper since the vorticity cannot represent flow rotation. Liutex is a new physical quantity like velocity, vorticity, temperature, pressure, which has been ignored by our founding fathers of fluid dynamics for centuries but is particularly important for vortex dynamics and turbulence research. Introduction of Liutex, RS decomposition of vorticity, and R-NR decomposition of velocity gradient tensor would open a new era for vortex dynamics and new turbulence research, likely new fluid dynamics.

Liutex has paved the foundation of vortex science and serious turbulence research quantitatively but not qualitatively. Further research includes mechanism of

Liutex generation, turbulence generation and sustenance, Liutex structure in turbulence, correlation between fluctuation and Liutex, correlation between turbulence energy spectrum and Liutex, large vortex formation, multiple level vortex formation, correlation between Reynolds stress and Liutex and S which is the anti-symmetric shear part of the vorticity.

Turbulence is an unsolved scientific problem for over a century, which is not only important to science but also to industrial applications in aerospace engineering, mechanical engineering, energy engineering, bio engineering and many others. Wallace [27] pointed out: "there has been remarkable progress in turbulent boundary layer research in the past 50 years, particularly in understanding the structural organization of the flow. Consensus exists that vortices drive momentum transport but not about the exact form of the vortices or how they are created and sustained." As computer power has been increased much, high order direct numerical simulation (DNS) becomes a major tool to study the physics of turbulence generation and sustenance. More and more people realize that the vortex is not only the building element of turbulence but also the driven force of the turbulence generation and sustenance. The mechanism of vortex formation, development, and sustenance becomes the first priority of the turbulence study. We propose to use high order DNS with large number of grids to study the mechanism of the vortex structure generation and try to give exact form of the vortices and how they are created and sustained in turbulent boundary layer based on the definition of Liutex and RS decomposition of vorticity. This research will be focused on the late natural boundary layer transition, which is the last stages how laminar flow becomes turbulent. This work is aimed at identifying vortex structure and vortex generation mechanisms including large vortex formation and small vortex generation.

LIUTEX – A NEW EIGENVECTOR BASED MATHEMATICAL DEFINITION FOR FLUID ROTATION

Four Principles

To reasonably define a vortex vector or Liutex, we use the following principles: local [17], Galilean invariant [14], unique, and systematical including the scalar, vector, and tensor definition.

Definition of Liutex

First, we must give a definition of local rotation axis for the vortex area.

Definition 1: A local rotation axis is defined as the direction of $\vec{r}$ and $\vec{v} = \alpha d\vec{r}$, where $d\vec{r}$ is a small section of the local rotation axis $\vec{r}$ and $d\vec{v}$ is the increment of velocity $\vec{v}$ along $d\vec{r}$ (details can be found in [29, 30]).

This definition means that, in the direction of the local rotation axis, there is no cross-velocity gradient. For example, if the z-axis is the rotation axis in a reference frame, the velocity can only increase or decrease along the z-axis, which means only the z-component of velocity can increase or decrease along the z-axis or, in other words, only $dw \neq 0$, but $du = 0$ and $dv = 0$. Accordingly, we can obtain the following theorem:

Theorem 1. The direction of the local rotation axis is the real eigenvector of the velocity gradient tensor $\nabla \vec{v}$.

Apparently, if $\vec{r} = [r_x, r_y, r_z]^T$ represents the direction of the local rotation axis, we have $d\vec{v} = \alpha d\vec{r}$. On the other hand, from the definition of the velocity gradient tensor

$$d\vec{v} = \nabla \vec{v} \cdot d\vec{r} \tag{1}$$

Therefore,

$$d\vec{v} = \nabla \vec{v} \cdot d\vec{r} = \alpha d\vec{r} \tag{2}$$

and

$$\nabla \vec{v} \cdot \vec{r} = \alpha \vec{r} \tag{3}$$

which means $\vec{r}$ is the real eigenvector of $\nabla \vec{v}$ and α is the real eigenvalue.

Definition 2: Liutex is defined as $\vec{R} = R\vec{r}$ where $\vec{r}$ is the local rotational axis and R is the strength of the local rotation.

Assume the XYZ-frame is a coordinate system with $\vec{r}$ as the Z-axis. If U, V, W are velocity components along the X, Y and Z axes, respectively, the matrix representation of the velocity gradient tensor in the XYZ-frame can be written as

$$\nabla \vec{V} = \begin{bmatrix} \dfrac{\partial U}{\partial X} & \dfrac{\partial U}{\partial Y} & 0 \\ \dfrac{\partial V}{\partial X} & \dfrac{\partial V}{\partial Y} & 0 \\ \dfrac{\partial W}{\partial X} & \dfrac{\partial W}{\partial Y} & \dfrac{\partial W}{\partial Z} \end{bmatrix} \tag{4}$$

Generally, the z-axis in the original xyz-frame is not parallel to the Z-axis, so the velocity gradient tensor in the original xyz-frame

$$\nabla \vec{v} = \begin{bmatrix} \dfrac{\partial u}{\partial x} & \dfrac{\partial u}{\partial y} & \dfrac{\partial u}{\partial z} \\[2mm] \dfrac{\partial v}{\partial x} & \dfrac{\partial v}{\partial y} & \dfrac{\partial v}{\partial z} \\[2mm] \dfrac{\partial w}{\partial x} & \dfrac{\partial w}{\partial y} & \dfrac{\partial w}{\partial z} \end{bmatrix} \tag{5}$$

cannot be the same as Eq. (4). There exists a corresponding transformation between $\nabla \vec{V}$ and $\nabla \vec{v}$

$$\nabla \vec{V} = Q \nabla \vec{v} Q^{-1} \tag{6}$$

where Q is a rotation matrix which is orthogonal.

$$Q^{-1} = Q^{\mathrm{T}} \tag{7}$$

If the direction of the Z-axis in the xyz-frame is given by $\vec{r} = \left[r_x, r_y, r_z \right]^T$,

$$\vec{r} = Q^T \begin{bmatrix} 0 \\ 0 \\ 1 \end{bmatrix} \tag{8}$$

and

$$Q\vec{r} = \begin{bmatrix} 0 \\ 0 \\ 1 \end{bmatrix} \tag{9}$$

represents the direction of the local rotation axis Z in the XYZ-frame.

Once the local rotation axis Z is obtained, the rotation strength is determined in the XY plane perpendicular to the local rotation axis Z. This can be achieved by a second coordinate rotation in the XY plane. When the XYZ-frame is rotated around the Z-axis by an angle θ, the velocity gradient tensor will become

$$\nabla \vec{V}_\theta = P \nabla \vec{V} P^{-1} \tag{10}$$

where $\boldsymbol{P}$ is the rotation matrix around the Z-axis and can be written as

$$\boldsymbol{P} = \begin{bmatrix} cos\theta & sin\theta & 0 \\ -sin\theta & cos\theta & 0 \\ 0 & 0 & 1 \end{bmatrix} \tag{11}$$

$$\boldsymbol{P}^{-1} = \boldsymbol{P}^T = \begin{bmatrix} cos\theta & -sin\theta & 0 \\ sin\theta & cos\theta & 0 \\ 0 & 0 & 1 \end{bmatrix} \tag{12}$$

The criterion to determine the existence of local fluid rotation in the XY plane is

$$g_{Zmin} = \beta^2 - \alpha^2 > 0 \tag{13}$$

$$\alpha = \frac{1}{2}\sqrt{\left(\frac{\partial V}{\partial Y} - \frac{\partial U}{\partial X}\right)^2 + \left(\frac{\partial V}{\partial X} + \frac{\partial U}{\partial Y}\right)^2} \tag{14}$$

$$\beta = \frac{1}{2}\left(\frac{\partial V}{\partial X} - \frac{\partial U}{\partial Y}\right) \tag{15}$$

And the fluid-rotational angular velocity in the XY plane is defined as follows

$$\omega_{rot} = \omega_{\theta min} = \begin{cases} \beta - \alpha, & g_{Zmin} > 0 \\ 0, & g_{Zmin} \leq 0 \end{cases} \tag{16}$$

Here, we assume $\beta > 0$. If $\beta < 0$, we can rotate the local rotation axis to the opposite direction to make β positive. The local rotation strength (the magnitude of Liutex) is defined as twice the fluid-rotational angular velocity

$$R = \begin{cases} 2(\beta - \alpha), & g_{Zmin} > 0 \\ 0, & g_{Zmin} \leq 0 \end{cases} \tag{17}$$

The factor 2 is related to using 1/2 in the expression for the 2-D vorticity tensor component. For details of the definition of Liutex vector direction and magnitude, one can refer to Refs. [29-31].

Wang *et al.*, [33] recently derived an explicit formula to calculate the magnitude of R and dramatically simplify the implementation. The magnitude R can be obtained as

$$R = \langle \boldsymbol{\omega}, \boldsymbol{r} \rangle - \sqrt{\langle \boldsymbol{\omega}, \boldsymbol{r} \rangle^2 - 4\lambda_{ci}^2} \tag{18}$$

where ω is the vorticity vector. This formulation not only provides an explicit formula for Liutex and but also shows the relation between Liutex and vorticity and λ_{ci}.

To be more explicit, the new vortex mathematical definition, *i.e.* Liutex can be simply described as follows:

$$\vec{R} = R\vec{r} \text{ and } \vec{r} \cdot (\nabla \times \vec{v}) > 0 \tag{19}$$

where $\vec{r}$ is the eigenvector of $\nabla\vec{v}$.

This is an important breakthrough in modern fluid dynamics in that the real eigenvector of the velocity gradient tensor in a vortex area is the local rotational axis. After 160 years of struggles after Helmholtz proposed his idea to use small vorticity tube (vortex filament) to represent a vortex in 1858 [34], people finally discover that the vortex has local rotational axis and the axis is the real eigenvector of the velocity gradient tensor [30], and the vortex magnitude is the angular speed of the rigid rotation of the fluids. This is the first time that a mathematical definition of the vortex is given, which is a natural phenomenon.

Calculation Procedure for Liutex

By relying on the eigenvector-based definition, the complete calculation procedure consists of the following steps:

a)   Compute the velocity gradient tensor $\nabla\vec{v}$ in the xyz-frame;
b)   Calculate the eigenvalues of the velocity gradient tensor $\nabla\vec{v}$ when the complex eigenvalues exist;
c)   Calculate the (normalized) real eigenvector $\vec{r} = [r_x, r_y, r_z]^T$ corresponding to the real eigenvalue λ_r, which is the direction of Liutex;
d)   Obtain R according to Eq. (18);
e)   Compute Liutex $\vec{R}$ by Eq. (19).

Vorticity Decomposition

According to the Cauchy-Stokes decomposition, in the rotated coordinate system XYZ, the velocity gradient tensor can be decomposed to a symmetric part and an anti-symmetric part:

$$
\begin{bmatrix}
\dfrac{\partial U}{\partial X} & \dfrac{\partial U}{\partial Y} & 0 \\[2mm]
\dfrac{\partial V}{\partial X} & \dfrac{\partial V}{\partial Y} & 0 \\[2mm]
\dfrac{\partial W}{\partial X} & \dfrac{\partial W}{\partial Y} & \dfrac{\partial W}{\partial Z}
\end{bmatrix}
=
\begin{bmatrix}
\dfrac{\partial U}{\partial X} & \dfrac{1}{2}\left(\dfrac{\partial V}{\partial X}+\dfrac{\partial U}{\partial Y}\right) & \dfrac{1}{2}\dfrac{\partial W}{\partial X} \\[2mm]
\dfrac{1}{2}\left(\dfrac{\partial V}{\partial X}+\dfrac{\partial U}{\partial Y}\right) & \dfrac{\partial V}{\partial Y} & \dfrac{1}{2}\dfrac{\partial W}{\partial Y} \\[2mm]
\dfrac{1}{2}\dfrac{\partial W}{\partial X} & \dfrac{1}{2}\dfrac{\partial W}{\partial Y} & \dfrac{\partial W}{\partial Z}
\end{bmatrix}
+
$$

$$
\begin{bmatrix}
0 & -\dfrac{1}{2}\left(\dfrac{\partial V}{\partial X}-\dfrac{\partial U}{\partial Y}\right) & -\dfrac{1}{2}\dfrac{\partial W}{\partial X} \\[2mm]
\dfrac{1}{2}\left(\dfrac{\partial V}{\partial X}-\dfrac{\partial U}{\partial Y}\right) & 0 & -\dfrac{1}{2}\dfrac{\partial W}{\partial Y} \\[2mm]
\dfrac{1}{2}\dfrac{\partial W}{\partial X} & \dfrac{1}{2}\dfrac{\partial W}{\partial Y} & 0
\end{bmatrix}
= A + B \tag{20}
$$

Theorem 2. A vorticity tensor B can be decomposed as a rotational tensor and a non-rotational tensor.

Proof. The velocity gradient tensor after the XYZ-frame rotating around the Z axis by an angle θ, the velocity gradient tensor becomes

$$
\nabla \vec{V}_R =
\begin{bmatrix}
\lambda_{cr} & -(\beta-\alpha) & 0 \\[2mm]
(\beta+\alpha) & \lambda_{cr} & 0 \\[2mm]
\dfrac{\partial W}{\partial X}\Big|_\theta & \dfrac{\partial W}{\partial X}\Big|_\theta & \lambda_r
\end{bmatrix}
= A + B \tag{21}
$$

$$
B =
\begin{bmatrix}
0 & -\beta & -\dfrac{1}{2}\dfrac{\partial W}{\partial X}\big|_\theta \\[2mm]
\beta & 0 & -\dfrac{1}{2}\dfrac{\partial W}{\partial Y}\big|_\theta \\[2mm]
\dfrac{1}{2}\dfrac{\partial W}{\partial X}\big|_\theta & \dfrac{1}{2}\dfrac{\partial W}{\partial Y}\big|_\theta & 0
\end{bmatrix}
=
\begin{bmatrix}
0 & -\dfrac{R}{2} & 0 \\[2mm]
\dfrac{R}{2} & 0 & 0 \\[2mm]
0 & 0 & 0
\end{bmatrix}
+
$$

$$
\begin{bmatrix}
0 & -\beta+\dfrac{R}{2} & -\dfrac{1}{2}\dfrac{\partial W}{\partial X}\big|_\theta \\[2mm]
\beta-\dfrac{R}{2} & 0 & -\dfrac{1}{2}\dfrac{\partial W}{\partial Y}\big|_\theta \\[2mm]
\dfrac{1}{2}\dfrac{\partial W}{\partial X}\big|_\theta & \dfrac{1}{2}\dfrac{\partial W}{\partial Y}\big|_\theta & 0
\end{bmatrix}
= C + D \tag{22}
$$

where C is a tensor corresponding to a rigid rotation with an angle speed of $\dfrac{R}{2}$ and D is the anti-symmetric shear without any rotation because

$$
\beta - \frac{R}{2} = \frac{\partial U}{\partial Y}\Big|_\theta \text{ and } g_z = \frac{\partial U}{\partial Y}\Big|_\theta \cdot \frac{\partial V}{\partial X}\Big|_\theta = \frac{\partial U}{\partial Y}\Big|_\theta \cdot 0 = 0 \text{ (See Ref. [29])}
$$

Theorem 3. A vorticity vector can be decomposed to a rotational Liutex vector and a non-rotational shear vector, *i.e.* $\nabla \times \vec{V} = \vec{R} + \vec{S}$.

Proof: Assume $d\vec{l}$ is an arbitrarily selected real vector,

$$2\boldsymbol{B} \cdot d\vec{l} = -d\vec{l} \times (\nabla \times \boldsymbol{V}) = 2\boldsymbol{C} \cdot d\vec{l} + 2\boldsymbol{D} \cdot d\vec{l} = -d\vec{l} \times \vec{R} - d\vec{l} \times \vec{S}$$

$$d\vec{l} \times (\nabla \times \boldsymbol{V}) = d\vec{l} \times \vec{R} + d\vec{l} \times \vec{S} = d\vec{l} \times (\vec{R} + \vec{S}) \tag{23}$$

Since $d\vec{l}$ is arbitrarily selected, we will have

$$\nabla \times \vec{V} = \vec{R} + \vec{S} \tag{24}$$

Velocity Gradient Tensor Decomposition Based on Liutex

The traditional velocity gradient decomposition, Cauchy-Stokes decomposition, divides the velocity gradient tensor to a symmetric tensor and an anti-symmetric part. The symmetric part is corresponding to fluid deformation and the anti-symmetric part is corresponding to fluid rotation. However, this is a misunderstanding since Helmholtz [34], because the anti-symmetric part (also called vorticity part) is not corresponding to fluid rotation. Therefore, as the fluid dynamics foundation which is the fluid velocity decomposition, the Cauchy-Stokes decomposition is not appropriate as our founding fathers ignored an, extremely important physical quantity for centuries, which is Liutex. Liutex is the foundation for vortex dynamics, turbulence research and doomed to be extremely important to fluid dynamics. Based on the definition of Liutex, an appropriate decomposition called R-NR decomposition has been proposed as a counterpart of Cauchy-Stokes decomposition. Although Liutex is defined as a vector, we can propose a tensor interpretation of Liutex. After the P-Rotation (assume $\beta > 0$), the velocity gradient tensor becomes

$$\nabla \vec{V}_{\theta min} = \begin{bmatrix} \frac{1}{2}\left(\frac{\partial U}{\partial X} + \frac{\partial V}{\partial Y}\right) & -(\beta - \alpha) & 0 \\ \beta + \alpha & \frac{1}{2}\left(\frac{\partial U}{\partial X} + \frac{\partial V}{\partial Y}\right) & 0 \\ \left.\frac{\partial W}{\partial X}\right|_{\theta min} & \left.\frac{\partial W}{\partial Y}\right|_{\theta min} & \left.\frac{\partial W}{\partial Z}\right|_{\theta min} \end{bmatrix} \tag{25}$$

If λ_{cr} represents the real part of the complex eigenvalues, λ_{ci} the imaginary part of the complex eigenvalues and λ_r the real eigenvalue, we can obtain

where $\lambda_r = \left.\frac{\partial W}{\partial Z}\right|_{\theta min}$, $\lambda_{cr} = \frac{1}{2}\left(\frac{\partial U}{\partial X} + \frac{\partial V}{\partial Y}\right)$, $\lambda_{ci} = \sqrt{\beta^2 - \alpha^2}$.

Eq. (25) can be decomposed into two parts

$$\nabla \vec{V}_{\theta min} = \begin{bmatrix} \lambda_{cr} & -R/2 & 0 \\ R/2 + \varsigma & \lambda_{cr} & 0 \\ \xi & \eta & \lambda_r \end{bmatrix} = R + NR \tag{26}$$

where

$$R = \begin{bmatrix} 0 & -R/2 & 0 \\ R/2 & 0 & 0 \\ 0 & 0 & 0 \end{bmatrix}, \; NR = \begin{bmatrix} \lambda_{cr} & 0 & 0 \\ \varsigma & \lambda_{cr} & 0 \\ \xi & \eta & \lambda_r \end{bmatrix}, \; \varsigma = 2\alpha, \; \xi = \frac{\partial w}{\partial x}\Big|_{\theta min}$$

and $\eta = \frac{\partial w}{\partial y}\Big|_{\theta min}$.

Since the local rotational strength (magnitude) can be regarded as the scalar version of Liutex and the direction of the local rotational axis with the magnitude can be regarded as the vector version, R in Eq. (26) can be regarded as the tensor interpretation of Liutex which exactly represents the local rigidly rotational part of the velocity gradient tensor and consistent with the scalar and vector interpretations of Liutex. Eq. (26) contains a non-rotational part, because NR has three real eigenvalues (multiple λ_{cr} and λ_r) and implies no local rotation. Note that Cauchy-Stokes decomposition is impropriate in physics because the anti-symmetric part is not corresponding to fluid rotation, it is also impropriate in mathematics since the real eigenvector of $\nabla \vec{V}$ is neither eigenvector of symmetric tensor A nor eigenvector of anti-symmetric tensor of B, but is exactly the eigenvector of R and NR.

$$\nabla \vec{V} \, \vec{r} = A \, \vec{r} + B \, \vec{r} \tag{27}$$

where $\vec{r}$ is not the eigenvector of A or B, but

$$\nabla \vec{V} \, \vec{r} = R \, \vec{r} + NR\vec{r} \tag{28}$$

where $\vec{r}$ is the eigenvector of R and NR.

Vortex Identification

We use Liutex to identify vortex structures in late boundary layer transition on a flat plate (Fig. **1**) which is simulated by a DNS code called DNSUTA [28]. The DNS was conducted with near 60 million grid points and over 400000 time steps at a free stream Mach number of 0.5. Being different from other vortex identification methods which only use iso-surfaces to detect the vortex structure, this new defined Liutex can use not only Liutex iso-surface but also Liutex vector and lines to identify the vortex structure. Fig. (**2**) demonstrates the comparison of

Liutex vector field and vorticity vector field for the vortex legs. It can be obviously observed that the Liutex vectors are nearly tangent to the Liutex iso-surface (Figs. **2a** and **2b**) while the vorticity lines seem to be nearly orthogonal to the Liutex iso-surface (Figs. **2c** and **2d**). Figs. (**3**) shows both the Liutex iso-surfaces and the Liutex vector fields for the first vortex ring, which also clearly shows that the Liutex vectors are nearly tangent to the Liutex iso-surfaces. Figs. (**4** and **5**) depict the comparison of Liutex lines and vorticity lines for the Lambda vortex and the ring-like vortex respectively, which clearly show that Liutex lines could be generated, developed, and ended inside the flow field, which is very different from vorticity lines. These examples clearly show that the vorticity cannot well describe the vortex structure, but the newly defined Liutex can do properly by using not only Liutex iso-surface but also Liutex vectors and Liutex lines.

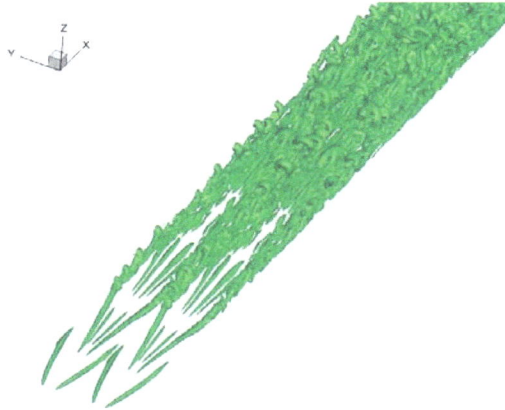

Fig. (1). Vortex structure of late boundary layer transition with iso-surfaces of |R|=0.05.

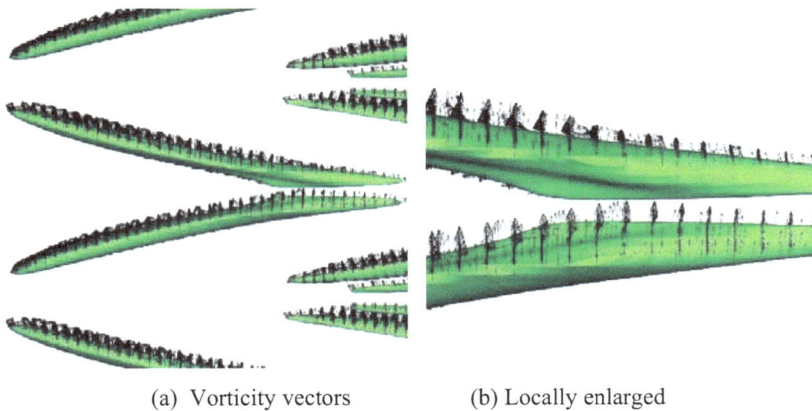

(a) Vorticity vectors   (b) Locally enlarged

Fig. 2 cont.....

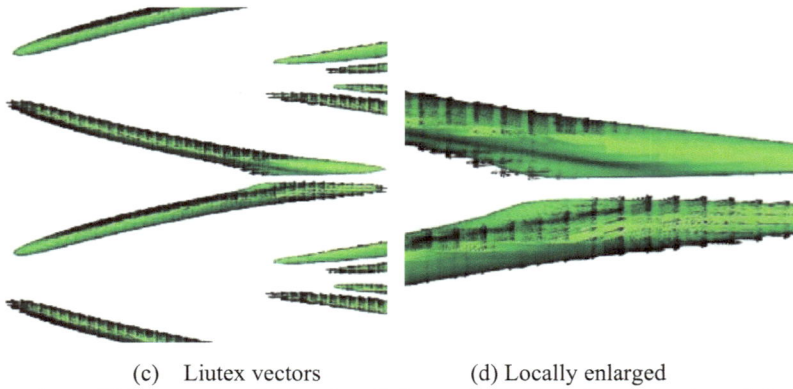

(c)   Liutex vectors   (d) Locally enlarged

Fig. (2). Comparison of Liutex vector field and vorticity vector field for the vortex legs: (**a**) vorticity vector field and the Liutex iso-surface, (**b**) the local enlargement of vorticity vector field near the head part, (**c**) Liutex vector field and the Liutex iso-surface, and (**d**) the local enlargement of Liutex vector field near the head part.

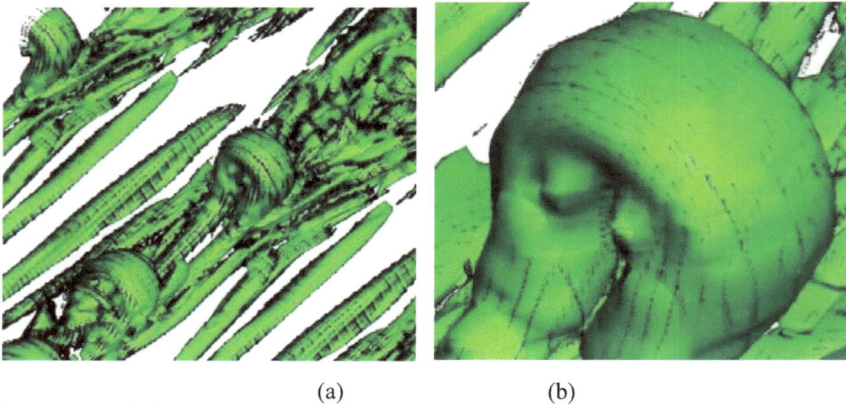

(a)   (b)

Fig. (3). Liutex vector field and the Liutex iso-surface for the first vortex ring: (**a**) global view and (**b**) local enlargement of the marked area.

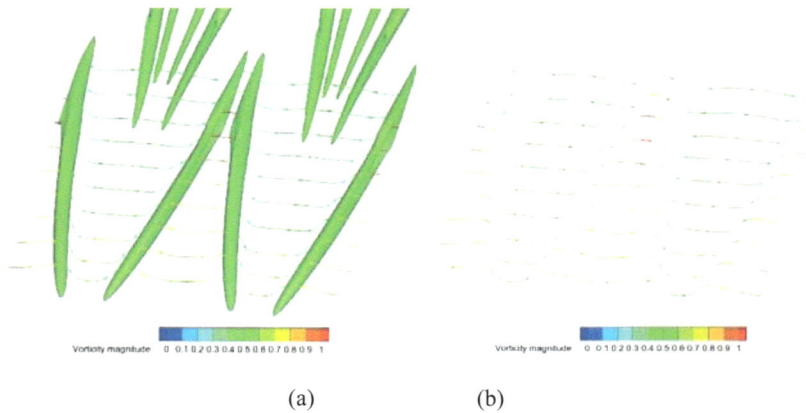

(a)   (b)

Fig. 4 cont.....

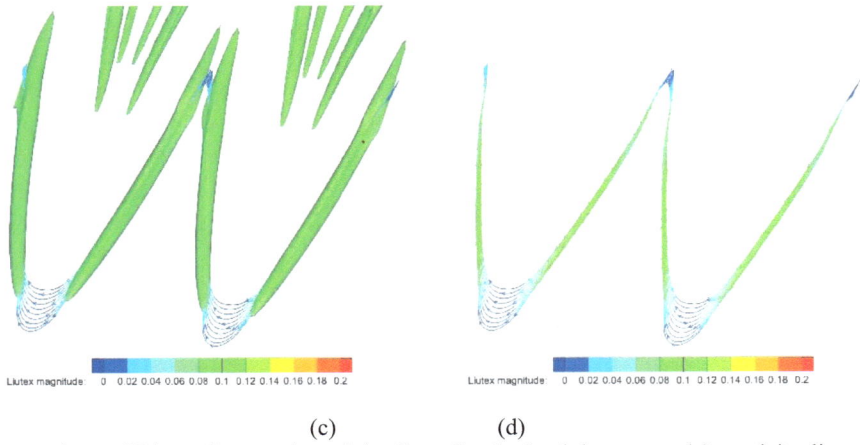

Fig. (4). Comparison of Liutex lines and vorticity lines for the lambda vortex: (**a**) vorticity lines with the Liutex iso-surface visible - not aligned, (**b**) vorticity lines with the Liutex iso-surface hidden, (**c**) Liutex lines with the Liutex iso-surface visible - aligned, and (**d**) Liutex lines with the Liutex iso-surface hidden.

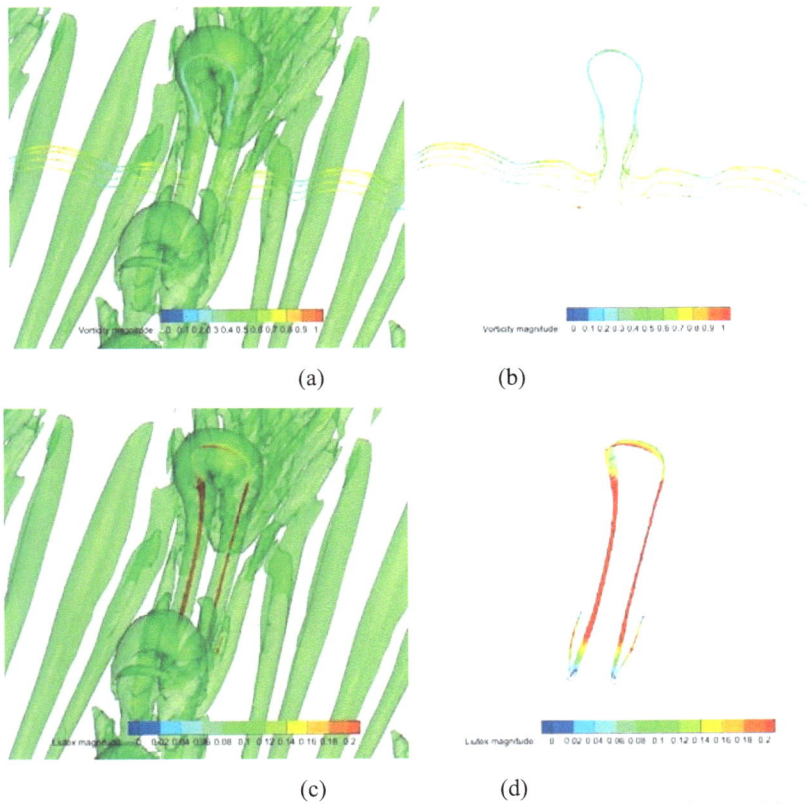

Fig. (5). Comparison of Liutex lines and vorticity lines for the ring-like vortex: (**a**) vorticity lines with the Liutex iso-surface visible – not aligned, (**b**) vorticity lines with the Liutex iso-surface hidden, (**c**) Liutex lines with the Liutex iso-surface visible - aligned, and (**d**) Liutex lines with the Liutex iso-surface hidden.

Fig. (**6**) shows the structures of Liutex iso-surfaces and lines and Fig. (**7**) demonstrates that vorticity lines are not aligned with vortices. As can be seen, vorticity lines can only represent the ring part of the hairpin vortex. In contrast, Liutex lines can provide a skeleton of the whole hairpin vortex. It is expected that Liutex lines will offer a new perspective to analyze the vortical structures. Fig. (**8**) shows the Liutex line structure in the late flow transition. Note that most vortex visualizations are performed by iso-surfaces of some kind of criteria, but Liutex can give not only iso-surfaces, but also Liutex vectors and Liutex lines. Vorticity lines just mislead the visualization of the vortex structure (see Figs. **2**-**7**).

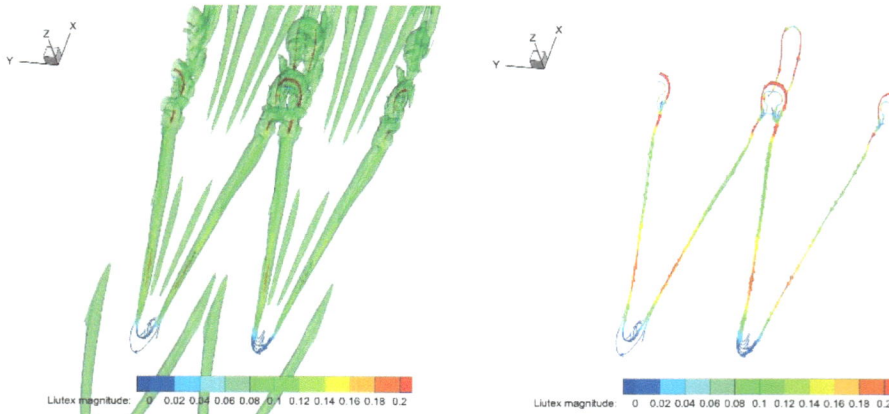

Fig. (6). Liutex lines for hairpin vortices.

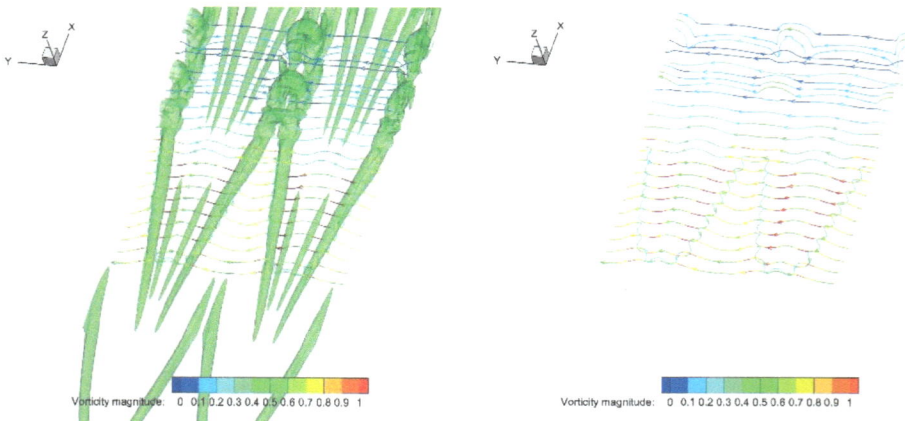

Fig. (7). Vorticity lines for hairpin vortices.

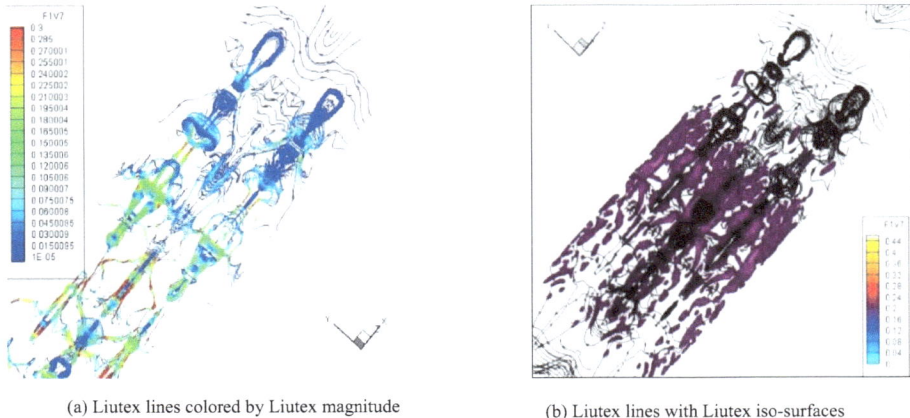

(a) Liutex lines colored by Liutex magnitude

(b) Liutex lines with Liutex iso-surfaces

Fig. (8) Liutex line structure in late boundary layer transition.

LIUTEX FOR COMPRESSIBLE FLOW

Since Liutex is a kinematical definition which is solely determined by the velocity gradient tensor, Liutex is unrelated to flow compressibility. Chapter 6 will address this point of view. Our high speed flow case like flow around the micro vortex generator (MVG) also demonstrates that Liutex will give the correct visualization of the vortex structure for both incompressible and compressible flows without formula change.

CONCLUSION

An eigenvector-based definition of Liutex is introduced. A systematic interpretation of scalar, vector and tensor versions of Liutex is presented to provide a unified characterization of the local fluid rotation. Several advantages can be summarized as follows:

(1)   Liutex is a mathematical vortex definition. The real eigenvector of the velocity gradient tensor is used to determine the direction of Liutex, which represents the axis of the local fluid rotation in the vortex area, and the rotational strength obtained in the plane perpendicular to the local axis is defined as the magnitude of Liutex.

(2)   Liutex is a unique, without an accurate and Galilean invariant definition for fluid rotation. Without accurate definition of vortex, we do not have vortex science. Since turbulence is built up and driven by vortices, we do not have serious theoretical turbulence research ever before because we have no definition of the vortex.

(3)   Liutex is an extremely important physical quantity of fluid flow which is ignored by our pioneering scholars of fluid dynamics for centuries. Liutex is critical for vortex science, turbulence research and fluid dynamics.

(4)   The vorticity cannot be applied to represent the vortex since vortex and vorticity are two vectors which are completely different in the direction and magnitude.

(5)   The vorticity can be decomposed to two parts: Liutex vector and anti-symmetric shear vector.

(6)   The velocity gradient tensor can be decomposed to two parts: a rotational tensor (Liutex tensor) and a non-rotational tensor.

(7)   Fluid motion can be decomposed as transition, deformation, and rotation. However, Cauchy-Stokes decomposition is not appropriate since the anti-symmetric tensor is corresponding to vorticity, but the vorticity cannot represent fluid rotation. In addition, the anti-symmetric part cannot represent the fluid deformation either.

(8)   The Cauchy-Stokes decomposition is not appropriate in mathematics since the real eigenvector of the velocity gradient tensor is inconsistent to the real eigenvector of A and B which are the symmetric and anti-symmetric parts, respectively.

CONSENT FOR PUBLICATION

Not applicable.

CONFLICT OF INTEREST

The authors confirm that this chapter contents have no conflict of interest.

ACKNOWLEDGEMENTS

Declared none.

REFERENCES

[1]   D. Küchemann, "Report on the IUTAM symposium on concentrated vortex motions in fluids", *J. Fluid Mech.,* vol. 21, pp. 1-20, 1965.
[http://dx.doi.org/10.1017/S0022112065000010]

[2]   A. Hussain, "Coherent structures and turbulence", *J. Fluid Mech.,* vol. 173, pp. 303-356, 1986.
[http://dx.doi.org/10.1017/S0022112086001192]

[3]   S.I. Green, *Fluid Vortices.* Kluwer Academic Publishers: Dordrecht, 1995. [http://dx.doi.org/10.1007/978-94-011-0249-0]

[4]   J. Wu, H. Ma, and M. Zhou, *Vorticity and vortices dynamics.* Springer-Verlag: Berlin, Heidelberg, 2006.

[http://dx.doi.org/10.1007/978-3-540-29028-5]

[5]   M. Chong, A. Perry, and B. Cantwell, "A general classification of three-dimensional flow fields", *Phys. Fluids A Fluid Dyn.,* vol. 2, pp. 765-777, 1990. [http://dx.doi.org/10.1063/1.857730]

[6]   P. Saffman, *Vortices dynamics.* Cambridge university press: Cambridge, 1992.

[7]   A. Majda, and A. Bertozzi, *Vorticity and Incompressible Flow.* Cambridge university press: Cambridge, 2001.
        [http://dx.doi.org/10.1017/CBO9780511613203]

[8]   H. Lamb, *Hydrodynamics.* Cambridge university press: Cambridge, 1932.

[9]   M. Nitsche, Vortex Dynamics.*Encyclopedia of Mathematics and Physics.* Academic Press: New York, 2006.
        [http://dx.doi.org/10.1016/B0-12-512666-2/00254-6]

[10]   S. Robinson, A review of vortex structures and associated coherent motions in turbulent boundary layers.*Structure of Turbulence and Drag Reduction.* Springer-Verlag: Berlin, Heidelberg, 1990.
        [http://dx.doi.org/10.1007/978-3-642-50971-1_2]

[11]   Y. Wang, Y. Yang, G. Yang, and C. Liu, "DNS study on vortex and vorticity in late boundary layer transition", *Commun. Comput. Phys.,* vol. 22, pp. 441-459, 2017.
        [http://dx.doi.org/10.4208/cicp.OA-2016-0183]

[12]   H. Lugt, The dilemma of defining a vortices.*Recent developments in theoretical and experimental fluid mechanics.* Springer-Verlag: Berlin, Heidelberg, 1979.
        [http://dx.doi.org/10.1007/978-3-642-67220-0_32]

[13]   S. Robinson, S. Kline, and P. Spalart, "A review of quasi-coherent structures in a numerically simulated turbulent boundary layer", *Tech. rep., NASA TM-102191,* 1989.

[14]   J. Jeong, and F. Hussain, "On the identification of a vortices", *J. Fluid Mech.,* vol. 285, pp. 69-94, 1995.
        [http://dx.doi.org/10.1017/S0022112095000462]

[15]   J. Hunt, A. Wray, and P. Moin, "Eddies, streams, and convergence zones in turbulent flows", *Center for Turbulence Research Proceedings of the Summer Program,* vol. 193, 1988

[16]   J. Zhou, R. Adrian, S. Balachandar, and T. Kendall, "Mechanisms for generating coherent packets of hairpin vortices in channel flow", *J. Fluid Mech.,* vol. 387, pp. 353-396, 1999.
        [http://dx.doi.org/10.1017/S002211209900467X]

[17]   P. Chakraborty, S. Balachandar, and R.J. Adrian, "On the relationships between local vortex identification schemes", *J. Fluid Mech.,* vol. 535, pp. 189-214, 2005.
        [http://dx.doi.org/10.1017/S0022112005004726]

[18]   C. Liu, Y. Wang, Y. Yang, and Z. Duan, "New Omega vortex identification method", *Sci. China Phys. Mech. Astron.,* vol. 59, 2016.684711 [http://dx.doi.org/10.1007/s11433-016-0022-6]

[19]   V. Kolář, "Vortex identification: New requirements and limitations", *Int. J. Heat Fluid Flow,* vol. 28, pp. 638-652, 2007.
        [http://dx.doi.org/10.1016/j.ijheatfluidflow.2007.03.004]

[20]   V. Kolář, J. Šístek, F. Cirak, and P. Moses, "Average corotation of line segments near a point and vortex identification", *AIAA J.,* vol. 51, pp. 2678-2694, 2013.
        [http://dx.doi.org/10.2514/1.J052330]

[21]   V. Kolář, P. Moses, and J. Šístek, "Local Corotation of Line Segments and Vortex Identification", *17th Australasian Fluid Mechanics Conference,* 2014 Auckland, New Zealand.

[22]   G. Haller, "A variational theory of hyperbolic Lagrangian coherent structures", *Physica D,* vol. 240, pp. 574-598, 2011.
        [http://dx.doi.org/10.1016/j.physd.2010.11.010]

[23]   G. Haller, and F. Geron-Vera, "Coherent Lagrangian vortices: The black holes of turbulence", *J. Fluid Mech.,* vol. 731, p. R4, 2013.
        [http://dx.doi.org/10.1017/jfm.2013.391]

[24]   Y. Zhang, K. Liu, H. Xian, and X. Du, "A review of methods for vortex identification in hydroturbines", *Renew. Sustain. Energy Rev.,* vol. 81, pp. 1269-1285, 2017. [http://dx.doi.org/10.1016/j.rser.2017.05.058]

[25]   Y. Zhang, X. Qiu, F. Chen, K. Liu, Y. Zhang, X. Dong, and C. Liu, "A selected review of vortex identification methods with applications", *J. Hydrodynam.,* vol. 30, no. 5, 2018.
        [http://dx.doi.org/10.1007/s42241-018-0112-8]

[26]   B. Epps, "Review of Vortex Identification Methods", *AIAA 2017-0989,* 2017.
        [http://dx.doi.org/10.2514/6.2017-0989]

[27]   J.M. Wallace, "Highlights from 50 years of turbulent boundary layer research", *J. Turbul.,* vol. 13, no. 53, pp. 1-70, 2013.

[28]   C. Liu, Y. Yan, and P. Lu, "Physics of turbulence generation and sustenance in a boundary layer", *Comput. Fluids,* vol. 102, pp. 353-384, 2014. [http://dx.doi.org/10.1016/j.compfluid.2014.06.032]

[29]   C. Liu, Y. Gao, S. Tian, and X. Dong, "Rortex—A new vortex vector definition and vorticity tensor and

vector decompositions", *Phys. Fluids,* vol. 30, 2018.035103
[http://dx.doi.org/10.1063/1.5023001]

[30]   Y. Gao, and C. Liu, "An eigenvector-based definition of Rortex and comparison with eigenvalue based vortex identification criteria", *Phys. Fluids,* vol. 30, 2018.085107 [http://dx.doi.org/10.1063/1.5040112]

[31]   S. Tian, Y. Gao, X. Dong, and C. Liu, "A definition of vortex vector and vortex", *Journal of Fluid Mechanics,* vol. 849, pp. 312-339, 2018.

[32]   C. Liu, Y. Gao, X. Dong, Y. Wang, J. Liu, Y. Zhang, X. Cai, and N. Gui, "Third generations of vortex identification methods: Omega and Liutex/Rortex based systems", *J. Hydrodynam.,* vol. 31, no. 2, pp. 1-19, 2019.
[http://dx.doi.org/10.1007/s42241-019-0022-4]

[33]   Y. Wang, Y. Gao, J. Liu, and C. Liu, "Explicit formula for the Liutex vector and physical meaning of vorticity based on the Liutex-Shear decomposition", *J. Hydrodynam.,* vol. 31, no. 3, pp. 464-474, 2019.
[http://dx.doi.org/10.1007/s42241-019-0032-2]

[34]   H. Helmholtz, "Über Integrale der hydrodynamischen Gleichungen, welche den Wirbelbewegungen entsprechen", *J. Reine Angew. Math.,* vol. 55, pp. 25-55, 1858. [J].

CHAPTER 2

Liutex and Its Calculation and Galilean Invariance

Yiqian Wang[1,*], Yisheng Gao[2] and Chaoqun Liu[3]

[1]School of Mathematical Science, Soochow University, Suzhou 215006, China
[2]College of Aerospace Engineering, Nanjing University of Aeronautics and Astronautics, Nanjing 210016, China
[3]Department of Mathematics, University of Texas at Arlington, Arlington, Texas 76019, USA

Abstract: The Liutex (previously known Rortex) method introduces a vortex vector field to mathematically and systematically describe vortices in flow fields. In this study, the previous calculation procedures of Liutex which includes a two-step reference coordinate rotation is revisited first. An explicit formula to calculate Liutex is then derived and the physical intuition and efficiency improvement brought by this formula are discussed. It is estimated that the computation time of Liutex vector from velocity gradient field can be reduced by 36.6% compared with that of the previous method. Besides, the Galilean invariance widely accepted as a preliminary check for a successful vortex identification method is discussed for Liutex vector.

Keywords: Angular velocity, Coherent structures, Galilean invariance, Liutex, Vortex identification, Velocity gradient tensor.

INTRODUCTION

Despite that turbulence is often considered random, instantaneous organized structures, or coherent structures exist in turbulence and play a significant role in turbulent momentum transport [1-2]. Two typical coherent structures found in near-wall turbulence are the hairpin vortex and low-speed streaks. Theordorsen [3], back in 1952, proposed a conceptual model of "horseshoe" or "hairpin" vortex to describe the regeneration cycle of turbulence. Actually, these hairpin shaped vortices are ubiquitously found in wall turbulence both from numerical simulations and experiments. Adrian [4] has stipulated that hairpins may autogenerate to form hairpin vortex packets which are presumed to be the prevalent coherent structures in wall bounded turbulence. In a transitional boundary layer flow, the development from the Λ-vortex to the hairpin vortex was carefully studied by Wang *et al*. [5] and its preponderance and statistical importance in turbulence and transition was further investigated by Eitel-Amor [6]. The second type of coherent structure considered

*****Corresponding author Yiqian Wang:** School of Mathematical Science, Soochow University, Suzhou 215006, China; Tel: +1-15950504139; E-mail: yiqianw@sina.com

Chaoqun Liu and Yisheng Gao (Eds.)

here is the near-wall streaks of alternating high and low streamwise momentum fluids which were first reported by Kline *et al.* [7] based on flow visualization in the viscous sublayer using hydrogen bubbles. Thereafter, substantial attention had been drawn to the streaks with the surrounding staggered quasi-streamwise vortices and the dynamics of the nonlinear self-sustain cycle concerning the streak instability that leads to the formation of quasi-longitudinal vortices was proposed [8]. Clearly, both the hairpins and the streaks are related to the notion of vortex, which has a clear physical intuition but hardly a mathematical definition. Δ, λ_{ci}, Q and Ω methods [9-12] which are classified into velocity-gradient-based Eulerian scalar vortex identification methods have been proved to be able to capture the vortical rotational strength to some extent. To give a more precise and unambiguous definition of a vortex, a vector field which includes information of both the direction and the magnitude named Liutex (previously named Rortex) [13-14] has been introduced recently. Then Gao and Liu [15] have improved the calculation method of Liutex by pointing out that the local rotational axis is actually the real eigenvector of the velocity gradient tensor provided that the other two corresponding eigenvalues are complex conjugates which serves as the sufficient and the necessary condition of local fluid rotation.

However, an explicit formula for Liutex vector has not been reached, making the derivation of governing equation more difficult and the appreciation of the physical meaning ambiguous. In the present study, a careful derivation of such an explicit formula is given after a revisit of the Liutex vector calculation procedure. Thereafter, the physical meaning and efficiency improvement of calculating the Liutex vector from this formula are discussed and also the Galilean invariance of Liutex is reassured according to the explicit formula.

REVIST OF THE LIUTEX VECTOR DEFINITION

The computation of Liutex vector includes the following three steps.

1) Obtain the directional information of Liutex.

Calculate the eigenvalues of the 3×3 matrix $\nabla \vec{v}$ (velocity gradient tensor) in the original xyz-frame. If all the three eigenvalues are real, there is no fluid rotation. Thus, the Liutex vector equals to zero. If $\nabla \vec{v}$ has two complex conjugated eigenvalues and one real eigenvalue, the corresponding real unit eigenvector $\vec{t}_r$ is the local rotational axis and thus Liutex direction. Then, make a coordinate rotation (Q rotation) that rotates the original z-axis to the local rotational axis $\vec{t}_r$ and obtain the new velocity gradient tensor $\nabla \vec{V}_Q$ in the resulting XYZ_Q frame by $\nabla \vec{V}_Q = Q\nabla \vec{v}Q^{\mathrm{T}}$.

2) Obtain the magnitude of Liutex.

After the Q rotation, $\nabla \vec{V}_Q$ in the XYZ_Q reference frame has the form of:

$$\nabla \vec{V}_Q = \begin{bmatrix} \dfrac{\partial U_Q}{\partial X_Q} & \dfrac{\partial U_Q}{\partial Y_Q} & 0 \\[2mm] \dfrac{\partial V_Q}{\partial X_Q} & \dfrac{\partial V_Q}{\partial Y_Q} & 0 \\[2mm] \dfrac{\partial W_Q}{\partial X_Q} & \dfrac{\partial W_Q}{\partial Y_Q} & \dfrac{\partial W_Q}{\partial Z_Q} \end{bmatrix} \tag{1}$$

Another rotation (P rotation) is then applied to rotate the reference frame around the Z_O-axis and the velocity gradient tensor $\nabla \vec{V}_P$ in the resulting XYZ_P coordinate can be expressed as:

$$\nabla \vec{V}_P = P\nabla \vec{V}_Q P^{-1} \tag{2}$$

where

$$P = \begin{bmatrix} cos\theta & sin\theta & 0 \\ -sin\theta & cos\theta & 0 \\ 0 & 0 & 1 \end{bmatrix} \tag{3}$$

The resulting $\partial U_P/\partial Y_P$ under rotation angle θ is the angular velocity at this azimuth angle θ, and could be obtained as:

$$\frac{\partial U_P}{\partial Y_P}\Big|_\theta = \alpha \sin(2\theta + \varphi) - \beta \tag{4}$$

with α and β determined by the elements of $\nabla \vec{V}_Q$:

$$\alpha = \frac{1}{2}\sqrt{\left(\frac{\partial V_Q}{\partial Y_Q} - \frac{\partial U_Q}{\partial X_Q}\right)^2 + \left(\frac{\partial V_Q}{\partial X_Q} + \frac{\partial U_Q}{\partial Y_Q}\right)^2} \tag{5}$$

$$\beta = \frac{1}{2}\left(\frac{\partial V_Q}{\partial X_Q} - \frac{\partial U_Q}{\partial Y_Q}\right) \tag{6}$$

while φ is a constant angle determined by $\nabla \vec{V}_Q$. The rotational strength or the magnitude of Liutex is then defined as twice the minimal absolute value of the off-diagonal components in the 2×2 upper left submatrix, *i.e.*, the minimal absolute value of $\partial U_P / \partial Y_P |_\theta$, which is given as:

$$R = \begin{cases} 2(\beta - \alpha), & \alpha^2 - \beta^2 < 0 \\ 0, & \alpha^2 - \beta^2 \geq 0 \end{cases} \tag{7}$$

and here it is assumed that $\beta > 0$ (if $\beta < 0$, first rotate the local rotation axis to the opposite direction to make β positive).

Finally, the Liutex vector is obtained: $\vec{R} = R\vec{t}_r$.

THE PHYSICAL MEANING OF LIUTEX, VORTICITY AND λ_{ci}

Consider the problem of vortex related rotational movements by studying how a surrounding point B moves about a point A with a velocity gradient $\nabla \vec{v}_A$. Suppose B is in a neighbourhood of A and the distance between point A and point B satisfies $L_{AB} < \delta$, with δ an arbitrary small value. If L_{AB} is small enough, the velocity at point B could be estimated as:

$$\vec{v}_B = \vec{v}_A + \nabla \vec{v}_A d\vec{s} \tag{8}$$

Here, $\vec{v}_A$ and $\vec{v}_B$ denote the velocity vector at the corresponding points while $d\vec{s}$ is the vector pointing from A to B. Because only the relative motion of B to A is cared here, let $\vec{v}_A = 0$ without loss of generality. In addition, the origin of the coordinates is selected to be point A so that $d\vec{s}$ can be represented by the coordinate components of point B, *i.e.*, $d\vec{s} = [x_B \ y_B \ z_B]^T$ and notice that $\vec{v}_B = d[x_B \ y_B \ z_B]^T / dt$. The Equation 8 becomes:

$$\frac{d}{dt}\begin{bmatrix} x_B \\ y_B \\ z_B \end{bmatrix} = \nabla \vec{v}_A \begin{bmatrix} x_B \\ y_B \\ z_B \end{bmatrix} \tag{9}$$

Equation 9 describes the motion of surrounding fluid (represented as point B) in a small enough neighbourhood region based on the velocity gradient $\nabla \vec{v}_A$. To make the linear system valid, (1) B must be close enough to A so that the linear estimation of $\vec{v}_B$ based on Equation 8 is plausible; (2) the time-span integrated should be sufficiently small so that the non-linear effect has not been accumulated. For the first-order ordinary differential equation system (Equation 9), the solution greatly depends on the eigenvalues and corresponding eigenvectors of the velocity gradient tensor $\nabla \vec{v}_A$. Because the elements of $\nabla \vec{v}_A$ are all real numbers, there are only two scenarios: (1) three real eigenvalues with corresponding eigenvectors; (2) two complex conjugate eigenvalues and one real with corresponding eigenvectors. In the first scenario, either stretch or compression exists in the direction of three eigenvectors without any rotational motion while in the second scenario the fluid will experience a rotation because of the complex conjugate eigenvalues. Actually, the condition that $\nabla \vec{v}_A$ has a pair of complex conjugate eigenvalues is the condition used by Liutex to determine whether a vortex region is detected. Gao and Liu [15] further pointed out that the rotational axis *i.e.*, the direction of Liutex vector, is actually given by the real eigenvector, along which only stretch and compression could be found. Because only vortex related phenomenon is concerned, it is mainly the second scenario that will be discussed herein, hence the solution to Equation 9 has the form of:

$$\begin{bmatrix} x \\ y \\ z \end{bmatrix} = T \begin{bmatrix} e^{(\lambda_{cr}+i\lambda_{ci})t} & 0 & 0 \\ 0 & e^{(\lambda_{cr}-i\lambda_{ci})t} & 0 \\ 0 & 0 & e^{\lambda_r t} \end{bmatrix} T^{-1} \begin{bmatrix} x_0 \\ y_0 \\ z_0 \end{bmatrix} \tag{10}$$

where λ_r and $\lambda_{1,2} = \lambda_{cr} \pm i\lambda_{ci}$ are the real and complex conjugate eigenvalues respectively while T is the eigenvector matrix with T^{-1} as its inverse. $[x_0 \ y_0 \ z_0]^T$ is used to denote the initial position of B, while $[x \ y \ z]^T$ is the coordinates of B as time evolves. From now on, for simplicity the subscript of B will be left out.

To gain a clearer understanding of the Liutex vector, the procedures of calculating Liutex vector by using a point in boundary layer transition is illustrated for the velocity gradient as below:

$$\nabla \vec{v} = \begin{bmatrix} 0.1102 & -0.4789 & 0.0586 \\ 0.1144 & 0.0455 & -0.4407 \\ -0.0108 & 0.1402 & -0.1669 \end{bmatrix} \tag{11}$$

the eigenvalues of which are $\lambda_r = -0.0458$ and $\lambda_{1,2} = \lambda_{cr} \pm i\lambda_{ci} = 0.0173 \pm 0.3123i$. Based on Equation 10, the problem can then be solved and visualized as in Fig. (**1a**). The position of point A is represented by the filled magenta circle, while the real eigenvector is the straight red line. The remaining four colored curves are pathlines originating from corresponding seeding points (initial positions)

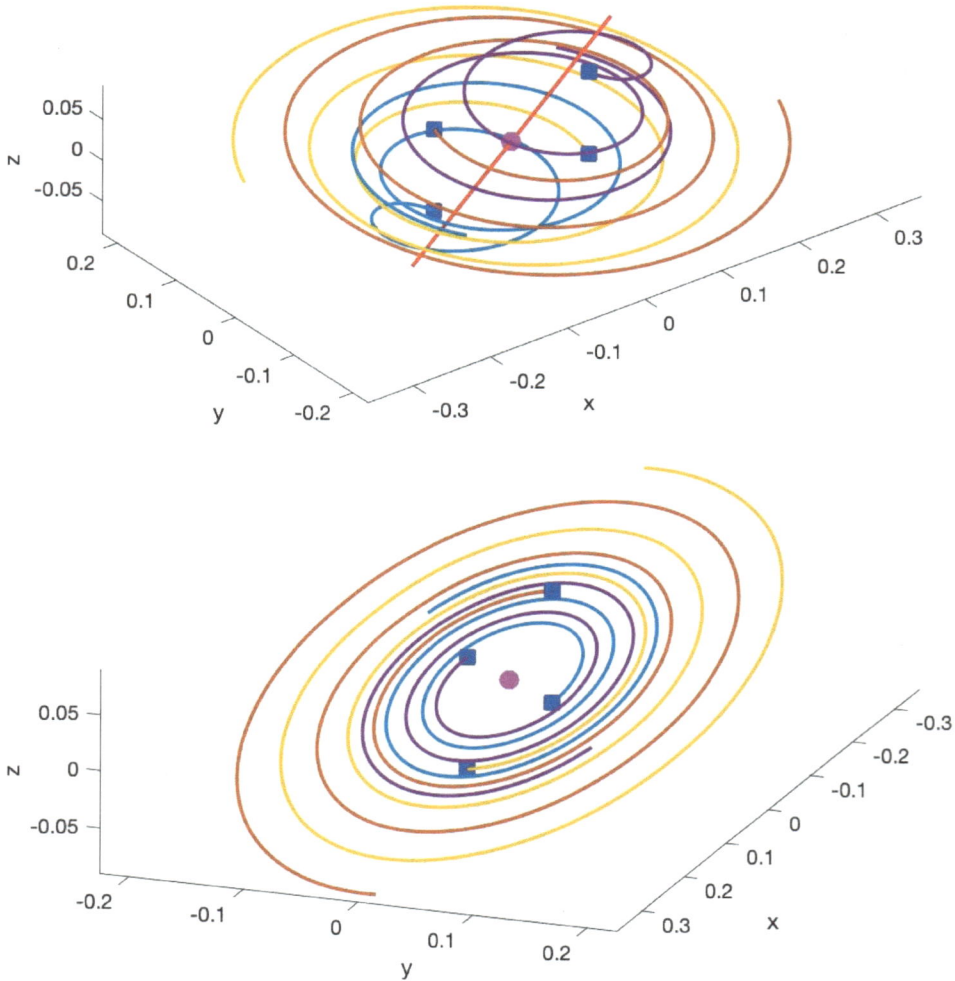

Fig. (1). (a) The fluid motion around point A, **(b)** same as in **(a)** except that the observation plane is now perpendicular to the real eigenvector.

denoted as blue squares. Because of the linearity, the coordinate in Fig. (**1**) can always be multiplied by an arbitrary small number to satisfy the condition that B should be sufficiently close to point A. In addition, the pathlines are identical to the streamlines owing to the linearity, and this makes the results meaningful despite the fact that the time span considered may be larger than the requirement to make Equation 9 valid. It can be seen from Fig. (**1a**) that the fluid particles in the neighborhood of point A are making spiral movement around the real eigenvector of $\nabla \vec{v}$, which is more obvious in Fig. (**1b**) with the observation plane being perpendicular to the real eigenvector. It should be also noted from Fig. (**1b**) that the trajectories are generally elliptic.

The second step of calculating Liutex vector involves firstly a Q rotation for the coordinates so that the real eigenvector $\vec{t}_r$ becomes the new Z_Q axis with the subscript of Q denoting the coordinates or quantities that are measured after Q rotation. For the considered point A, the new velocity gradient tensor becomes:

$$\nabla \vec{V}_Q = \begin{bmatrix} -0.0389 & 0.1892 & 0 \\ -0.5320 & 0.0735 & 0 \\ -0.0094 & 0.4172 & -0.0458 \end{bmatrix} \tag{12}$$

Based on Equation 10, the rotational direction is now the positive Z_Q axis, and the trajectories viewed from positive Z_Q axis are shown in Fig. (**2a**). The purpose of the Q rotation is clear that the new real eigenvector becomes $[0\ 0\ 1]^T$. However, what will the shapes of these trajectories become after a second P rotation by finding the minimum $\partial U_P / \partial Y_P$ as a function of in-plane rotation angle θ? After the P rotation, the velocity gradient tensor becomes:

$$\nabla \vec{V}_P = \begin{bmatrix} 0.0173 & 0.1802 & 0 \\ -0.5410 & 0.0173 & 0 \\ 0.0751 & -0.4105 & -0.0458 \end{bmatrix} \tag{13}$$

And the trajectories viewed from positive Z_P is shown in Fig. (**2b**). We can clearly see that the P rotation actually rotates the coordinates so that the major axis and minor axis of the elliptic shapes formed by these trajectories coincide with the new Y_P and X_P-axes respectively. In addition, from the revisit of Liutex vector calculation above, $\partial U_P / \partial Y_P = 0.1802$ represents the minimum angular velocity of Y_P-axis and $2 \times \partial U_P / \partial Y_P = 0.3604$ is thus the magnitude of Liutex vector.

Based on Equation 4, the angular velocity at the new Y_P-axis after an angle θ P-rotation is actually a function of $\sin(2\theta)$, with θ defined as the in-plane rotational angle about the Z_Q-axis. Hereafter, the angle θ is redefined backward, *i.e.*, θ represents the clockwise rotation angle from the optimum case *i.e.*, with the Y_P-axis and X_P-axis coinciding with the major axis and minor axis of the elliptic trajectories. The situation is depicted in Fig. (**2b**), and thus the angular velocity in the trajectories around point A can be given as:

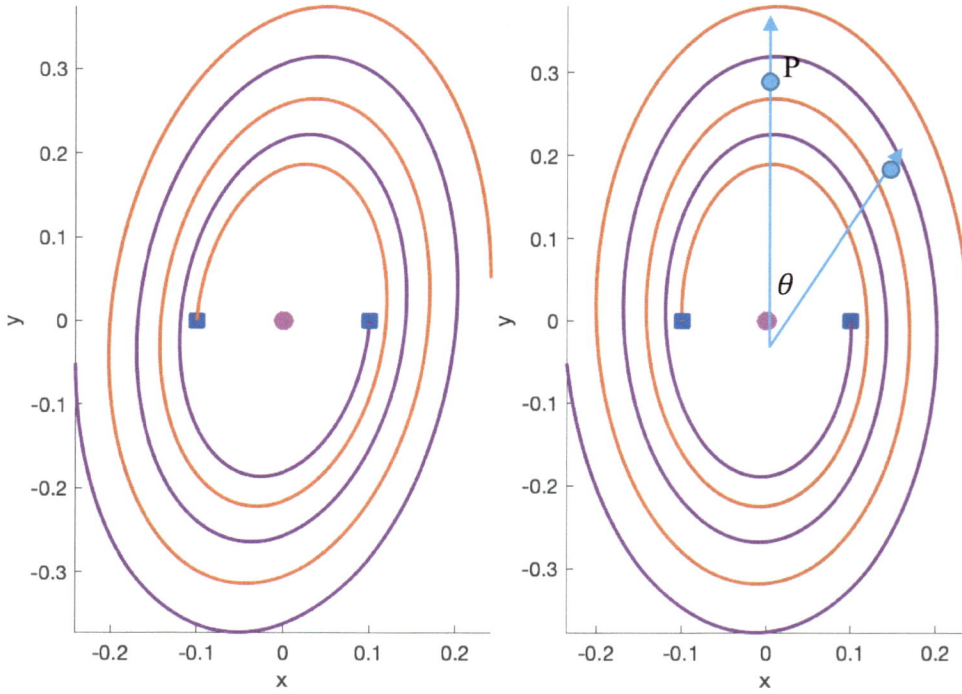

Fig. (2). Fluid motion after Q rotation (**a**) and after P rotation (**b**).

$$\dot{\theta} = \frac{(\dot{\theta}_{max} - \dot{\theta}_{min})}{2} \sin\left(2\theta - \frac{\pi}{2}\right) + \frac{\dot{\theta}_{max} + \dot{\theta}_{min}}{2} \qquad (14)$$

where $\dot{\theta}_{max}$ and $\dot{\theta}_{min}$ defined as below are the maximum and minimum angular velocity in the minor axis and major axis of the elliptic trajectories respectively shown in Fig. (**2b**) as point P1 and P2.

$$\dot{\theta}_{min} = \frac{\partial U_P}{\partial Y_P} \qquad (15)$$

$$\dot{\theta}_{max} = -\frac{\partial V_P}{\partial X_P} \qquad (16)$$

Therefore, the spatial mean angular velocity:

$$\langle \dot{\theta} \rangle = \frac{\dot{\theta}_{max} + \dot{\theta}_{min}}{2} = \frac{\partial U_P / \partial Y_P - \partial V_P / \partial X_P}{2} = \frac{1}{2}\omega_{Z_P} \qquad (17)$$

i.e., ω_{Z_P} (the vorticity in the Z_P direction) or $\langle \vec{\omega}, \vec{r} \rangle$ (the vorticity in the real eigenvector direction) equals $2\langle \dot{\theta} \rangle$ (twice the spatial mean angular velocity). $\langle \vec{\omega}, \vec{r} \rangle$ represents the dot product of vorticity tensor $\vec{\omega}$ and $\vec{r}$ both in the original coordinate. Thus, one possible physical meaning for the vorticity along the real eigenvector $\vec{r}$ is twice the spatial mean angular velocity. According to Equation 10, ω_{X_P} and ω_{Y_P} represent the modulation effect on the spiral or circular motion in the plane perpendicular to the rotational axis, which are not as significant compared with ω_{Z_P} to determine the rotational motion.

On the other hand, Equation 10 means the surrounding particles of A rotate with a "period" of $T_p = 2\pi/\lambda_{ci}$ ignoring the compression or stretch determined by the real eigenvalue λ_r and the movement away from or towards the rotation axis determined by λ_{cr}, *i.e.,* the real part of the complex eigenvalues. Under these assumptions, the trajectories can form closed curves which make the movements periodic. That means in a time-span $T_p = 2\pi/\lambda_{ci}$, the surrounding fluid particles can rotate an angle of 2π. Therefore, the time-average angular velocity is $2\pi/T_p = \lambda_{ci}$. However, as stated above, the time-span considered must be small enough for Equation 9 to be valid which means a full circular motion may or may not be made. Despite the argument, the pathlines and the streamlines are identical for linear systems. Thus, the time-averaged angular velocity λ_{ci} still has an implication of the rotational motion of the instantaneous flow field. Hereafter, λ_{ci} will be denoted as the pseudo time-average angular velocity to be more precise.

From above discussion, $\langle \vec{\omega}, \vec{r} \rangle$, *i.e.,* the vorticity in the direction of real eigenvector, could be interpreted as twice the spatial mean angular velocity and λ_{ci} can be viewed as the pseudo time-average angular velocity. Obviously, some relation between the spatial and time mean values can be found. Firstly, from Equation 14 one can obtain:

$$\frac{1}{2} \omega_{Z_P} = \frac{1}{2\pi} \int_0^{2\pi} \dot{\theta} d\theta = \frac{\dot{\theta}_{max} + \dot{\theta}_{min}}{2} \tag{18}$$

The pseudo time-period T_p is also related to $\dot{\theta}$ as:

$$T_p = \int_0^{2\pi} \frac{d\theta}{\dot{\theta}} \tag{19}$$

with $\dot{\theta}$ given as $\dot{\theta} = 0.5 \times \left(\dot{\theta}_{max} - \dot{\theta}_{min} \right) \sin(2\theta - \pi/2) + 0.5 \times \left(\dot{\theta}_{max} + \dot{\theta}_{min} \right)$. Then the integral in Equation 19 can be solved and finally obtain:

$$\lambda_{ci} = \frac{2\pi}{T} = \sqrt{\dot{\theta}_{min}\dot{\theta}_{max}} \tag{20}$$

In the framework of Liutex, the Liutex magnitude is defined as twice the minimum angular velocity, *i.e.*,

$$R = 2\dot{\theta}_{min} \tag{21}$$

Therefore, the three quantities of ω_{Z_P}(vorticity in the direction of real eigenvector), λ_{ci} (the imaginary part of the complex eigenvalue) and R (the magnitude of Liutex vector) are all related to the minimum and maximum angular velocity in the plane perpendicular to the real eigenvector. Obviously, the three quantities are all connected to the rotational motion of the fluid particles around A. However, the first two quantities, as averaged values, cannot uniquely determine the rotational motion, while the minimum angular velocity can be interpreted as the rigid-body rotation part of the fluid motion. In summary, R is a better quantity to describe the fluid motion in the neighborhood of point A.

DERIVATION OF THE EXPLICIT FORMULA TO CALCULATE LIUTEX

Derivation - Approach 1

Given that $\nabla\vec{v}$ has two complex conjugate eigenvalues $\lambda_r \pm \lambda_{ci}i$, $\nabla\vec{v}$ and its eigenvalues and eigenvectors must satisfy:

$$\nabla\vec{v}[\vec{t}_r \quad \vec{t}_{cr} \quad \vec{t}_{ci}] = [\vec{t}_r \quad \vec{t}_{cr} \quad \vec{t}_{ci}]\begin{bmatrix} \lambda_r & 0 & 0 \\ 0 & \lambda_{cr} & \lambda_{ci} \\ 0 & -\lambda_{ci} & \lambda_{cr} \end{bmatrix} \tag{22}$$

where $\vec{t}_r$ is the real eigenvector corresponding to the real eigenvalue λ_r while $\vec{t}_{cr}$ and $\vec{t}_{ci}$ are formed by the real and imaginary part of the eigenvectors corresponding to $\lambda_r \pm \lambda_{ci}i$, respectively. It can be inferred from Equation 10 that only stretching or compression exist in the direction of $\vec{t}_r$ while the other two directions are coupled with rotational motion. However, $\vec{t}_r$, $\vec{t}_{cr}$ and $\vec{t}_{ci}$ are not necessarily orthogonal to each other. As introduced above, our aim is to find the behaviour of $\nabla\vec{v}$ to the vectors in the plane perpendicular to the real eigenvector $\vec{t}_r$. Thus, Schmidt orthogonalization is used here to do the job.

Denote $k = \langle\vec{t}_r, \vec{t}_{cr}\rangle$, and thus:

$$\vec{t}'_{cr} = \vec{t}_{cr} - \langle\vec{t}_r, \vec{t}_{cr}\rangle\vec{t}_r = \vec{t}_{cr} - k\vec{t}_r \tag{23}$$

$$\left|\vec{t}_{cr}'\right|^2 = \langle \vec{t}_{cr}', \vec{t}_{cr}' \rangle = \langle \vec{t}_{cr}, \vec{t}_{cr} \rangle - \langle \vec{t}_r, \vec{t}_{cr} \rangle^2 \tag{24}$$

Note that, the new $\vec{t}_{cr}'$ is orthogonal to $\vec{t}_r$, i.e., $\langle \vec{t}_r, \vec{t}_{cr}' \rangle = 0$. Denote $k_1 = \langle \vec{t}_r, \vec{t}_{ci} \rangle$, and $k_2 = \langle \vec{t}_{cr}', \vec{t}_{ci} \rangle / \langle \vec{t}_{cr}', \vec{t}_{cr}' \rangle$, and let:

$$\vec{t}_{ci}' = \vec{t}_{ci} - k_1 \vec{t}_r - k_2 \vec{t}_{cr}' \tag{25}$$

Expand the expressions, we get:

$$k_2 = \frac{\langle \vec{t}_{cr}, \vec{t}_{ci} \rangle - \langle \vec{t}_r, \vec{t}_{cr} \rangle \langle \vec{t}_r, \vec{t}_{ci} \rangle}{\langle \vec{t}_{cr}, \vec{t}_{cr} \rangle - \langle \vec{t}_r, \vec{t}_{cr} \rangle^2} \tag{26}$$

$$\left|\vec{t}_{ci}'\right| = \sqrt{\langle \vec{t}_{ci}, \vec{t}_{ci} \rangle - \langle \vec{t}_r, \vec{t}_{ci} \rangle^2 - \frac{\left(\langle \vec{t}_{cr}, \vec{t}_{ci} \rangle - \langle \vec{t}_r, \vec{t}_{cr} \rangle \langle \vec{t}_r, \vec{t}_{ci} \rangle\right)^2}{\langle \vec{t}_{cr}, \vec{t}_{cr} \rangle - \langle \vec{t}_r, \vec{t}_{cr} \rangle^2}} \tag{27}$$

Clearly, $\vec{t}_r$, $\vec{t}_{cr}'$ and $\vec{t}_{ci}'$ are now orthogonal to each other, but $\left|\vec{t}_{cr}'\right| \neq 1$, $\left|\vec{t}_{ci}'\right| \neq 1$ while $\left|\vec{t}_r\right| = 1$. In matrix form, it reads:

$$\begin{bmatrix} \vec{t}_r & \vec{t}_{cr}' & \vec{t}_{ci}' \end{bmatrix} = \begin{bmatrix} \vec{t}_r & \vec{t}_{cr} & \vec{t}_{ci} \end{bmatrix} \begin{bmatrix} 1 & -k & 0 \\ 0 & 1 & 0 \\ 0 & 0 & 1 \end{bmatrix} \begin{bmatrix} 1 & 0 & -k_1 \\ 0 & 1 & -k_2 \\ 0 & 0 & 1 \end{bmatrix} \tag{28}$$

Now normalize $\vec{t}_{cr}'$ and $\vec{t}_{ci}'$, one gets:

$$\vec{t}_{cr}'' = \frac{\vec{t}_{cr}'}{\left|\vec{t}_{cr}'\right|} \tag{29}$$

$$\vec{t}_{ci}'' = \frac{\vec{t}_{ci}'}{\left|\vec{t}_{ci}'\right|} \tag{30}$$

Thus,

$$\begin{bmatrix} \vec{t}_r & \vec{t}_{cr}'' & \vec{t}_{ci}'' \end{bmatrix} = \begin{bmatrix} \vec{t}_r & \vec{t}_{cr} & \vec{t}_{ci} \end{bmatrix} \begin{bmatrix} 1 & -k & 0 \\ 0 & 1 & 0 \\ 0 & 0 & 1 \end{bmatrix} \begin{bmatrix} 1 & 0 & -k_1 \\ 0 & 1 & -k_2 \\ 0 & 0 & 1 \end{bmatrix} \begin{bmatrix} 1 & 0 & 0 \\ 0 & 1/\left|\vec{t}_{cr}'\right| & 0 \\ 0 & 0 & 1/\left|\vec{t}_{ci}'\right| \end{bmatrix} \tag{31}$$

Now $\vec{t}_{cr}''$ and $\vec{t}_{ci}''$ are orthogonal in the plane normal to real eigenvector $\vec{t}_r$.

Denote

$$
T = \begin{bmatrix} 1 & -k & 0 \\ 0 & 1 & 0 \\ 0 & 0 & 1 \end{bmatrix} \begin{bmatrix} 1 & 0 & -k_1 \\ 0 & 1 & -k_2 \\ 0 & 0 & 1 \end{bmatrix} \begin{bmatrix} 1 & 0 & 0 \\ 0 & 1/|\vec{t}'_{cr}| & 0 \\ 0 & 0 & 1/|\vec{t}'_{ci}| \end{bmatrix} = \begin{bmatrix} 1 & -\dfrac{k}{|\vec{t}'_{cr}|} & \dfrac{-k_1+kk_2}{|\vec{t}'_{ci}|} \\ 0 & \dfrac{1}{|\vec{t}'_{cr}|} & -\dfrac{k_2}{|\vec{t}'_{ci}|} \\ 0 & 0 & \dfrac{1}{|\vec{t}'_{ci}|} \end{bmatrix} \tag{32}
$$

$$
T^{-1} = \begin{bmatrix} 1 & 0 & 0 \\ 0 & |\vec{t}'_{cr}| & 0 \\ 0 & 0 & |\vec{t}'_{ci}| \end{bmatrix} \begin{bmatrix} 1 & 0 & k_1 \\ 0 & 1 & k_2 \\ 0 & 0 & 1 \end{bmatrix} \begin{bmatrix} 1 & k & 0 \\ 0 & 1 & 0 \\ 0 & 0 & 1 \end{bmatrix} = \begin{bmatrix} 1 & k & k_1 \\ 0 & |\vec{t}'_{cr}| & k_2|\vec{t}'_{cr}| \\ 0 & 0 & |\vec{t}'_{ci}| \end{bmatrix} \tag{33}
$$

Thus,

$$
\begin{bmatrix} \vec{t}_r & \vec{t}''_{cr} & \vec{t}''_{ci} \end{bmatrix} = \begin{bmatrix} \vec{t}_r & \vec{t}_{cr} & \vec{t}_{ci} \end{bmatrix} T \tag{34}
$$

$$
\begin{bmatrix} \vec{t}_r & \vec{t}_{cr} & \vec{t}_{ci} \end{bmatrix} = \begin{bmatrix} \vec{t}_r & \vec{t}''_{cr} & \vec{t}''_{ci} \end{bmatrix} T^{-1} \tag{35}
$$

Insert the above equations into the Equation 22, we get:

$$
\nabla \vec{v} \begin{bmatrix} \vec{t}_r & \vec{t}''_{cr} & \vec{t}''_{ci} \end{bmatrix} = \begin{bmatrix} \vec{t}_r & \vec{t}''_{cr} & \vec{t}''_{ci} \end{bmatrix} T^{-1} \begin{bmatrix} \lambda_r & 0 & 0 \\ 0 & \lambda_{cr} & \lambda_{ci} \\ 0 & -\lambda_{ci} & \lambda_{cr} \end{bmatrix} T \equiv
$$

$$
\begin{bmatrix} \vec{t}_r & \vec{t}''_{cr} & \vec{t}''_{ci} \end{bmatrix} H \tag{36}
$$

with

$$
H = \begin{bmatrix} \lambda_r & \dfrac{-k\lambda_r+k\lambda_{cr}-k_1\lambda_{ci}}{|\vec{t}'_{cr}|} & \dfrac{(kk_2-k_1)(\lambda_r-\lambda_{cr})+(k_1k_2+k)\lambda_{ci}}{|\vec{t}'_{ci}|} \\ 0 & \lambda_{cr}-k_2\lambda_{ci} & \lambda_{ci}\dfrac{|\vec{t}'_{cr}|}{|\vec{t}'_{ci}|}(1+k_2^2) \\ 0 & -\lambda_{ci}\dfrac{|\vec{t}'_{ci}|}{|\vec{t}'_{cr}|} & \lambda_{cr}+k_2\lambda_{ci} \end{bmatrix} \tag{37}
$$

Now, although $\vec{t}''_{cr}$ and $\vec{t}''_{ci}$ are normalized, orthogonal vector in the plane normal to real eigenvector v_r, but they are not necessary coincides with the major or the minor axis. An in-plane rotation can be introduced as multiplication of:

$$r(\theta) = \begin{bmatrix} 1 & 0 & 0 \\ 0 & cos(\theta) & -sin(\theta) \\ 0 & sin(\theta) & cos(\theta) \end{bmatrix} \tag{38}$$

i.e.,

$$\begin{bmatrix} \vec{t}_r & \vec{t}_{cr}''' & \vec{t}_{ci}''' \end{bmatrix} = \begin{bmatrix} \vec{t}_r & \vec{t}_{cr}'' & \vec{t}_{ci}'' \end{bmatrix} r(\theta) \tag{39}$$

Thus:

$$\nabla \vec{v} \begin{bmatrix} \vec{t}_r & \vec{t}_{cr}''' & \vec{t}_{ci}''' \end{bmatrix} = \begin{bmatrix} \vec{t}_r & \vec{t}_{cr}''' & \vec{t}_{ci}''' \end{bmatrix} r(\theta)^{-1} T^{-1} \begin{bmatrix} \lambda_r & 0 & 0 \\ 0 & \lambda_{cr} & \lambda_{ci} \\ 0 & -\lambda_{ci} & \lambda_{cr} \end{bmatrix} Tr(\theta) \tag{40}$$

Now, the above equation describes the motion along $\vec{t}_{cr}'''$ and $\vec{t}_{ci}'''$ which are arbitrary orthogonal directions in the plane normal to real eigenvector $\vec{t}_r$. Actually, $\vec{t}_r$ is decoupled from the other two directions, while $\vec{t}_{cr}'''$ and $\vec{t}_{ci}'''$ are coupled. In addition, the right multiplication of matrices, *i.e.*,

$$r(\theta)^{-1} T^{-1} \begin{bmatrix} \lambda_r & 0 & 0 \\ 0 & \lambda_{cr} & \lambda_{ci} \\ 0 & -\lambda_{ci} & \lambda_{cr} \end{bmatrix} Tr(\theta) \equiv TT \tag{41}$$

$$TT(1,1) = \lambda_r \tag{42}$$

$$TT(2,1) = TT(3,1) = 0 \tag{43}$$

$$TT(1,2) = \frac{-k\lambda_r + k\lambda_{cr} - k_1\lambda_{ci}}{|\vec{t}_{cr}'|} cos(\theta) + \frac{(kk_2 - k_1)(\lambda_r - \lambda_{cr}) + (k_1 k_2 + k)\lambda_{ci}}{|\vec{t}_{ci}'|} sin(\theta) \tag{44}$$

$$TT(1,3) = -\frac{-k\lambda_r + k\lambda_{cr} - k_1\lambda_{ci}}{|\vec{t}_{cr}'|} sin(\theta) + \frac{(kk_2 - k_1)(\lambda_r - \lambda_{cr}) + (k_1 k_2 + k)\lambda_{ci}}{|\vec{t}_{ci}'|} cos(\theta) \tag{45}$$

$$TT(2,2) = \lambda_{cr} + \lambda_{ci} \left\{ -k_2 cos(2\theta) + \frac{1}{2} \left[\frac{|\vec{t}_{cr}'|}{|\vec{t}_{ci}'|} (1 + k_2^2) - \frac{|\vec{t}_{ci}'|}{|\vec{t}_{cr}'|} \right] sin(2\theta) \right\} \tag{46}$$

$$TT(3,3) = \lambda_{cr} - \lambda_{ci} \left\{ -k_2 cos(2\theta) + \frac{1}{2} \left[\frac{|\vec{t}_{cr}'|}{|\vec{t}_{ci}'|} (1 + k_2^2) - \frac{|\vec{t}_{ci}'|}{|\vec{t}_{cr}'|} \right] sin(2\theta) \right\} \tag{47}$$

$$TT(2,3) = \lambda_{ci} \left\{ k_2 sin(2\theta) + \frac{1}{2} \left[\frac{|\vec{t}_{cr}'|}{|\vec{t}_{ci}'|} (1 + k_2^2) - \frac{|\vec{t}_{ci}'|}{|\vec{t}_{cr}'|} \right] cos(2\theta) + \frac{1}{2} \left[\frac{|\vec{t}_{cr}'|}{|\vec{t}_{ci}'|} (1 + k_2^2) + \frac{|\vec{t}_{ci}'|}{|\vec{t}_{cr}'|} \right] \right\} \tag{48}$$

$$TT(2,3) = \lambda_{ci}\left\{k_2 sin(2\theta) + \frac{1}{2}\left[\frac{|\vec{t}_{cr}'|}{|\vec{t}_{ci}'|}(1 + k_2^2) - \frac{|\vec{t}_{ci}'|}{|\vec{t}_{cr}'|}\right]cos(2\theta) - \frac{1}{2}\left[\frac{|\vec{t}_{cr}'|}{|\vec{t}_{ci}'|}(1 + k_2^2) + \frac{|\vec{t}_{ci}'|}{|\vec{t}_{cr}'|}\right]\right\}(49)$$

TT is now the matrix to describe the behaviour of velocity gradient tensor $\nabla\vec{v}$ along $\vec{t}_r$, $\vec{t}_{cr}'''$, and $\vec{t}_{ci}'''$, which are orthogonal to each other and have been standardized. In addition, $\vec{t}_{cr}'''$ and $\vec{t}_{ci}'''$ are arbitrary in the plane normal to v_r due to the introduction of θ. Particularly, as in the original framework of Liutex, the minimum absolute value of $TT(2,3)$ (or $TT(3,2)$) can be used to define the magnitude of Liutex vector. *i.e.*,

$$|R| = \lambda_{ci}\left\{\frac{1}{2}\left[\frac{|\vec{t}_{cr}'|}{|\vec{t}_{ci}'|}(1 + k_2^2) + \frac{|\vec{t}_{ci}'|}{|\vec{t}_{cr}'|}\right] - \sqrt{k_2^2 + \frac{1}{4}\left[\frac{|\vec{t}_{cr}'|}{|\vec{t}_{ci}'|}(1 + k_2^2) - \frac{|\vec{t}_{ci}'|}{|\vec{t}_{cr}'|}\right]^2}\right\}\qquad(50)$$

Combined with expressions for k_2, $|\vec{t}_{cr}'|$ and $|\vec{t}_{ci}'|$,

$$|R| = \lambda_{ci}\left\{\frac{\langle\vec{t}_{ci},\vec{t}_{ci}\rangle - \langle\vec{t}_r,\vec{t}_{ci}\rangle^2 + \langle\vec{t}_{cr},\vec{t}_{cr}\rangle - \langle\vec{t}_r,\vec{t}_{cr}\rangle^2}{\sqrt{(\langle\vec{t}_{ci},\vec{t}_{ci}\rangle - \langle\vec{t}_r,\vec{t}_{ci}\rangle^2)(\langle\vec{t}_{cr},\vec{t}_{cr}\rangle - \langle\vec{t}_r,\vec{t}_{cr}\rangle^2) - (\langle\vec{t}_{cr},\vec{t}_{ci}\rangle - \langle\vec{t}_r,\vec{t}_{cr}\rangle\langle\vec{t}_r,\vec{t}_{ci}\rangle)^2}} - \right.$$
$$\left.\frac{\sqrt{(\langle\vec{t}_{ci},\vec{t}_{ci}\rangle - \langle\vec{t}_r,\vec{t}_{ci}\rangle^2 - \langle\vec{t}_{cr},\vec{t}_{cr}\rangle + \langle\vec{t}_r,\vec{t}_{cr}\rangle^2)^2 + 4(\langle\vec{t}_{cr},\vec{t}_{ci}\rangle - \langle\vec{t}_r,\vec{t}_{cr}\rangle\langle\vec{t}_r,\vec{t}_{ci}\rangle)^2}}{\sqrt{(\langle\vec{t}_{ci},\vec{t}_{ci}\rangle - \langle\vec{t}_r,\vec{t}_{ci}\rangle^2)(\langle\vec{t}_{cr},\vec{t}_{cr}\rangle - \langle\vec{t}_r,\vec{t}_{cr}\rangle^2) - (\langle\vec{t}_{cr},\vec{t}_{ci}\rangle - \langle\vec{t}_r,\vec{t}_{cr}\rangle\langle\vec{t}_r,\vec{t}_{ci}\rangle)^2}}\right\}\qquad(51)$$

Lemma 1: The denominator in Equation 51 is actually the absolute value of determinant of matrix $[\vec{t}_r \quad \vec{t}_{cr} \quad \vec{t}_{ci}]$, *i.e.*,

$$\sqrt{(\langle\vec{t}_{ci},\vec{t}_{ci}\rangle - \langle\vec{t}_r,\vec{t}_{ci}\rangle^2)(\langle\vec{t}_{cr},\vec{t}_{cr}\rangle - \langle\vec{t}_r,\vec{t}_{cr}\rangle^2) - (\langle\vec{t}_{cr},\vec{t}_{ci}\rangle - \langle\vec{t}_r,\vec{t}_{cr}\rangle\langle\vec{t}_r,\vec{t}_{ci}\rangle)^2}$$

$$= |det([\vec{t}_r \quad \vec{t}_{cr} \quad \vec{t}_{ci}])|\qquad(52)$$

Prove: Just expend the expressions, it can be proved. Remember to use the identity $\langle\vec{t}_r,\vec{t}_r\rangle = 1$.

Based on the physical meaning of determinant, the denominator is actually the volume of a parallelepiped formed by vectors $\vec{t}_r$, $\vec{t}_{cr}$ and $\vec{t}_{ci}$, and also equals

$$|det([\vec{t}_r \quad \vec{t}_{cr} \quad \vec{t}_{ci}])| = |\vec{t}_r \cdot (\vec{t}_{cr} \times \vec{t}_{ci})| = |\vec{t}_{cr} \cdot (\vec{t}_r \times \vec{t}_{ci})| = |\vec{t}_{ci} \cdot (\vec{t}_r \times \vec{t}_{cr})|$$

Lemma 2: The first part in the equation, *i.e.*,

$$\lambda_{ci} \frac{\langle \vec{t}_{ci},\vec{t}_{ci}\rangle - \langle \vec{t}_r,\vec{t}_{ci}\rangle^2 + \langle \vec{t}_{cr},\vec{t}_{cr}\rangle - \langle \vec{t}_r,\vec{t}_{cr}\rangle^2}{\sqrt{\left(\langle \vec{t}_{ci},\vec{t}_{ci}\rangle - \langle \vec{t}_r,\vec{t}_{ci}\rangle^2\right)\left(\langle \vec{t}_{cr},\vec{t}_{cr}\rangle - \langle \vec{t}_r,\vec{t}_{cr}\rangle^2\right) - \left(\langle \vec{t}_{cr},\vec{t}_{ci}\rangle - \langle \vec{t}_r,\vec{t}_{cr}\rangle\langle \vec{t}_r,\vec{t}_{ci}\rangle\right)^2}} \tag{53}$$

is the magnitude of projected vorticity vector $\vec{\omega} = \left(\omega_x, \omega_y, \omega_z\right)^T$ onto the real eigenvector $\vec{t}_r$, *i.e.* $|\langle \omega,\vec{t}_r\rangle|$. Here, the direction of real eigenvector is chosen to ensure that $\langle \vec{\omega},\vec{t}_r\rangle$ to be positive as suggested by Liu.

Prove:

To simplify the deduction, $\vec{a}$, $\vec{b}$ and $\vec{c}$ are used to denote $\vec{t}_r$, $\vec{t}_{cr}$ and $\vec{t}_{ci}$. And the components are:

$$\vec{a} = \begin{bmatrix} a_1 \\ a_2 \\ a_3 \end{bmatrix} \tag{54}$$

$$\vec{b} = \begin{bmatrix} b_1 \\ b_2 \\ b_3 \end{bmatrix} \tag{55}$$

$$\vec{c} = \begin{bmatrix} c_1 \\ c_2 \\ c_3 \end{bmatrix} \tag{56}$$

Equation 1 now becomes:

$$\nabla \vec{v}\begin{bmatrix} \vec{a} & \vec{b} & \vec{c} \end{bmatrix} = \begin{bmatrix} \vec{a} & \vec{b} & \vec{c} \end{bmatrix}\begin{bmatrix} \lambda_r & 0 & 0 \\ 0 & \lambda_{cr} & \lambda_{ci} \\ 0 & -\lambda_{ci} & \lambda_{cr} \end{bmatrix} \tag{57}$$

We denote $\Psi = \begin{bmatrix} \vec{a} & \vec{b} & \vec{c} \end{bmatrix}$, and the determinant of Ψ is $det(\Psi)$, the inverse of Ψ is:

$$\Psi^{-1} = \frac{1}{det(\Psi)}\begin{bmatrix} b_2 c_3 - b_3 c_2 & -b_1 c_3 + b_3 c_1 & b_1 c_2 - b_2 c_1 \\ -a_2 c_3 + a_3 c_2 & a_1 c_3 - a_3 c_1 & -a_1 c_2 + a_2 c_1 \\ a_2 b_3 - a_3 b_2 & -a_1 b_3 + a_3 b_1 & a_1 b_2 - a_2 b_1 \end{bmatrix} \tag{58}$$

Thus,

$$\nabla \vec{v} = \begin{bmatrix} u_x & u_y & u_z \\ v_x & v_y & v_z \\ w_x & w_y & w_z \end{bmatrix} = \Psi \begin{bmatrix} \lambda_r & 0 & 0 \\ 0 & \lambda_{cr} & \lambda_{ci} \\ 0 & -\lambda_{ci} & \lambda_{cr} \end{bmatrix} \Psi^{-1} \tag{59}$$

Insert the expressions,

$$u_x = \frac{1}{det(\Psi)} \{ a_1 (b_2 c_3 - b_3 c_2) \lambda_r + [b_1 (-a_2 c_3 + a_3 c_2) + c_1 (a_2 b_3 - a_3 b_2)] \lambda_{cr}$$
$$+ [-c_1 (-a_2 c_3 + a_3 c_2) + b_1 (a_2 b_3 - a_3 b_2)] \lambda_{ci} \}$$

$$v_y = \frac{1}{det(\Psi)} \{ a_2 (-b_1 c_3 + b_3 c_1) \lambda_r + [b_2 (a_1 c_3 - a_3 c_1) + c_2 (-a_1 b_3 + a_3 b_1)] \lambda_{cr}$$
$$+ [-c_2 (a_1 c_3 - a_3 c_1) + b_2 (-a_1 b_3 + a_3 b_1)] \lambda_{ci} \}$$

$$w_z = \frac{1}{det(\Psi)} \{ a_3 (b_1 c_2 - b_2 c_1) \lambda_r + [b_3 (-a_1 c_2 + a_2 c_1) + c_3 (a_1 b_2 - a_2 b_1)] \lambda_{cr}$$
$$+ [-c_3 (-a_1 c_2 + a_2 c_1) + b_3 (a_1 b_2 - a_2 b_1)] \lambda_{ci} \}$$

$$u_y = \frac{1}{det(\Psi)} \{ a_1 (-b_1 c_3 + b_3 c_1)(\lambda_r - \lambda_{cr}) + [-c_1 (a_1 c_3 - a_3 c_1) + b_1 (-a_1 b_3 + a_3 b_1)] \lambda_{ci} \}$$

$$v_x = \frac{1}{det(\Psi)} \{ a_2 (b_2 c_3 - b_3 c_2)(\lambda_r - \lambda_{cr}) + [-c_2 (-a_2 c_3 + a_3 c_2) + b_2 (a_2 b_3 - a_3 b_2)] \lambda_{ci} \}$$

$$u_z = \frac{1}{det(\Psi)} \{ a_1 (b_1 c_2 - b_2 c_1)(\lambda_r - \lambda_{cr}) + [-c_1 (-a_1 c_2 + a_2 c_1) + b_1 (a_1 b_2 - a_2 b_1)] \lambda_{ci} \}$$

$$w_x = \frac{1}{det(\Psi)} \{ a_3 (b_2 c_3 - b_3 c_2)(\lambda_r - \lambda_{cr}) + [-c_3 (-a_2 c_3 + a_3 c_2) + b_3 (a_2 b_3 - a_3 b_2)] \lambda_{ci} \}$$

$$v_z = \frac{1}{det(\Psi)} \{ a_2 (b_1 c_2 - b_2 c_1)(\lambda_r - \lambda_{cr}) + [-c_2 (-a_1 c_2 + a_2 c_1) + b_2 (a_1 b_2 - a_2 b_1)] \lambda_{ci} \}$$

$$w_y = \frac{1}{det(\Psi)} \{ a_3 (-b_1 c_3 + b_3 c_1)(\lambda_r - \lambda_{cr}) + [-c_3 (a_1 c_3 - a_3 c_1) + b_3 (-a_1 b_3 + a_3 b_1)] \lambda_{ci} \}$$

The vorticity vector:

$$\vec{\omega} = \begin{bmatrix} \omega_x \\ \omega_y \\ \omega_z \end{bmatrix} = \begin{bmatrix} w_y - v_z \\ u_z - w_x \\ v_x - u_y \end{bmatrix} \tag{60}$$

The first part of vorticity $\vec{\omega}_1$:

$$\vec{\omega}_1 = \frac{(\lambda_r - \lambda_{cr})}{det(\Psi)} \begin{bmatrix} a_3(-b_1c_3 + b_3c_1) - a_2(b_1c_2 - b_2c_1) \\ a_1(b_1c_2 - b_2c_1) - a_3(b_2c_3 - b_3c_2) \\ a_2(b_2c_3 - b_3c_2) - a_1(-b_1c_3 + b_3c_1) \end{bmatrix} \tag{61}$$

It is easy to show $\vec{\omega}_1$ is orthogonal to $\vec{a}$ by showing that:

$$\begin{bmatrix} a_3(-b_1c_3 + b_3c_1) - a_2(b_1c_2 - b_2c_1) \\ a_1(b_1c_2 - b_2c_1) - a_3(b_2c_3 - b_3c_2) \\ a_2(b_2c_3 - b_3c_2) - a_1(-b_1c_3 + b_3c_1) \end{bmatrix} \cdot \begin{bmatrix} a_1 \\ a_2 \\ a_3 \end{bmatrix} = 0 \tag{62}$$

The second part of vorticity:

$$\vec{\omega}_2 = \frac{\lambda_{ci}}{det(\Psi)} \begin{bmatrix} -a_1c_3^2 + a_3c_1c_3 - a_1b_3^2 + a_3b_1b_3 - a_1c_2^2 + a_2c_1c_2 - a_1b_2^2 + a_2b_1b_2 \\ -a_2c_1^2 + a_1c_1c_2 - a_2b_1^2 + a_1b_1b_2 - a_2c_3^2 + a_3c_2c_3 - a_2b_3^2 + a_3b_2b_3 \\ -a_3c_2^2 + a_2c_2c_3 - a_3b_2^2 + a_2b_2b_3 - a_3c_1^2 + a_1c_1c_3 - a_3b_1^2 + a_1b_1b_3 \end{bmatrix} \tag{63}$$

Thus,

$$\begin{aligned} \langle \vec{\omega}_2, \vec{a} \rangle = \frac{\lambda_{ci}}{det(\Psi)} \{ &-a_1^2b_2^2 - a_1^2b_3^2 - a_1^2c_2^2 - a_1^2c_3^2 - a_2^2b_1^2 - a_2^2b_3^2 - a_2^2c_1^2 \\ &- a_2^2c_3^2 - a_3^2b_1^2 - a_3^2b_2^2 - a_3^2c_1^2 - a_3^2c_2^2 \\ &+ 2(a_1a_2b_1b_2 + a_1a_2c_1c_2 + a_1a_3b_1b_3 + a_1a_3c_1c_3 + a_2a_3b_2b_3 \\ &+ a_2a_3c_2c_3) \} \end{aligned}$$

Notice that:

$$\langle \vec{t}_{ci}, \vec{t}_{ci} \rangle - \langle \vec{t}_r, \vec{t}_{ci} \rangle^2 + \langle \vec{t}_{cr}, \vec{t}_{cr} \rangle - \langle \vec{t}_r, \vec{t}_{cr} \rangle^2$$

$$= \langle \vec{c}, \vec{c} \rangle - \langle \vec{a}, \vec{c} \rangle^2 + \langle \vec{b}, \vec{b} \rangle - \langle \vec{a}, \vec{b} \rangle^2$$

$$= \langle \vec{a}, \vec{a} \rangle \langle \vec{c}, \vec{c} \rangle - \langle \vec{a}, \vec{c} \rangle^2 + \langle \vec{a}, \vec{a} \rangle \langle \vec{b}, \vec{b} \rangle - \langle \vec{a}, \vec{b} \rangle^2$$

$$= (a_1^2 + a_2^2 + a_3^2)(c_1^2 + c_2^2 + c_3^2) - (a_1c_1 + a_2c_2 + a_3c_3)^2$$
$$+ (a_1^2 + a_2^2 + a_3^2)(b_1^2 + b_2^2 + b_3^2) - (a_1b_1 + a_2b_2 + a_3b_3)^2$$

$$= (a_1^2 + a_2^2 + a_3^2)(c_1^2 + c_2^2 + c_3^2) - (a_1c_1 + a_2c_2 + a_3c_3)^2$$
$$+ (a_1^2 + a_2^2 + a_3^2)(b_1^2 + b_2^2 + b_3^2) - (a_1b_1 + a_2b_2 + a_3b_3)^2$$

$$= a_1^2b_2^2 + a_1^2b_3^2 + a_1^2c_2^2 + a_1^2c_3^2 + a_2^2b_1^2 + a_2^2b_3^2 + a_2^2c_1^2 + a_2^2c_3^2 + a_3^2b_1^2 + a_3^2b_2^2 + a_3^2c_1^2$$
$$+ a_3^2c_2^2$$
$$- 2(a_1a_2b_1b_2 + a_1a_2c_1c_2 + a_1a_3b_1b_3 + a_1a_3c_1c_3 + a_2a_3b_2b_3$$
$$+ a_2a_3c_2c_3)$$

$$= (a_1b_2 - a_2b_1)^2 + (a_1b_3 - a_3b_1)^2 + (a_2b_3 - a_3b_2)^2 + (a_1c_2 - a_2c_1)^2 + (a_1c_3 - a_3c_1)^2$$
$$+ (a_2c_3 - a_3c_2)^2$$

Thus,

$$\langle \vec{\omega}, \vec{a} \rangle = \langle \vec{\omega}_2, \vec{a} \rangle = -\frac{\lambda_{ci}}{det(\Psi)} \left(\langle \vec{c}, \vec{c} \rangle - \langle \vec{a}, \vec{c} \rangle^2 + \langle \vec{b}, \vec{b} \rangle - \langle \vec{a}, \vec{b} \rangle^2 \right) \qquad (64)$$

The expression in the above parentheses is non-negative.

Prove: denote the angle between $\vec{a}$ and $\vec{c}$ as α, the angle between $\vec{a}$ and $\vec{b}$ as β:

$$\langle \vec{c}, \vec{c} \rangle - \langle \vec{a}, \vec{c} \rangle^2 + \langle \vec{b}, \vec{b} \rangle - \langle \vec{a}, \vec{b} \rangle^2$$

$$= \langle \vec{a}, \vec{a} \rangle \langle \vec{c}, \vec{c} \rangle - \langle \vec{a}, \vec{c} \rangle^2 + \langle \vec{a}, \vec{a} \rangle \langle \vec{b}, \vec{b} \rangle - \langle \vec{a}, \vec{b} \rangle^2$$

$$= |\vec{a}|^2 |\vec{c}|^2 - [|\vec{a}||\vec{c}|cos(\alpha)]^2 + |\vec{a}|^2 |\vec{b}|^2 - \left[|\vec{a}||\vec{b}|cos(\beta) \right]^2$$

$$= |\vec{a}|^2 |\vec{c}|^2 sin^2(\alpha) + |\vec{a}|^2 |\vec{b}|^2 sin^2(\beta) \geq 0$$

It is required $\langle \vec{\omega}, \vec{a} \rangle$ is positive, thus $det(\Psi)$ has to be negative, and

$$|\langle \vec{\omega}, \vec{a} \rangle| = \frac{\lambda_{ci}}{|det(\Psi)|} \left(\langle \vec{c}, \vec{c} \rangle - \langle \vec{a}, \vec{c} \rangle^2 + \langle \vec{b}, \vec{b} \rangle - \langle \vec{a}, \vec{b} \rangle^2 \right) \qquad (65)$$

$$\frac{1}{2} |\langle \vec{\omega}, \vec{a} \rangle| = \frac{\lambda_{ci}}{2|det(\Psi)|} \left(\langle \vec{c}, \vec{c} \rangle - \langle \vec{a}, \vec{c} \rangle^2 + \langle \vec{b}, \vec{b} \rangle - \langle \vec{a}, \vec{b} \rangle^2 \right) \qquad (66)$$

Thus, Lemma 2 has been proved.

Lemma 3. The second part of the equation

$$\lambda_{ci} \frac{\sqrt{\left(\langle\vec{t}_{ci},\vec{t}_{ci}\rangle - \langle\vec{t}_r,\vec{t}_{ci}\rangle^2 - \langle\vec{t}_{cr},\vec{t}_{cr}\rangle + \langle\vec{t}_r,\vec{t}_{cr}\rangle^2\right)^2 + 4\left(\langle\vec{t}_{cr},\vec{t}_{ci}\rangle - \langle\vec{t}_r,\vec{t}_{cr}\rangle\langle\vec{t}_r,\vec{t}_{ci}\rangle\right)^2}}{\sqrt{\left(\langle\vec{t}_{ci},\vec{t}_{ci}\rangle - \langle\vec{t}_r,\vec{t}_{ci}\rangle^2\right)\left(\langle\vec{t}_{cr},\vec{t}_{cr}\rangle - \langle\vec{t}_r,\vec{t}_{cr}\rangle^2\right) - \left(\langle\vec{t}_{cr},\vec{t}_{ci}\rangle - \langle\vec{t}_r,\vec{t}_{cr}\rangle\langle\vec{t}_r,\vec{t}_{ci}\rangle\right)^2}}$$

$$= \sqrt{\lambda_{ci}^2 \frac{\left(\langle\vec{t}_{ci},\vec{t}_{ci}\rangle - \langle\vec{t}_r,\vec{t}_{ci}\rangle^2 + \langle\vec{t}_{cr},\vec{t}_{cr}\rangle - \langle\vec{t}_r,\vec{t}_{cr}\rangle^2\right)^2}{\left(\langle\vec{t}_{ci},\vec{t}_{ci}\rangle - \langle\vec{t}_r,\vec{t}_{ci}\rangle^2\right)\left(\langle\vec{t}_{cr},\vec{t}_{cr}\rangle - \langle\vec{t}_r,\vec{t}_{cr}\rangle^2\right) - \left(\langle\vec{t}_{cr},\vec{t}_{ci}\rangle - \langle\vec{t}_r,\vec{t}_{cr}\rangle\langle\vec{t}_r,\vec{t}_{ci}\rangle\right)^2} - 4\lambda_{ci}^2}$$

$$= \sqrt{|\langle\omega,a\rangle|^2 - 4\lambda_{ci}^2} \tag{67}$$

Lemma 4. Finally, the equation to calculate magnitude of Liutex becomes very simple:

$$|\vec{R}| = |\langle\vec{\omega},\vec{a}\rangle| - \sqrt{|\langle\omega,\vec{a}\rangle|^2 - 4\lambda_{ci}^2} \tag{68}$$

Thus, the Liutex vector can be given as:

$$\vec{R} = |\vec{R}|\vec{a} = \left\{|\langle\vec{\omega},\vec{a}\rangle| - \sqrt{|\langle\vec{\omega},\vec{a}\rangle|^2 - 4\lambda_{ci}^2}\right\}\vec{a} \tag{69}$$

or

$$\vec{R} = |\vec{R}|\vec{t}_r = \left\{|\langle\vec{\omega},\vec{t}_r\rangle| - \sqrt{|\langle\vec{\omega},\vec{t}_r\rangle|^2 - 4\lambda_{ci}^2}\right\}\vec{t}_r \tag{70}$$

In the above equation, $\vec{\omega}$ is the vorticity vector and $\vec{t}_r$ (or $\vec{a}$) is real eigenvector (required that $\langle\vec{\omega},\vec{t}_r\rangle \geq 0$), λ_{ci} is the imaginary part of the complex eigenvalues. The first part is clearly the vorticity parallel to the real eigenvector while the second part has to be the pure shear.

Derivation – Approach 2

From the discussion in the previous section, it is noticed that from Equation 18,

$$\dot{\theta}_{min} + \dot{\theta}_{max} = \langle\vec{\omega},\vec{t}_r\rangle \tag{71}$$

In addition, from Equation 20,

$$\dot{\theta}_{min}\dot{\theta}_{max} = \lambda_{ci}^2 \tag{72}$$

By combining above two equations, one can get:

$$|\vec{R}| = 2\dot{\theta}_{min} = \langle \vec{\omega}, \vec{t}_r \rangle - \sqrt{\langle \vec{\omega}, \vec{t}_r \rangle^2 - 4\lambda_{ci}^2} \tag{73}$$

Thus, the Liutex vector can be given as

$$\vec{R} = R\vec{t}_r = \left\{ \langle \vec{\omega}, \vec{t}_r \rangle - \sqrt{\langle \vec{\omega}, \vec{t}_r \rangle^2 - 4\lambda_{ci}^2} \right\} \vec{t}_r \tag{74}$$

Which is identical to Equation 70. According to Equation 74, once the velocity gradient tensor has been obtained along with its eigenvalues and eigenvectors, the Liutex vector can be easily calculated. It is for the first time such an explicit expression of is systematically derived in terms of eigenvalues and eigenvector of velocity gradient tensor and vorticity vector. Besides, physical understandings of the terms in Equation 74 can be provided as (1) $\vec{t}_r$, the real eigenvector, is the rotational axis, *i.e.*, the Liutex direction; (2) $\langle \vec{\omega}, \vec{t}_r \rangle$ is the magnitude of vorticity along the direction of $\vec{t}_r$; (3) $sqrt(\langle \vec{\omega}, \vec{t}_r \rangle^2 - 4\lambda_{ci}^2)$ represents the pure shear part of the fluid motion. That means the explicit expression itself is actually the Liutex (Rortex)-shear decomposition, *i.e.*, vorticity can be further decomposed into a rotational part R and a shear part S.

Besides the physical intuition above, Equation 74 can also further improve the computational efficiency of Liutex vector. Gao and Liu [15] have substantially decreased the computational time by pointing out the real eigenvector of velocity gradient tensor is actually the rotational axis. Equation 74 can help shortening the time for calculating the Liutex magnitude. Here, the efficiency improvement brought by Equation 74 is compared with the version of Gao and Liu [15]. The Fortran program used for computing Liutex vector consists of four parts: (1) reading mesh and solution files $(t - read)$, (2) evaluating velocity gradient tensor $(t - vg)$, (3) obtaining the Liutex vector $(t - L)$ and (4) writing results to files $(t - write)$. According to Equation 74, the only modification made to the code is in the third part for calculating Liutex vector. The previous and present methods are applied to the boundary layer transition DNS (Direct Numerical Simulation) data which are generated by the code DNSUTA. Table **1** gives the corresponding computational times and all these computations are done on a MacBook Pro (2017) laptop with 2.9 GHz CPU and 16 GB memory.

Considering the modified part of the code for calculating Liutex vector from velocity gradient tensor, the time consumed is substantially decreased from $2.76s$ to $1.75s$ for one time-step data. The increase of efficiency can be measured as $(2.76 - 1.75)/2.76 \approx 36.6\%$, which is rather significant. If the whole procedure including reading and writing files is considered, however, the efficiency increase becomes $(2.76 - 1.75)/(4.63 + 1.66 + 2.76 + 3.74) \approx 7.9\%$. It is argued, however, that the efficiency improvement provided by Equation 74 could still be very important if the RS (Liutex-Shear) decomposition is integrated into Navier-Stokes equation solvers as modelling inputs of turbulence model, for example, which might require the solving of Liutex vector every time step.

Table 1. Comparison of computational times.

	t − read	t − vg	Previous t − L	Present t − L	t − write
Time(s)	4.63	1.66	2.76	1.75	3.74

GALILEAN INVARIANCE OF LIUTEX VECTOR

The Galilean transformation of two coordinate systems can be represented by:

$$\begin{bmatrix} x' \\ y' \\ z' \end{bmatrix} = Q_c \begin{bmatrix} x \\ y \\ z \end{bmatrix} + \vec{c}_1 t + \vec{c}_2 \qquad (75)$$

where x, y and z are the coordinates of original system while x', y' and z' are the coordinates of the transformed system. Q_c is a constant rotation matrix and satisfies $Q_c^{-1} = Q_c^T$ representing the spatial rotation. $\vec{c}_1$ denotes a constant relative motion while $\vec{c}_2$ is a fixed translation in space. A quantity is said to be Galilean invariant if it doesn't change under constant spatial rotation, relative motion and translation in both space and time according to Equation 75. The necessity of requiring vortex identification method Galilean invariant has been accepted by many researchers. For example, definitions based on spiralling or closed streamlines suffer from the fact that the streamline topology can dramatically change under Galilean transformations. A reasonable definition should be therefore Galilean invariant. For the Liutex method, its Galilean invariance has been proved by Wang *et al.* [16]

based on the calculation procedure of Gao *et al.* [15] with examples of the Burgers vortex, the Sullivan vortex and a transitional boundary layer flow. However, the proof given in Wang *et al.* [16] was tedious. By referring to Equation 74, the Liutex vector depends on the vorticity vector, eigenvalues and eigenvectors of the velocity gradient tensor, which are all Galilean invariants. Thus, the Galilean invariance of Liutex vector is naturally proved.

CONCLUSIONS

With the newly introduced Liutex vector, the instantaneous rotational motion induced by the velocity gradient tensor has been studied carefully. It is summarized that the vorticity in the direction of real eigenvector is actually twice the spatial average angular velocity, while the imaginary part of complex eigenvalue is the pseudo average angular velocity and the magnitude of Liutex vector is twice the minimum angular velocity. Then, it is argued the minimum value is a better choice than the other two mean quantities to uniquely represent the rotational motion of the fluid. In addition, an explicit formula for Liutex vector is for the first time derived and the efficiency improvement by this formula is then discussed. Finally, the Galilean invariance of Liutex vector is verified by referring to this explicit formula.

LIST OF SYMBOLS

xyz: the original reference frame

XYZ_Q: the reference frame after Q rotation

XYZ_P: the reference frame after P rotation

$\nabla \vec{v}$:   velocity gradient tensor in xyz frame

$\nabla \vec{V}_Q$:   velocity gradient tensor in XYZ_Q frame

$\nabla \vec{V}_P$:   velocity gradient tensor in XYZ_P frame

$\vec{R}$:   the Liutex vector

R: the magnitude of the Liutex vector

$\vec{r}$:   the direction of the Liutex vector

λ_r: the real eigenvalue of velocity gradient tensor

$\lambda_{cr} \pm i\lambda_{ci}$:   the complex conjugate eigenvalues of velocity gradient tensor

ω_{Z_P}:   vorticity in the direction of Z_P

θ: a rotation angle in the plane normal to $\vec{r}$

$\dot{\theta}$:   angular velocity

CONSENT FOR PUBLICATION

Not applicable.

CONFLICT OF INTEREST

The author(s) confirm that this chapter contents have no conflict of interest.

ACKNOWLEDGEMENTS

This work is supported by the National Natural Science Foundation of China (Grant No. 11702159) and China Post-Doctoral Science Foundation (Grant No. 2017M610876). This work is accomplished by using Code DNSUTA released by Dr. Chaoqun Liu at University of Texas at Arlington in 2009.

REFERENCES

[1]   S.K. Robinson, "Coherent Motions in the Turbulence Boundary Layer", *Annu. Rev. Fluid Mech.,* vol. 23, pp. 601-639, 1991.
        [http://dx.doi.org/10.1146/annurev.fl.23.010191.003125]
[2]   J. Jiménez, "Coherent structures in wall-bounded turbulence", *J. Fluid Mech.,* vol. 842, p. P1, 2018.
        [http://dx.doi.org/10.1017/jfm.2018.144]
[3]   T. Theodorsen, "Mechanism of turbulence", *Proceedings of the Second Midwestern Conference on Fluid Mechanics,* 1952pp. 1-18
[4]   R.J. Adrian, "Hairpin vortex organization in wall turbulencea", *Phys. Fluids,* vol. 19, no. 4, 2007.
        [http://dx.doi.org/10.1063/1.2717527]
[5]   Y.Q. Wang, H. Al-Dujaly, Y.H. Yan, N. Zhao, and C.Q. Liu, "Physics of multiple level hairpin vortex structures in turbulence", *Sci. China Phys. Mech. Astron.,* vol. 59, no. 2, pp. 1-11, 2016.
        [http://dx.doi.org/10.1007/s11433-015-5757-5]
[6]   G. Eitel-Amor, "R. órlú, P. Schlatter, and O. Flores, "Hairpin vortices in turbulent boundary layers", *Phys. Fluids,* vol. 27, no. 2, 2015. [http://dx.doi.org/10.1063/1.4907783]
[7]   "W. C. Reynolds, "The structure of turbulent boundary layers", *J. Fluid Mech.,* vol. 30, no. 04, pp. 741-773, 1967.
        [http://dx.doi.org/10.1017/S0022112067001740]
[8]   W. Schoppa, and F. Hussain, "Coherent structure generation in near-wall turbulence", *J. Fluid Mech.,*

vol. 453, pp. 57-108, 2002.
[http://dx.doi.org/10.1017/S002211200100667X]

[9]   M.S. Chong, A.E. Perry, and B.J. Cantwell, "A general classification of three-dimensional flow fields", *Phys. Fluids A Fluid Dyn.,* vol. 2, no. 5, pp. 765-777, 1990. [http://dx.doi.org/10.1063/1.857730]

[10]   J.C.R. Hunt, A.A. Wray, and P. Moin, "Eddies, streams, and convergence zones in turbulent flows", *Proceedings of the Summer Program,* 1988

[11]   J. Jeong, and F. Hussain, "On the identification of a vortex", *J. Fluid Mech.,* vol. 285, pp. 69-94, 1995. [http://dx.doi.org/10.1017/S0022112095000462]

[12]   C.Q. Liu, Y.Q. Wang, Y. Yang, and Z.W. Duan, "New omega vortex identification method", *Sci. China Phys. Mech. Astron.,* vol. 59, no. 8, 2016. [http://dx.doi.org/10.1007/s11433-016-0022-6]

[13]   C. Liu, Y. Gao, S. Tian, and X. Dong, "Rortex - A new vortex vector definition and vorticity tensor and vector decompositions", *Phys. Fluids,* vol. 30, no. 3, 2018. [http://dx.doi.org/10.1063/1.5023001]

[14]   S. Tian, Y. Gao, X. Dong, and C. Liu, "A Definition of Vortex Vector and Vortex", *J. Fluid Mech.,* vol. 849, pp. 312-339, 2018. [http://dx.doi.org/10.1017/jfm.2018.406]

[15]   Y. Gao, and C. Liu, "Rortex and comparison with eigenvalue-based vortex identification criteria", *Phys. Fluids,* vol. 30, no. 085107, 2018. [http://dx.doi.org/10.1063/1.5040112]

[16]   Y. Wang, Y. Gao, and C. Liu, "Galilean invariance of Rortex", *Phys. Fluids,* vol. 30, no. 111701, 2018.

New Omega Vortex Identification Method Based on Determined Epsilon

Xiangrui Dong[1,2], Yisheng Gao[3] and Chaoqun Liu[2,*]

[1]*Institute of Energy and Power Engineering, University of Shanghai for Science and Technology, Shanghai 200093, China*

[2]*Department of Mathematics, University of Texas at Arlington, Arlington, Texas 76019, USA*

[3]*College of Aerospace Engineering, Nanjing University of Aeronautics and Astronautics, Nanjing 210016, China*

Abstract: A new Omega (Ω) method with ε determination is introduced to represent the ratio of vorticity square over the sum of vorticity squared and deformation squared, for vortex identification. the advantages of the new Ω method can be summarized as follows: (1) Ω, as a ratio of the vorticity squared over the sum of the vorticity squared and deformation squared, is a normalized and case-independent function which satisfies $\Omega \in [0,1]$; (2) Compared with the other vortex visualization methods, which require a wide threshold to capture the vortex structures, Ω can always be set as 0.52 to capture vortex for different cases and time steps; (3) ε is defined as a function without any adjustment on its coefficient for all cases; (4) The Ω method can capture both strong and weak vortices simultaneously. In addition, Ω is quite robust with no obvious change in vortex visualization.

Keywords: Case-independent, Deformation, Omega method, Vortex identification, Vorticity.

INTRODUCTION

The definition and identification of vortex has been a longstanding issue in fluid dynamics. Robinson *et al.* [1] proposed a rather accurate definition: a vortex exists when instantaneous streamlines mapped onto a plane normal to the vortex core exhibit a roughly circular or spiral pattern, when viewed from a reference frame moving with the center of the vortex core. Several vortex identification methods based on the velocity gradient tensor ∇V have been widely used to investigate the

---

*Corresponding author Chaoqun Liu: Department of Mathematics, University of Texas at Arlington, Arlington, Texas 76019, USA; Tel: +1-8172725151; Fax: +1-8172725802; E-mail: cliu@uta.edu

vortex structures in turbulent flows. Perry and Chong [2] suggested a $\tilde{\Delta}$-method with an idea that the vortices exist where eigenvalues of velocity gradient tensor ∇V are complex, which implies the streamline pattern is spiral or closed viewed from a reference frame moving with the point. This method was further developed by Zhou *et al.* [3] and called λ_{ci}. They suggested employing iso-surfaces of imaginary part of the complex eigenvalue to capture vortices. A famous Q-criterion was introduced by Hunt *et al.* [4], in which an eddy is defined as the region with positive second invariant Q of the velocity gradient tensor. Another well-known scheme is the λ_2 method, introduced by Jeong and Hussain [5]. They suggested the usage of second eigenvalue of the symmetric tensor $\mathbf{S}^2 + \mathbf{\Omega}^2$ trying to capture the pressure minimum in a plane normal to the vortex axis. All these methods have achieved some success. However, a threshold is required case by case and time by time, which means different thresholds will lead to different vortex structures. Zhang *et al.* [6] discussed various vortex identification methods in their study, and pointed out that those identification methods are too sensitive to the chosen threshold, making them inadequate for the quantitative analysis of the vortex. The improper thresholds may be able to only capture strong vortices but lose weak ones.

DEFINITION OF A VORTEX BY NEW OMEGA METHOD

A new vortex identification method, called Omega (Ω), first proposed by Liu *et al.* [7], appears to overcome the above-mentioned weaknesses. Recently, Zhang *et al.* [6] applied this new Ω method into the analysis of the reversible pump turbine and indicated that this new omega method is quite suitable for the analysis of complex flow of hydro-turbines, especially for the unsteady flow cases. Tao *et al.* [8] also utilized Ω to investigate the wake flow from moving bodies in their study, and they pointed out that comparing to other vortex identification methods, the new Ω method has a clear physical meaning and vortex is formed when the vorticity is strong but deformation is weak. Ω method was also used by other researchers [9-11] to compare with the existing vortex identification methods. Thus, the definition of Ω is introduced in the following section.

A parameter Ω is introduced to represent the ratio of vorticity over the whole velocity gradient inside a vortex core. According to Helmholtz velocity decomposition, the velocity gradient tensor ∇V can be decomposed into a symmetric tensor and an anti-symmetric tensor,

$$\nabla V = \frac{1}{2}(\nabla V + \nabla V^T) + \frac{1}{2}(\nabla V - \nabla V^T) = \mathbf{A} + \mathbf{B} \qquad (1)$$

where $\mathbf{A}$ is symmetric part which represents deformation and $\mathbf{B}$ is anti-symmetric part which is related to the whole vorticity. Now the ratio Ω is defined as a ratio of vorticity squared over the sum of vorticity squared and deformation squared, which shows vortex is formed when the vorticity is strong but deformation is weak,

$$\Omega = \frac{\|\mathbf{B}\|_F^2}{\|\mathbf{A}\|_F^2 + \|\mathbf{B}\|_F^2} = \frac{b}{a+b} \tag{2}$$

where $\|\cdot\|_F$ is the Frobenius norm. a and b is given below,

$$a = \text{trace}(\mathbf{A}^T\mathbf{A}) = \sum_{i=1}^{3}\sum_{j=1}^{3}\left(\mathbf{A}_{ij}^2\right) \tag{3}$$

$$b = \text{trace}(\mathbf{B}^T\mathbf{B}) = \sum_{i=1}^{3}\sum_{j=1}^{3}\left(\mathbf{B}_{ij}^2\right) \tag{4}$$

There is no doubt that $\Omega \in [0, 1]$, since both a and b are not negative. In fact vortex is a measurement of fluid stiffness. If $\Omega = 1$, fluid will behave as a solid rotation. If $\Omega = 0.5$, fluid has strong shear without rotation. Fluid is different from solid and is a mixture of vorticity and deformation. $\Omega > 0.5$ represents the region where vorticity overtakes deformation ($b > a$), which is defined as vortex. Although Ω is non-dimensional and satisfies $\Omega \in [0, 1]$, some serious noises (clouds) may appear inside the flow domain if both term a and b in equation (2) are in close proximity to zero due to the systematic computational errors. These noises can be reduced or even removed by introducing a proper positive number, ε, in the denominator of Ω. Therefore, in application, we pick,

$$\Omega = \frac{b}{a+b+\varepsilon} \tag{5}$$

ε is introduced in fact to remove noises caused by the consequence of division by zero. Apparently, as a positive number, ε is dependent upon the dimension of the physical variables and needs to be adjusted to a proper number case by case and time by time. A linear correlation is found between ε and the maximum of $b - a$. The Epsilon ε is defined as a function of $(b - a)_{\max}$, which is a fixed parameter at each time step in each case. In this study, ε is proposed as follows,

$$\varepsilon = 0.001 * (b - a)_{\max} = 0.002 * Q_{\max} \tag{6}$$

It should be noted that equation (6) is an empirical formula based on a large number of test results from different cases. The term $(b - a)_{\max}$ represents the maximum of the difference of vorticity squared and deformation squared, and is easy to obtain as a fixed number at each time step in a certain case. The adjustment of ε in any case in vortex visualization is therefore not necessary after ε is determined in equation (6). Unlike solid body which can have rotation without any deformation, vortex in fluid flow is always a mixture of vorticity and deformation. It should be noted that there are still difficulties to give a rigorous definition for "vortex", thus, the Omega criterion just shows the vorticity overtakes the deformation ($\Omega > 0.5$) when the "vortex" is formed. Therefore, the Ω can be set as 0.52 to capture the vortices properly in all the cases at different time steps.

APPLICATION OF OMEGA METHOD WITH DETERMINED EPSILON

In this section, a number of computational examples from several research groups are presented and analyzed. Firstly, the universality of the above Epsilon determination is verified. Different Epsilon levels are tested at different time steps. Then, the advantages of the new Ω method are compared with other vortex identification methods.

Case 1: Boundary Layer Transition

A boundary layer transition with $Ma = 0.5$ on a flat plate based on high order direct numerical simulation (DNS) is applied as Case 1 to test the Epsilon function in Omega method as well as the advantages of iso-surface of $\Omega = 0.52$ compared with Q creterion. The grid system is $n_{streamwise} \times n_{spanwise} \times n_{normal} = 1920 \times 128 \times 241$. Firstly we give a comparison of different ε levels. Fig. (**1**) shows the vortex structures in boundary layer transition at $t = 8.2T$ captured by iso-surfaces of $\Omega = 0.52$ with (a) $\varepsilon = 1.28 \times 10^{-4}$, (b) $\varepsilon = 1.28 \times 10^{-3}$ and (c) $\varepsilon = 1.28 \times 10^{-2}$, where $\varepsilon = 1.28 \times 10^{-3}$ in Fig. (**1b**) is based on the new Epsilon determination in equation (6), and T is the period of T-S wave in Case1. As is clearly seen, with a smaller ε, there are too many clouds (noises) in Fig. (**1a**) which is marked in a red square; however, some significant vortex structures cannot be shown with a larger ε. For instance, in the downstream region marked in a red square in Fig. (**1c**), the vortex rings disappear compared with the one in Fig. (**1b**). Therefore, for this case, the ε based on the Epsilon function is proper to be used in vortex identification given by $\Omega = 0.52$.

Then, the Epsilon function is also tested in different time steps of Case 1. Fig. (**2**) shows the iso-surfaces of $\Omega = 0.52$ with determined ε based on equation (6) at different time steps. The newly determined Epsilon in each time step is listed in Table **1** accordingly. As is shown, the typical vortex structures are well captured at different time steps without noises appearing in the space. Note that we do not need to adjust ε to a proper value for different time steps. ε can be calculated through the term of $(b - a)_{\max}$.

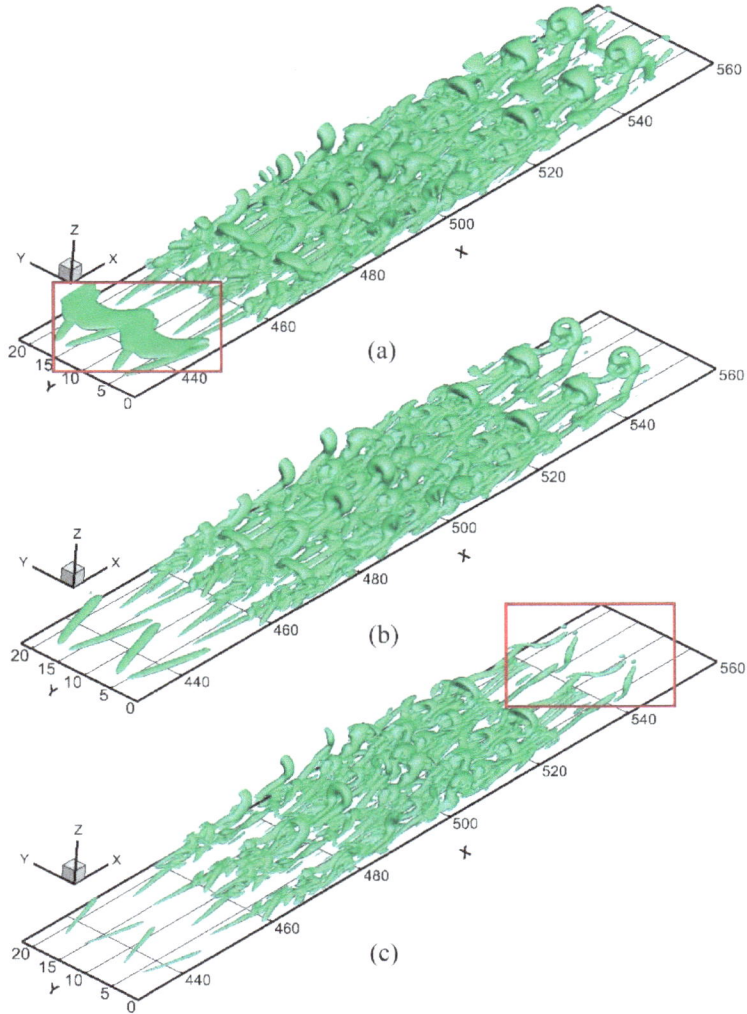

Fig. (1). Boundary layer transition vortex structures at $t = 8.2T$ captured by iso-surfaces of $\Omega = 0.52$ with different ε levels (the threshold of Ω is from 0 to 1 for the whole flow field) (a) $\varepsilon = 1.28 \times 10^{-4}$ (b) $\varepsilon = 1.28 \times 10^{-3}$ (c) $\varepsilon = 1.28 \times 10^{-2}$.

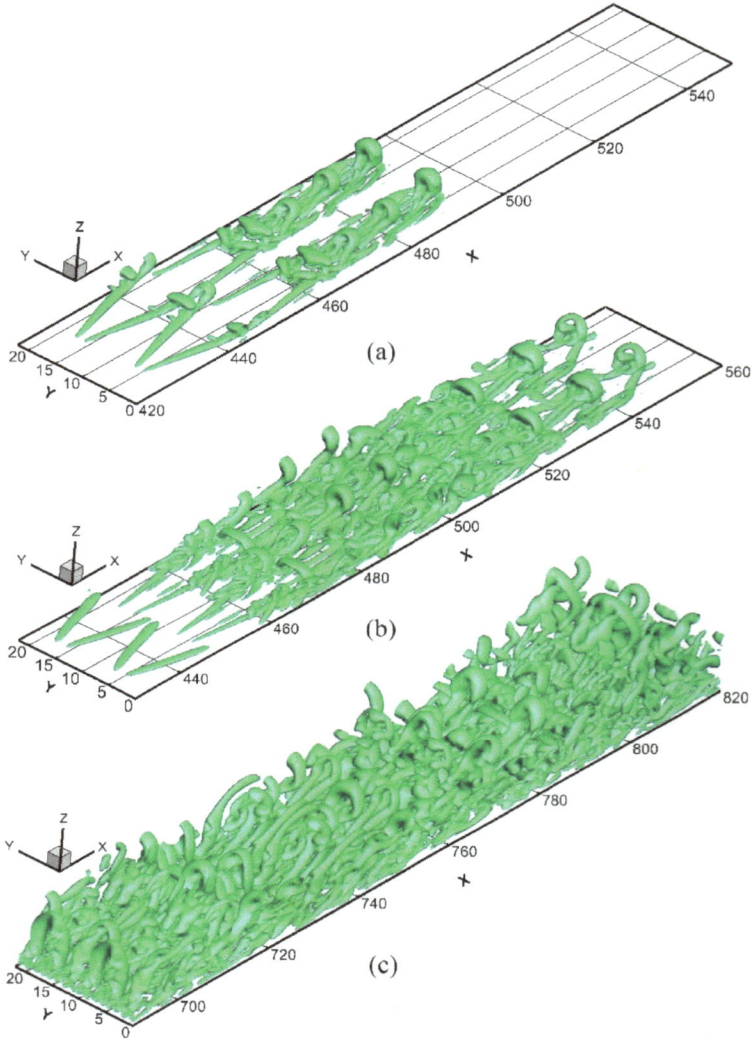

Fig. (2). Boundary layer transition vortex structures captured by iso-surface of $\Omega = 0.52$ at different times (a) $t = 7.0T$ (b) $t = 8.2T$ (c) $t = 18.0T$.

Table 1. Epsilon determination at different time steps in Case 1.

Time step	$t = 7.0T$	$t = 8.2T$	$t = 18.0T$
$(b-a)_{max}$	0.62	1.28	0.79
ε	0.62×10^{-3}	1.28×10^{-3}	0.79×10^{-3}

On the other hand, the advantages of the Ω method are introduced in the following. Fig. (**3**) shows the vortex structures in boundary layer transition at $t = 8.2\text{T}$ captured by the iso-surfaces of $Q = 0.001$, $Q = 0.003$ and $Q = 0.006$. The threshold of Q is from -0.8 to 0.6 for the whole flow field. Comparing Figs. (**3a, b, c**), it can be observed that a smaller magnitude Q can capture weaker vortices including the first vortex ring in downstream which is squared in red in Fig. (**3a**). However, the "clouds" above Λvortex, which smears the main structures, is also observed by this smaller magnitude Q. While, a larger magnitude threshold can lead to clear representation of the strong vortices while the weak vortices are all skipped. As can be seen in red square in Fig. (**3c**), compared with Figs. (**3a, b**), the iso-surface of $Q = 0.006$ is not able to capture the first ring in downstream and cause faked "vortex breakdown" since weak vortices may be missed by larger Q. Thus, it is quite hard to choose a proper threshold for iso-surface visualization of Q criterion, on the purpose of capturing the clear main structures without losing too much small vortices. Compared with Figs. (**3a-c**), Fig. (**4**) gives the vortex structures captured by the iso-surfaces of $\Omega = 0.52$, $\Omega = 0.53$ and $\Omega = 0.54$ ($\Omega \in [0, 1]$ for the whole flow field, ε is 1.28×10^{-3} at $t = 8.2\text{T}$). It is seen that the main vortical structures including Λ-vortex in upstream and the first ring in downstream are well captured by these three Ω levels, and are not very sensitive with the change of Ω. This indicates that Ω is quite robust with no obvious change in vortex visualization. The reason is that Ω is a ratio while Q is related to the vortex strength. Therefore, the Ω can represent vortices with much larger vortex strength latitude.

Furthermore, it can also be observed that Fig. (**4a**) shows the same first vortex rings of downstream as the ones in Fig. (**3a**), however, the Λ-vortex in upstream with weaker strength is lightly smeared by some non-physical vortex structures (clouds) in Fig. (**3a**). On the other hand, compared with Fig. (**4b**), the vortex rings downstream of the flow especially their heads cannot be observed in Fig. (**3b**), although same Λ-vortex structures can be captured. Thus, Ω has a capability of well capturing both strong and weak vortex structures.

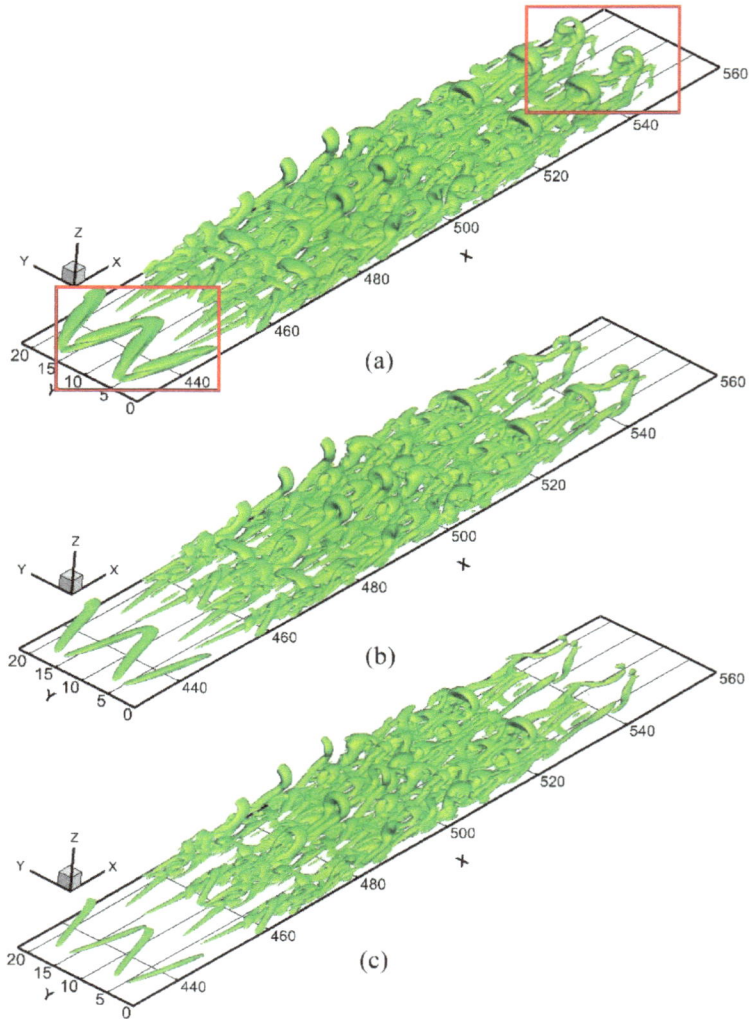

Fig. (3). Boundary layer transition vortex structures at $t = 8.2T$ by different Q iso-surfaces (the threshold of Q is from -0.8 to 0.6 for the whole flow field) (a) $Q = 0.001$ (b) $Q = 0.003$ (c) $Q = 0.006$.

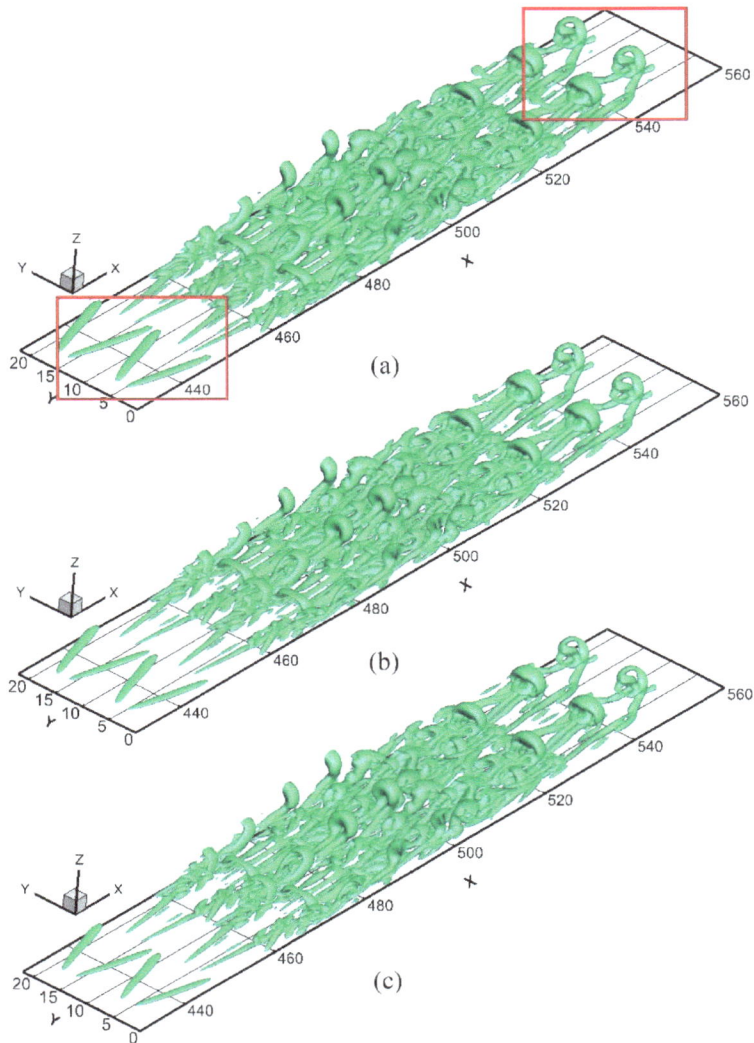

Fig. (4). Boundary layer transition vortex structures at $t = 8.2T$ by different Ω iso-surfaces with $\varepsilon = 1.28 \times 10^{-3}$ (the threshold of Ω is from 0 to 1 for the whole flow field) (a) $\Omega = 0.52$ (b) $\Omega = 0.53$ (c) $\Omega = 0.54$.

According to the review of vortex identification methods by Epps [12], Ω method can be considered to be a dimensionless criterion related to Q, and is similar to the kinematic vorticity number [13] as a measure of the rotational quality of the vorticity. The review also points out that $\Omega = 0.5$ is equivalent to $Q > 0$, which is partly inaccurate. Actually Ω and Q have essential differences.

On the one hand, Q is a special situation of Ω when $\Omega = 0.5$ and $\varepsilon = Q_{th}$, where Q_{th} is a certain number picked from a wildly changing thresholds by experience. For instance, if we pick $\Omega = b/(a + b + Q_{th}) = 0.5$, then we can obtain $b - a = Q_{th}$, thus, both $\Omega = 0.5$ and $Q = Q_{th}$ will give same results. Q_{th} is a threshold to adjust by experience in Q criterion, while ε in Ω method is fixed as $0.001 * (b - a)_{max}$ to remove the noises. It is easy to obtain the value of Ω from a given Q_{th}, but not the other way around. For example, if we pick $\Omega = 0.6$, we can obtain $Q = 1/2(b - a) = 1/2\varepsilon + 1/6b$, where b is the local vorticity square norm. It is in general impossible to find an equivalent Q_{th} corresponding to $\Omega = 0.6$, since Q_{th} has to be adjusted at every grid point.

On the other hand, Ω method can capture both strong and weak vortex structures simultaneously by setting $\Omega = 0.52$ but Q criterion may not. Ω denotes a ratio ranging from 0 to 1, which represents the rotational quality of the vorticity. For instance, for weak vortices, Q or $1/2(b - a)$ may be very small. The weak vortices must be lost when choosing a larger Q_{th} if clearly strong vortices are shown in a case. However, no matter how weak these vortices are, they are always well shown in the region where the vorticity overtakes the deformation. Otherwise, if we reduce Q_{th} to show the weak vortices, the strong vortices may be smeared and cannot be visualized clearly.

Case 2: SWBLIs Controlled by MVG

The SWBLI (shock wave and boundary layer interaction) in a supersonic ramp flow with MVG control at $Ma = 2.5$, which is simulated by the implicitly implemented LES (ILES) method, is selected as Case 2. The body-fitted grid system of the computational domain is $n_{streamwise} \times n_{normal} \times n_{spanwise} = 1600 \times 192 \times 137$. Similarly, Fig. (5) shows the iso-surfaces of $\Omega = 0.52$ with the determined ε at different times. The term $(b - a)_{max}$ and the ε at each time step are listed in Table 2, respectively. At different time steps, it can be observed that the strong vortex structures like vortex rings are well captured without any clouds in the upper space of the vortex rings.

Table 2. Epsilon determination at different time steps in Case 2.

Time step	$t = 1284T^*$	$t = 1984T^*$
$(b - a)_{max}$	7848	5368
ε	7.848	5.368

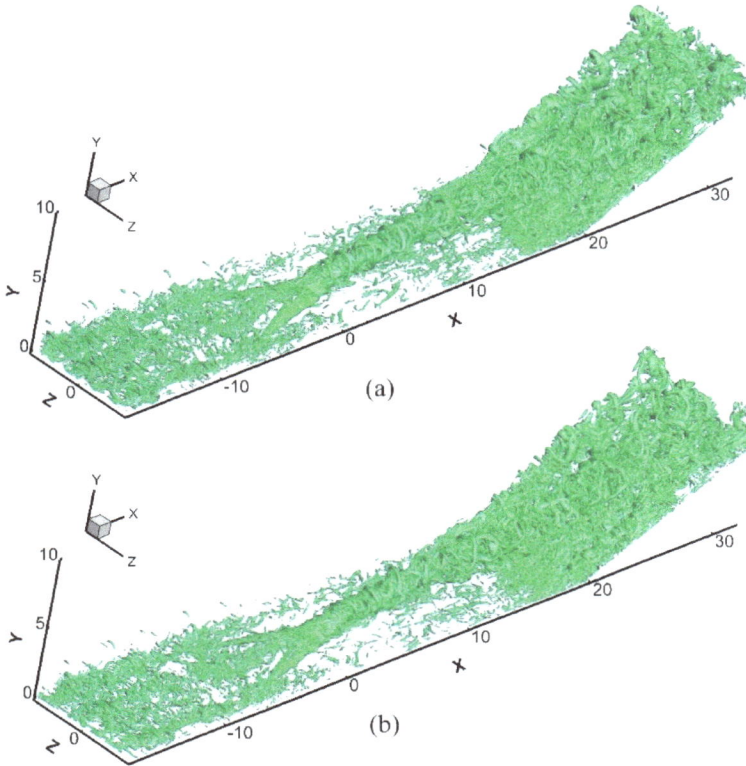

Fig. (5). Vortex structures behind MVG captured by the iso-surface of $\Omega = 0.52$ at different times (a) $t = 1284T^*$ (b) $t = 1984T^*$.

Fig. (**6**) gives the iso-surfaces of Ω and Q with different levels to show the vortex structures behind MVG at $t = 1284T^*$, where T* is the characteristic time in Case 2. The range of Ω and Q over the whole flow field in this case are $[0, 1]$ and $[-1595.49,\ 1520.79]$, respectively. Thus, the iso-surfaces of Ω and Q are chosen as $\Omega = 0.51$, $\Omega = 0.52$, $\Omega = 0.54$ and $Q = 1.0$, $Q = 1.5$ and $Q = 8.0$ for comparison. In Figs. (**6b, e**), the iso-surfaces of $\Omega = 0.52$ and $Q = 1.5$ present a similar chain of large-scale vortex rings behind MVG. However, some non-physical vortex structures (marked by red square in Figs. (**6d-f**), which may be caused by the trigonal geometry and the complex grid of MVG, occur when using the Q criterion; these non-physical vortex structures still exist even a high threshold is applied in Fig (**6f**). In contrast, the Ω iso-surfaces in Figs. (**6a-c**) do not have this problem. In addition, different levels of the Ω value are compared by Figs. (**6a-c**). As is shown, Ω is quite robust with no obvious change in vortex visualization when Ω goes from 0.51 to 0.54 ($\Omega \in [0, 1]$ and $\Omega > 0.5$ for the vortex region).

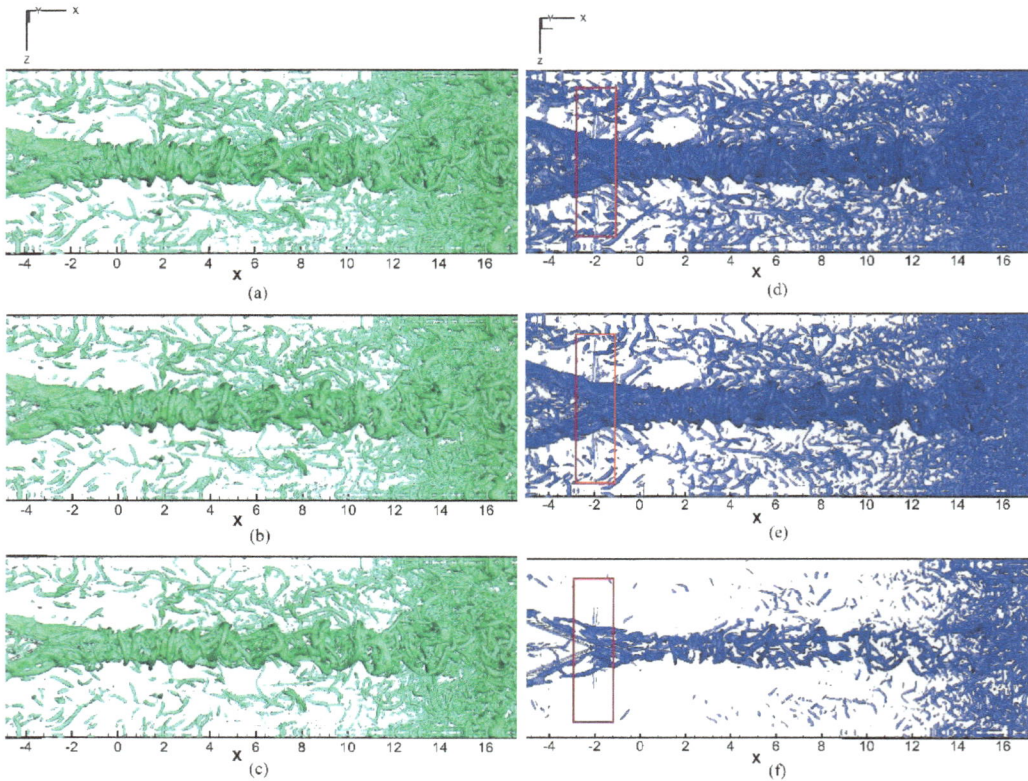

Fig. (6). Vortex structures behind MVG at $t = 1284T^*$ captured by the iso-surfaces of Ω and Q (a) $\Omega = 0.51$ (b) $\Omega = 0.52$ (c) $\Omega = 0.54$ (d) $Q = 1.0$ (e) $Q = 1.5$ (f) $Q = 8.0$ (the threshold of Q is from -1595.5 to 1520.8 for the whole flow field).

Case 3: Channel Flow with $\mathrm{Re}_\tau = 950$

The new Ω method is tested in a turbulent channel flow with $\mathrm{Re}_\tau = 950$ (Case 3), and the comparison with Q criteria is also shown here. The computational domain is $2\pi \times \pi \times 2$ with $768 \times 768 \times 385$ grid points in the streamwise, spanwise and normal direction, respectively. Half domain in the normal direction is used due to the symmetry condition. The comparison between Ω and Q criteria on capturing the vortex structures in this channel flow is given by (Fig. 7). Fig. (**7a**) shows the iso-surface of $\Omega = 0.52$, with $\varepsilon = 2.97$. Figs. (**7b, c**) give a zoom view of the channel flow captured by $\Omega = 0.6$ and $Q = 5$. As is seen, the same strong vortex structures captured by Ω and Q display in the blue circle. However, it is found that some weak vortices which are circled in red are 'broken down' or even lost by using Q. Thus, both strong and weak vortices can be well captured by iso-surface of $\Omega = 0.52$.

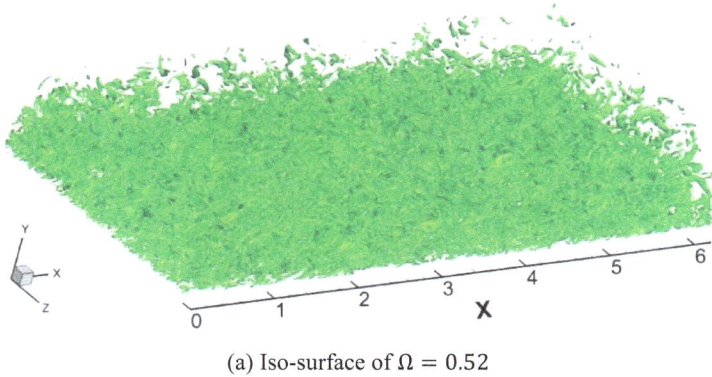

(a) Iso-surface of $\Omega = 0.52$

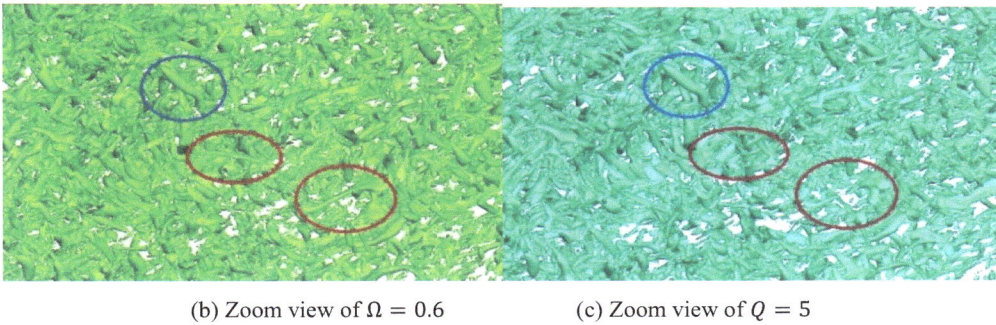

(b) Zoom view of $\Omega = 0.6$ (c) Zoom view of $Q = 5$

Fig. (7). Different iso-surfaces in Case 3.

CONCLUSIONS

A new Omega vortex identification method with ε determination is introduced to represent the ratio of vorticity square over the sum of vorticity square and deformation square. The new Ω has the following advantages:

(1) The newly determined ε in the denominator of Ω is easy to perform, since ε does not need to be adjusted from an arbitrary threshold but can be calculated as a proper and fixed number depending on the term $(b - a)_{\mathrm{max}}$ or Q_{max}. After the determination of ε, a proper iso-surface boundary of vortex structures can be well captured by $\Omega = 0.52$ for each case and each time step.

(2) It is difficult to obtain the accurate surface of vortex structures by Q, λ_2 or other criteria since they have widely varying thresholds. An improper Q and λ_2 threshold may lead to a situation that strong vortex can be captured while weak vortices are lost or weak vortices are captured but strong vortices are smeared. However, Ω is a ratio and has a clear physical meaning. It indicates a region where vorticity

overtakes deformation in a flow field. Ω can always be set as 0.52 for capturing both strong and weak vortices simultaneously.

(3) Ω method is quite robust without obvious change in vortex visualization.

CONSENT FOR PUBLICATION

Not applicable.

CONFLICT OF INTEREST

The author(s) confirm that this chapter contents have no conflict of interest.

ACKNOWLEDGEMENTS

This work was supported by Department of Mathematics at University of Texas at Arlington. The authors are grateful to Texas Advanced Computing Center (TACC) for the computation hours provided. This work was accomplished by using Code DNSUTA released by Dr. Chaoqun Liu at University of Texas at Arlington in 2009.

REFERENCES

[1]   S K Robinson, "The kinematics of turbulent boundary layer structure", *NASA STI/Recon Technical Report N,* p. NASA TM-103859, 1991.

[2]   M.S. Chong, A.E. Perry, and B.J. Cantwell, "A general classification of three- dimensional flow fields", *Phys. Fluids A Fluid Dyn.,* vol. 2, no. 5, pp. 765-777, 1990. [http://dx.doi.org/10.1063/1.857730]

[3]   J. Zhou, R.J. Adrian, and S. Balachandar, "Mechanisms for generating coherent packets of hairpin vortices in channel flow", *J. Fluid Mech.,* vol. 387, pp. 353-396, 1999. [http://dx.doi.org/10.1017/S002211209900467X]

[4]   J.C.R. Hunt, A.A. Wray, and P. Moin, "Eddies, streams, and convergence zones in turbulent flows", *Center for Turbulence Research: Proceedings of the Summer Program,* vol. N89-24555, 1988

[5]   J. Jeong, and F. Hussain, "On the identification of a vortex", *J. Fluid Mech.,* vol. 285, pp. 69-94, 1995. [http://dx.doi.org/10.1017/S0022112095000462]

[6]   Y. Zhang, K. Liu, and H. Xian, "A review of methods for vortex identification in hydroturbines", *Renew. Sustain. Energy Rev.,* vol. 81, pp. 1269-1285, 2018. [http://dx.doi.org/10.1016/j.rser.2017.05.058]

[7]   C.Q. Liu, Y.Q. Wang, and Y. Yang, "New omega vortex identification method", *Sci. China Phys. Mech. Astron.,* vol. 59, no. 8, 2016.684711 [http://dx.doi.org/10.1007/s11433-016-0022-6]

[8]   Y. Tao, K. Inthavong, and P. Petersen, "Numerical simulation and experimental verification of wake flows induced by moving manikins", *4ᵗʰ International Conference on Building Energy, Environment,* 2018 Melbourne, Australia

[9]   X. Dong, Y. Wang, and X. Chen, "Determination of epsilon for omega vortex identification method", *J. Hydrodynam.,* vol. 30, no. 4, pp. 541-548, 2018. [http://dx.doi.org/10.1007/s42241-018-0066-x]

[10]   E. Abdel-Raouf, M A R. Sharif, and J. Baker, "Impulsively started, steady and pulsated annular inflows", *Fluid Dyn. Res.,* vol. 49, no. 2, 2017.025511. [http://dx.doi.org/10.1088/1873-7005/aa5add]

[11]   J. Li, G. Dong, and H. Zhang, "A new kind of hairpin-like vortical structure induced by cross- interaction of sinuous streaks in turbulent channel", *Chin. Phys. B,* vol. 27, no. 8, 2018.084701 [http://dx.doi.org/10.1088/1674-1056/27/8/084701]

[12]   B. Epps, "Review of vortex identification methods", *55ᵗʰ AIAA Aerospace Sciences Meeting,* 2017p. 0989

[13]   C. Truesdell, "Two measures of vorticity", *Journal of Rational Mechanics and Analysis,* vol. 2, pp. 173-217, 1953.

Stability Analysis on Shear Flow and Vortices in Late Boundary Layer Transition

Jie Tang[*]

Department of Mathematics, University of Texas at Arlington, Arlington, Texas 76019, USA

Abstract： Turbulence is still an unsolved scientific problem, which has been regarded as "the most important unsolved problem of classical physics". Liu proposed a new mechanism about turbulence generation and sustenance after decades of research on turbulence and transition. One of them is the transitional flow instability. Liu believes that inside the flow field, shear (dominant in laminar) is unstable while rotation (dominant in turbulence) is relatively stable. This inherent property of flow creates the trend that non-rotational vorticity must transfer to rotational vorticity and causes the flow transition. To verify this new idea, this chapter analyzed the linear stability on two-dimensional shear flow and quasi-rotational flow. Chebyshev collocation spectral method is applied to solve Orr–Sommerfeld equation. Several typical parallel shear flows are tested as the basic-state flows in the equation. The instability of shear flow is demonstrated by the existence of positive eigenvalues associated with disturbance modes (eigenfunctions), *i.e.* the growth of these linear modes. Quasi-rotation flow is considered under cylindrical coordinates. An eigenvalue perturbation equation is derived to study the stability problem with symmetric flows. Shifted Chebyshev polynomial with Gauss collocation points is used to solve the equation. To investigate the stability of vortices in flow transition, a ring-like vortex and a leg-like vortex over time from our Direct Numerical Simulation (DNS) data are tracked. The result shows that, with the development over time, both ring-like vortex and leg-like vortex become more stable as Omega becomes close to 1.

Keywords: Shear flow, Stability analysis, Transition, Turbulence, Vortices.

INTRODUCTION

A Short History Review of Research on Flow Transition and Turbulence Generation

In fluid flow, the process of a laminar flow becoming turbulent is a fundamental scientific phenomenon, known as laminar-turbulent transition. Laminar flow describes the fluid flows in parallel layers, with no disruption between the layers [1]. Turbulent flow is characterized by eddies or small packets of fluid particles

*Corresponding author Jie Tang: Department of Mathematics, University of Texas at Arlington, Arlington, Texas 76019, USA; Tel: 8179081296; E-mail: jietanguta@gmail.com

which result in lateral mixing [2]. Laminar-turbulent transition is an extraordinarily complicated process which at present is still far from fully understood. Nevertheless, as the result of many decades of intensive research, classical comprehensive theories of physical mechanisms of the transition phenomenon have been proposed [3-5].

Boundary layer is a very important concept in transition theory. It is a thin layer of viscous fluid close to the solid surface of a wall in contact with a moving stream [6]. The flow velocity varies from zero at the wall up to approximate free stream velocity at the boundary. The fundamental concept of the boundary layer was suggested by L. Prandtl [7] in 1904. Modern research on fluid transition is most often studied in the context of boundary layers due to their ubiquity in real flows and their importance in many fluid-dynamic processes [8].

In a thin boundary layer, the velocity gradient is significant, and consequently the viscous shear stresses defined by is large, where μ is the dynamic viscosity, $u = u(y)$ describes the profile of the boundary layer longitudinal velocity component, y is the normal-to-wall direction. In other words, in a thin boundary layer, laminar flow is dominant with shear layers.

$$\tau = \mu \frac{du}{dy}$$

(1.1)

Computation of the boundary layer parameters is based on the solution of equations obtained from the Navier–Stokes equations for viscous fluid motion. Navier-Stokes equations describe the conservation of mass, momentum, and energy.

For boundary-layer flows, two main classes of transition are known [9-11] depending on the character of environmental disturbances. The first of them is usually observed when environmental disturbances are rather small. It is regarded as natural transition and has fundamental and practical importance in problems involving moving vehicles in air and water. The second class of transition, usually called bypass, is observed when high enough levels of environmental perturbations are present.

Classical theory on natural transition can be described by four stages: receptivity, linear instability, non-linear growth and vortex breakdown as shown in Fig. (**1-1**) [5].

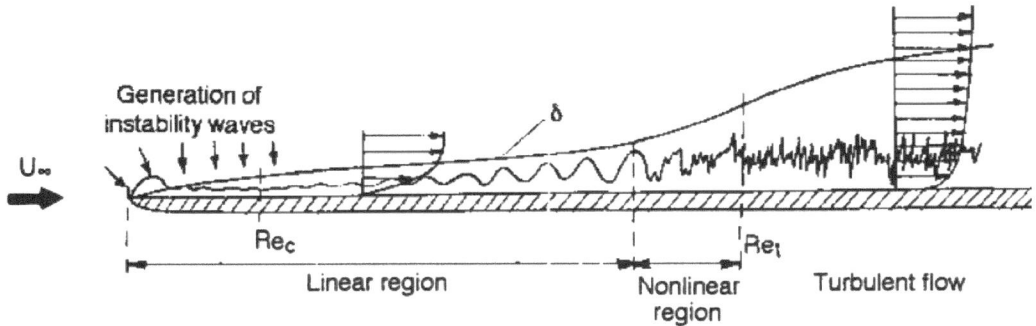

Fig. (1-1). Qualitative sketch of the process of turbulence onset in a boundary layer. δ is the thickness of the boundary layer, Re represents the Reynolds number and U_∞ is the income free stream [5].

The initial stage of the natural transition process is known as the receptivity phase and consists of the transformation of environmental disturbances into small perturbations (*i.e.* instability waves, usually called Tollmien-Schlichting waves) within the boundary layer. This aspect of the transition process was clearly formulated for the first time by Morkovin [10] in 1968. Many experimental and theoretical work of this process appeared in the 1970s [11-16]. Details of the subsequent rapid development of investigations on receptivity can be found in a number of books and review papers [17-21].

The second stage of transition corresponds to the linearly propagation of small-amplitude instability waves in the boundary. This stage is described by linear hydrodynamic stability theory, also called linear stability theory. Tollmien [22] started the research on linear stability theory in 1929. In the following century, it becomes the most developed branch of the transition problem with a lot of research achievements for two-dimensional and three-dimensional flows. For example, Schlichting [23], Lin [24], Herbert [25] and many others.

When the growth of linear instability waves reaches considerable values, the flow enters a phase of three-dimensional nonlinear growth, then the turbulent flow formed (so-called vortex breakdown). They are the last two stages. Although the region of nonlinear growth has been studied for more than half century, there are still many questions unanswered [26-31]. For example, the mechanism of vortices generation and deformation, the formation of turbulence and turbulence coherent structure.

Liu's New Theory on Boundary Layer Transition

Liu proposed a new comprehensive mechanism about turbulence generation and sustenance in a boundary layer [32] after decades of research on turbulence and transition [33-49]. Many new observations are made and new mechanisms are revealed in late boundary layer transition [32] including:

(1) Mechanism of spanwise vorticity rollup.
(2) Mechanism of transfer from flow shear to rotation.
(3) Mechanism of spanwise vortex tube formation and role of the linear unstable modes.
(4) Mechanism of K -vortex root formation.
(5) Mechanism of first ring-like vortex formation.
(6) Mechanism of multiple vortex ring formation.
(7) Mechanism of second sweep formation.
(8) Mechanism of high shear layer formation.
(9) Mechanism of positive spike formation.
(10) Mechanism of secondary and tertiary vortex formation.
(11) Mechanism of U-shaped vortex formation.
(12) Mechanism of small length vortices generation.
(13) Mechanism of multiple level high shear layer formation.
(14) Mechanism of energy transfer paths from the large length scale to the small ones.
(15) Mechanism of symmetry loss and flow chaos.
(16) Mechanism of thickening of turbulence boundary layer.
(17) Mechanism of high surface friction of turbulent flow.

This chapter focuses on Liu's second proposal: Mechanism of transfer from flow shear to rotation. More precisely, the mechanism of transfer from non-rotational vorticity to rotational vorticity.

In boundary layer, laminar flow is dominant by shear because of large velocity gradient. Turbulent flow in general consisted of eddies or small packets of fluid particles, is dominant by rotation. In classical stability theory, laminar is regarded as a stable state while turbulence is an unstable state with disorder, chaotic and random fluid layers. However, Dr. Liu has an opposite opinion, he believes that "Shear layer Instability" is the "mother of turbulence", and rotation is more stable than shear when the Reynolds number is large enough. This inherent property of flow creates the trend that non-rotational vorticity must transfer to rotational vorticity, and causes the occurrence of transition.

Liu also pointed out the very commonly confusion of vorticity and vortex in fluid dynamics in his paper [50,51]. Vorticity has rigorous mathematical definition (curl of velocity), but no clear physical meaning. On the other hand, vortex has clear physical meaning (rotation) but no rigorous mathematical definition. For a long time, many researchers and textbooks treat them as a same thing. Liu [50] gave the detailed DNS observations on the difference between vorticity and vortex, including:

(1) Vorticity tube is not vortex
(2) Vortex is not the congregation of vorticity.
(3) Vortex is never attached to the wall.

The classical description of "vortex breakdown" is also not accurate since rotation cannot break to pieces.

Based on Liu's discovery, some current stability analysis of vortices base on velocity-vorticity equation and vorticity profiles of base flow are questionable.

The Purpose of the Current Chapter

To verify Liu's new idea, this paper first uses linear stability theory and Orr-Sommerfeld equation to analyze the stability of shear. Second, quasi-rotation velocity profiles extracted from lambda vortex of our DNS case are used to analyze the stability of rotation on boundary layer transition.

Organization of this Chapter

This chapter is organized as flows. Section I is the Introduction. The numerical method, Chebyshev spectral method used in this book chapter is introduced in section 2. Section 3 studies the instability of 2D shear flow. Section 4 derives an eigenvalue perturbation equation with rotation flow under cylindrical coordinates. Finally, section 5 shows the numerical results of stability analysis on ring- and leg-like vortices.

CHEBYSHEV SPECTRAL METHODS

Introduction to Spectral Methods

Spectral methods are an important development of the class of discretization schemes for differential equations, known generically as the method of weighted residuals (MWR) [52]. The key elements of MWR are the trial functions (also called the expansion or approximating functions) and the test functions (also known as weight functions).

The general scheme of MWR is first to get a truncated series expansion of the solution by the trial functions. Then the residual can be produced by using the truncated expansion in the differential equation. Last, a desired truncated series expansion is achieved by minimizing the residual with respect to a suitable norm, defined as an orthogonality condition with respect to each of the test functions.

The choice of trial functions is one of the features which distinguishes spectral methods from finite-element and finite-difference method. The trial functions for spectral methods are infinitely differentiable global functions, while the trial functions for finite-element method or finite-difference method are specified locally in each element or cell. The most frequently used trial functions are trigonometric polynomials, Chebyshev polynomials and Legendre polynomials.

The choice of test functions leads to three most commonly used spectral methods schemes: Galerkin, collocation, and tau. In the Galerkin approach, the test functions are the same as the trial functions. It requires that the integral of the residual times each test function should be zero. In the collocation approach the test functions are translated Dirac Delta functions centered at special points, namely collocation points. This approach requires the residual to be zero at the collocation points. The tau approach is similar to the Galerkin approach with a supplementary set of equations used to apply the boundary conditions.

Chebyshev Polynomials of First Kind

The Fourier method is the most commonly used spectral method with trigonometric polynomials as trial functions. It is appropriate for periodic problems but is not adapted to non-periodic problems because of the existence of the Gibbs phenomenon at the boundaries. In this book chapter, the cases are all non-periodic problems, so better-suited trial functions like Chebyshev polynomials constitute a proper alternative to the Fourier method. Especially in a bounded domain, the use of Chebyshev polynomials has been advantageous. The stability calculations shown in this book chapter have been obtained by Chebyshev discretization of the Cartesian and cylindrical coordinates.

Let us first consider the definition and some properties of the Chebyshev polynomials of the first kind [53].

Definition (Chebyshev polynomial of the first kind $T_n(x)$). The Chebyshev polynomial of the first kind of order n is defined as follows:

$$T_n(x) = \cos[n\cos^{-1}(x)], \qquad x \in [-1,1], \qquad n = 0,1,2,\cdots \qquad (2.1)$$

From this definition, the following property is evident by setting $x = \cos\theta$:

$$T_n(\cos\theta) = \cos(n\theta), \qquad \theta \in [0,\pi], \qquad n = 0,1,2,\cdots \qquad (2.2)$$

Properties of the Chebyshev polynomials $T_n(x)$

The polynomials $T_n(x), n \geq 1$, satisfy the following properties, which follow straightforwardly from (2.1):

The Chebyshev polynomials $T_n(x)$ satisfy the following three-term recurrence relations:

$$T_{n+1}(x) = 2xT_n(x) - T_{n-1}(x), \qquad n = 1,2,3,\cdots, \qquad (2.3)$$

with starting values $T_0(x) = 1, T_1(x) = x$.

Explicit expressions for the first seven Chebyshev polynomials are

$$T_o(x) = 1,$$
$$T_1(x) = x,$$
$$T_2(x) = 2x^2 - 1,$$
$$T_3(x) = 4x^3 - 3x,$$
$$T_4(x) = 8x^4 - 8x^2 + 1,$$
$$T_5(x) = 16x^5 - 20x^3 + 5x,$$
$$T_6(x) = 32x^6 - 48x^4 + 18x - 1. \qquad (2.4)$$

The graphs of these Chebyshev polynomials are plotted in Fig. (**2-1**).

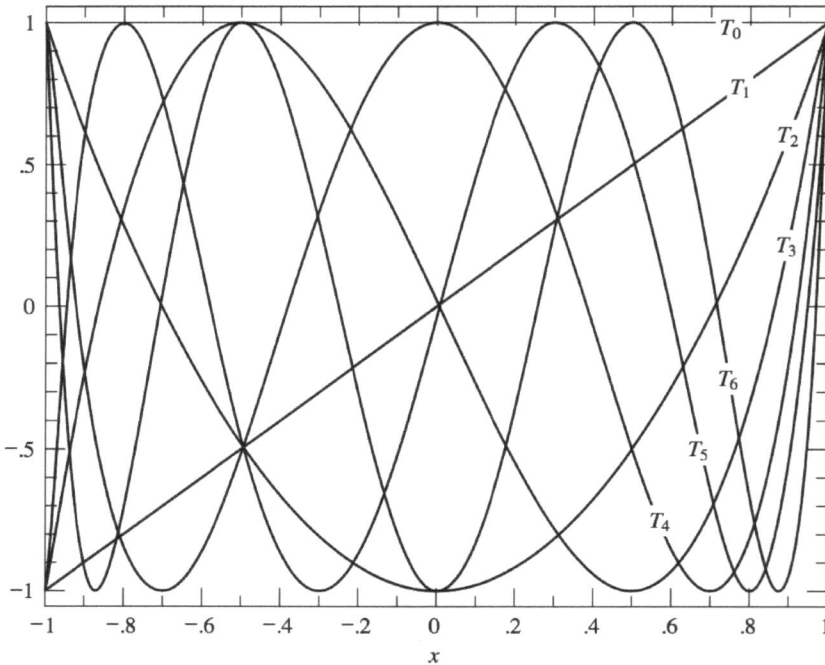

Fig. (2-1). Chebyshev polynomials $T_0(x)$ through $T_6(x)$ (from William H. Press *et al.* "Numerical Recipes in C", https://www.cec.uchile.cl/cinetica/pcordero/MC_libros/NumericalRecipesinC.pdf, pp.191).

The leading coefficient (of x^n) in $T_n(x)$ is 2^{n-1} and $T_n(-x) = (-1)^n T_n(x)$.

$T_n(x)$ has n zeros which lie in the interval $(-1,1)$. They are given by

$$x_k = \cos\left(\frac{2k+1}{2n}\pi\right), \qquad k = 0,1,\cdots,n-1. \qquad (2.5)$$

These points are called Chebyshev nodes or Gauss points.

$T_n(x)$ has $n+1$ extrema in the interval $[-1,1]$ and they are given by

$$x_k' = \cos\frac{k\pi}{n}, \qquad k = 0,1,\cdots,n. \qquad (2.6)$$

At these points, the values of the polynomials are $T_n(x_k') = (-1)^k$. They are called Gauss-Lobatto points.

The differentiation of T_n gives

$$T_n'(x) = \frac{d}{d\theta}(cosn\theta)\frac{d\theta}{dx} = n\frac{sink\theta}{sin\theta}, \qquad n = 0,1,2,\cdots. \qquad (2.7)$$

By the application of trigonometrical formulas, the recurrence relation on the derivative is:

$$\frac{T_{n+1}'(x)}{n+1} - \frac{T_{n-1}'(x)}{n-1} = 2T_n(x), \qquad n = 1,2,3,\cdots, \qquad (2.8)$$

with $T_0'(x) = 0, T_1'(x) = 1$.

Orthogonality relation

$$\int_{-1}^{1} T_r(x)T_s(x)(1-x^2)^{-\frac{1}{2}}dx = N_r\delta_{rs}, \qquad (2.9)$$

with $N_0 = \pi$ and $N_r = \frac{1}{2}\pi$ if $r \neq 0$.

This property means that the set of Chebyshev polynomials $\{T_n(x)\}$ is an orthogonal set with respect to the weight function $w(x) = (1-x^2)^{-\frac{1}{2}}$ in the interval (-1,1).

Discrete orthogonality relation

With the zeros of $T_{n+1}(x)$ as nodes (Chebyshev nodes): Let $n > 0$, $r,s \leq n$, and $x_j = \cos\left(\frac{\left(j+\frac{1}{2}\right)\pi}{n+1}\right)$. Then,

$$\sum_{j=0}^{n} T_r(x_j)T_s(x_j) = K_r\delta_{rs}, \qquad (2.10)$$

where $K_0 = n+1$ and $K_r = \frac{1}{2}(n+1)$ when $1 \leq r \leq n$.

With the extrema of $T_n(x)$ as nodes (Gauss-Lobatto points): Let $n > 0$, $r,s \leq n$, and $x_j = \cos\left(\frac{j\pi}{n}\right)$. Then,

$$\sum_{j=0}^{n} {''} T_r(x_j) T_s(x_j) = K_r \delta_{rs},$$ (2.11)

where $K_0 = K_n = n$ and $K_r = \frac{1}{2}n$ when $1 \le r \le n - 1$.

The double prime indicates that the terms with suffixes $j = 0$ and $j = n$ are to be halved.

Chebyshev Collocation Approach

Chebyshev collocation (*i.e.* interpolation) is a useful technique to approximate a given function. The commonly used collocation points are the Gauss-Lobatto points. The advantage of Gauss-Lobatto points is that both the boundary points are included.

Consider the Chebyshev approximation of the function $u(x)$ defined for $x \in [-1,1]$:

$$u_N(x) = \sum_{n=0}^{N} a_n T_n(x).$$ (2.12)

The technique consists of setting to zero the residual $R_N = u - u_N$ at the collocation points $x_i = \frac{cos\pi i}{N}, i = 0, \cdots, N$, let

$$u(x_i) = u_N(x_i) = \sum_{n=0}^{N} a_n T_n(x_i), \qquad i = 0 \cdots, N.$$ (2.13)

When discretizing ordinary or partial differential equations, derivatives of the solution are need as well. These derivatives must be expressed in terms of Chebyshev polynomials and the following recurrence relation between Chebyshev polynomials are their derivatives is used.

$$T_0^{(k)}(x_i) = 0,$$
$$T_1^{(k)}(x_i) = T_0^{(k-1)}(x_i),$$
$$T_2^{(k)}(x_i) = 4T_1^{(k-1)}(x_i),$$

$$T_n^{(k)}(x_i) = 2nT_{n-1}^{(k-1)}(x_i) + \frac{n}{n-2}T_{n-2}^{(k)}(x_i), \qquad n = 3,4,\cdots. \qquad (2.14)$$

with the superscript $k \geq 1$ denoting the order of differentiation.

Convergence of Chebyshev Spectral Method

An important difference between finite-difference approximations to the eigenvalues and eigenfunctions equation like Orr-Sommerfeld equation and the Chebyshev approximations is their order of accuracy. Finite-difference approximations give only a finite order of accuracy in the sense that errors behave like $(\Delta x)^p$ for some finite p when the grid scale Δx approaches zero. On the other hand, if the basic-state velocity profile $\bar{u}(y)$ is infinitely differentiable, the Chebyshev polynomial approximations used here are of infinite order in the sense that error decrease more rapidly than any power of $\frac{1}{N}$ as $N \to \infty$.

The latter statement is verified by Orszag [54]. If $\bar{u}(y)$ is infinitely differentiable, all the eigenfunction $v(y)$ of the Orr-Sommerfeld equation are infinitely differentiable for $y \in [-1,1]$ (with one-sided derivatives at the end-points). $T_n(y)$ denotes the nth-degree Chebyshev polynomial of the first kind. Recall Eq. (2.2), defined $y = \cos\theta$:

$$T_n(\cos\theta) = \cos n\theta$$

For all non-negative integers n. It is possible to expand $v(y)$ in the interval $y \in [-1,1]$ as

$$v(y) = \sum_{n=0}^{\infty} a_n T_n(y) \qquad (2.15)$$

where

$$a_n = \frac{2}{\pi c_n} \int_{-1}^{1} v(y)T_n(y)(1-y^2)^{-\frac{1}{2}}dy \qquad (2.16)$$

with $c_0 = 2, c_n = 1$. The rapidity of convergence of Eq. (2.3) for $|y| \leq 1$ is easily demonstrated by observing that

$$f(\theta) = v(\cos\theta) \qquad (2.17)$$

is an infinitely differentiable, even, periodic function of θ.

Consequently, the theory of Fourier series ensures that $f(\theta)$ possesses a Fourier cosine expansion

$$f(\theta) = \sum_{n=0}^{\infty} a_n cosn\theta \qquad (2.18)$$

with the property that the error after N terms decreases more rapidly than any power of $\frac{1}{N}$ as $N \to \infty$. The expansion (2.6) is precisely Eq. (2.3) for $y = cos\theta$.

Moreover, the error associated with the Chebyshev approximation is $O\left(\frac{1}{N^m}\right)$ where N refers to the truncation and m is connected to the number of continuous derivatives (if finite) of function under consideration.

Advantages of Chebyshev Nodes

In polynomial interpolation, Chebyshev nodes provide the resulting interpolation polynomial minimizes the effect of Runge's phenomenon.

Given a function $f \in C^N[-1,1]$ and Chebyshev nodes $y_0, \dots, y_{N-1}$, for each $y \in [-1,1]$, a number $\xi(y)$ exists in $(-1,1)$ with

$$f(y) - P_{N-1}(y) = \frac{f^{(N)}(\xi(y))}{N!} \prod_{j=0}^{N-1}(y - y_j) \qquad (2.19)$$

where $P_{N-1}(y)$ is the Lagrange interpolating polynomial.

Notice that $\prod_{j=0}^{N-1}(y - y_j)$ is the monic Chebyshev polynomial, that is,

$$\prod_{j=0}^{N-1}(y - y_j) = \frac{T_N(y)}{2^{N-1}}. \qquad (2.20)$$

Recall $T_N(y) = cos(n \, cos^{-1}(y))$, we have

$$\left| \prod_{j=0}^{N-1} (y - y_j) \right| \leq \frac{1}{2^{N-1}} \tag{2.21}$$

Then

$$\max_{y \in [-1,1]} |f(y) - P_{N-1}(y)| \leq \frac{1}{2^{N-1}N!} \max_{y \in [-1,1]} |f^{(N)}(y)| \tag{2.22}$$

From Eq. (2.22), it is obviously that with high order polynomial interpolation, the error is very small.

DIRECT NUMERICAL SIMULATION (DNS) CASE SET UP AND CODE VALIDATION

To investigate the linear stability problem of vortices in real-world case, our high order DNS with near 60 million grid points and about 400,000 time steps are used to visualize and track the generation of vortices on boundary layer flow transition.

Case Set-up

The computation domain of our DNS case is shown in Fig. (**3-1**). The mesh includes $1920 \times 128 \times 241$ points in streamwise (x), spanwise (y), and normal-to-wall (z) directions respectively. The grid is uniform in the streamwise and spanwise directions, while stretched in the normal direction. The first grid interval is carefully chosen to make sure the grid is fine enough to capture all the small scales ($Z^+ = 0.43$).

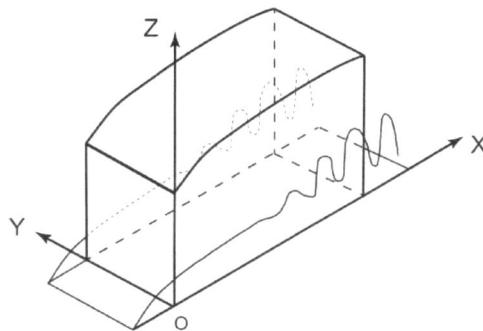

Fig. (3-1). Computational Domain.

The parallel computation is accomplished through the Message Passing Interface (MPI) together with the streamwise direction domain decomposition (shown in Fig. (3-2)). The flow conditions, including Reynolds number, Mach number, *etc.* are listed in Table **3-1**.

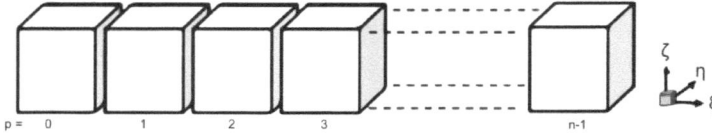

Fig. (3-2). Domain decomposition for MPI.

Table 3-1 Flow parameters.

M_∞	Re	x_{in}	Lx
0.5	1000	$300.79\delta_{in}$	$798.03\delta_{in}$
Ly	Lz_{in}	T_w	T_∞
$22\delta_{in}$	$40\delta_{in}$	273.15K	273.15K

where,

M_∞ = Mach number

Re = Reynolds number, define as $\frac{\rho_\infty U_\infty \delta_{in}}{\mu_\infty}$

δ_{in} = inflow displacement thickness

T_w = wall temperature

T_∞ = free stream temperature

Lz_{in} = height at inflow boundary

Lz_{out} = height at outflow boundary

Lx= length of computational domain along x direction

Ly= length of computational domain along y direction

x_{in} = distance between leading edge of flat plate and upstream boundary of computational domain

Governing Equation in Generalized Curvilinear Coordinates

The governing equations with three-dimensional compressible flow in Cartesian coordinates are shown in section 2. In this section, we give the expansion in Curvilinear Coordinates (ξ, η, ζ).

The compressible Navier-Stokes equations can be written by a vector form:

$$\frac{\partial Q}{\partial t} + \frac{\partial E}{\partial x} + \frac{\partial F}{\partial y} + \frac{\partial G}{\partial z} = \frac{\partial E_v}{\partial x} + \frac{\partial F_v}{\partial y} + \frac{\partial G_v}{\partial z} \tag{3.1}$$

where

$$Q = \begin{bmatrix} \rho \\ \rho u \\ \rho v \\ \rho w \\ e \end{bmatrix} \quad E = \begin{bmatrix} \rho u \\ \rho u^2 + p \\ \rho uv \\ \rho uw \\ (e + p)u \end{bmatrix} \quad F = \begin{bmatrix} \rho v \\ \rho vu \\ \rho v^2 + p \\ \rho vw \\ (e + p)v \end{bmatrix} \quad G = \begin{bmatrix} \rho w \\ \rho wu \\ \rho wv \\ \rho w^2 + p \\ (e + p)w \end{bmatrix} \tag{3.2}$$

$$E_v = \frac{1}{Re} \begin{bmatrix} 0 \\ \tau_{xx} \\ \tau_{xy} \\ \tau_{xz} \\ u\tau_{xx} + v\tau_{xy} + w\tau_{xz} + q_x \end{bmatrix} \tag{3.3}$$

$$F_v = \frac{1}{Re} \begin{bmatrix} 0 \\ \tau_{yx} \\ \tau_{yy} \\ \tau_{yz} \\ u\tau_{yx} + v\tau_{yy} + w\tau_{yz} + q_y \end{bmatrix} \tag{3.4}$$

$$G_v = \frac{1}{Re} \begin{bmatrix} 0 \\ \tau_{zx} \\ \tau_{zy} \\ \tau_{zz} \\ u\tau_{zx} + v\tau_{zy} + w\tau_{zz} + q_z \end{bmatrix} \tag{3.5}$$

$$q_x = \frac{\mu}{(\gamma - 1)M_\infty^2 Pr} \frac{\partial T}{\partial x} \tag{3.6}$$

$$q_y = \frac{\mu}{(\gamma - 1)M_\infty^2 Pr} \frac{\partial T}{\partial y} \tag{3.7}$$

$$q_z = \frac{\mu}{(\gamma - 1)M_\infty^2 Pr} \frac{\partial T}{\partial z} \tag{3.8}$$

$$p = \frac{1}{\gamma M_\infty^2} \rho T \tag{3.9}$$

$$\tau = \mu \begin{bmatrix} \frac{4}{3}\frac{\partial u}{\partial x} - \frac{2}{3}\left(\frac{\partial v}{\partial y} + \frac{\partial w}{\partial z}\right) & \frac{\partial u}{\partial y} + \frac{\partial v}{\partial x} & \frac{\partial u}{\partial z} + \frac{\partial w}{\partial x} \\ \frac{\partial u}{\partial y} + \frac{\partial v}{\partial x} & \frac{4}{3}\frac{\partial u}{\partial x} - \frac{2}{3}\left(\frac{\partial w}{\partial z} + \frac{\partial u}{\partial x}\right) & \frac{\partial v}{\partial z} + \frac{\partial w}{\partial y} \\ \frac{\partial u}{\partial z} + \frac{\partial w}{\partial x} & \frac{\partial v}{\partial z} + \frac{\partial w}{\partial y} & \frac{4}{3}\frac{\partial u}{\partial x} - \frac{2}{3}\left(\frac{\partial u}{\partial x} + \frac{\partial v}{\partial y}\right) \end{bmatrix} \tag{3.10}$$

Assume that the position frame of reference is fixed in time, *i.e.* the generalized coordinates do not change with time. Then, we can define the curvilinear coordinates in relation to the Cartesian coordinates as

$$\begin{cases} \xi = \xi(x, y, z) \\ \eta = \eta(x, y, z) \\ \zeta = \zeta(x, y, z) \end{cases} \tag{3.11}$$

Thus, the Navier-Stokes equations can be transformed to the system using generalized coordinates:

$$\frac{\partial \widehat{Q}}{\partial \tau} + \frac{\partial \widehat{E}}{\partial \xi} + \frac{\partial \widehat{F}}{\partial \eta} + \frac{\partial \widehat{G}}{\partial \zeta} = \frac{\partial \widehat{E}_v}{\partial \xi} + \frac{\partial \widehat{F}_v}{\partial \eta} + \frac{\partial \widehat{G}_v}{\partial \zeta} \tag{3.12}$$

with $\widehat{Q} = J^{-1}Q$ and

$$\widehat{E} = J^{-1}\left(\xi_x E + \xi_y F + \xi_z G\right) \tag{3.13}$$

$$\widehat{F} = J^{-1}\left(\eta_x E + \eta_y F + \eta_z G\right) \tag{3.14}$$

$$\widehat{G} = J^{-1}\left(\zeta_x E + \zeta_y F + \zeta_z G\right) \tag{3.15}$$

$$\widehat{E}_v = J^{-1}\left(\xi_x E_v + \xi_y F_v + \xi_z G_v\right) \tag{3.16}$$

$$\widehat{F}_v = J^{-1}\left(\eta_x E_v + \eta_y F_v + \eta_z G_v\right) \tag{3.17}$$

$$\widehat{G}_v = J^{-1}\left(\zeta_x E_v + \zeta_y F_v + \zeta_z G_v\right) \tag{3.18}$$

$$J^{-1} = \det\left(\frac{\partial(x, y, z)}{\partial(\xi, \eta, \zeta)}\right) \qquad (3.19)$$

Numerical Methods

A sixth order compact scheme [56] is used for the spatial discretization in the streamwise and normal-to-wall directions. The scheme is used for internal points $j = 3, \cdots, N - 2$ as follows:

$$\frac{1}{3}f'_{j-1} + f'_j + \frac{1}{3}f'_{j+1} = \frac{1}{h}\left(-\frac{1}{36}f_{j-2} - \frac{7}{9}f_{j-1} + \frac{7}{9}f_{j+1} + \frac{1}{36}f_{j+2}\right) + O(h^6) \quad (3.20)$$

where f'_j is the first derivative at the internal point j. The fourth order compact scheme is used at point $j = 2, N - 1$, and the third order one-sided compact scheme is used at the boundary points $j = 1, N$.

In the spanwise direction, the pseudo-spectral method is used for the periodic conditions. To eliminate the spurious numerical oscillations caused by central difference schemes, a high-order spatial scheme is used instead of artificial dissipation. An implicit sixth-order compact scheme for space filtering is applied to the primitive variables u, v, w, ρ, p after a specified number of time steps.

The governing equations are solved explicitly by a 3rd order Total Variation Diminishing (TVD) Runge-Kutta (RK) scheme for time marching:

$$Q^{(0)} = Q^n$$
$$Q^{(1)} = Q^{(0)} + \Delta t R^{(0)}$$
$$Q^{(2)} = \frac{3}{4}Q^{(0)} + \frac{1}{4}Q^{(1)} + \frac{1}{4}\Delta t R^{(1)}$$
$$Q^{n+1} = \frac{1}{4}Q^{(0)} + \frac{2}{3}Q^{(2)} + \frac{2}{3}\Delta t R^{(2)} \qquad (3.21)$$

CFL ≤ 1 is required to ensure the stability.

The adiabatic and the non-slipping conditions are enforced at the wall boundary on the flat plate. On the far field and the outflow boundaries, the non-reflecting boundary conditions are applied.

Blasius solution with enforced disturbance is introduced into inlet as a laminar base inflow. The disturbance includes a two-dimensional T-S wave and a pair of conjugate three-dimensional T-S waves. The inflow has a form:

$$q = q_{\text{lam}} + A_{2d}q'_{2d}e^{i(\alpha_{2d}x - \omega t)} + A_{3d}q'_{3d}e^{i(\alpha_{3d}x \pm \beta y - \omega t)}$$

with q represents the vector (u,v,w,p,T), q_{lam} is the Blasius solution for a two-dimensional laminar flat plat boundary layer. The streamwise wavenumber, spanwise wavenumber, frequency and amplitude are given respectively as follows:

$$\alpha_{2d} = 0.29919 - i5.09586 \times 10^{-3}$$

$$\beta = \pm 0.5712$$

$$\omega = 0.114027$$

$$A_{2d} = 0.03$$

$$A_{3d} = 0.01$$

The T-S wave parameters are obtained by solving the compressible boundary layer stability equations.

Code Validation

The code "DNSUTA" was developed at the University of Texas at Arlington and carefully validated by NASA Langley and UTA researchers [40, 55]. Only a short description of the validation would be addressed here, and readers are encouraged to refer to these papers for details. A more detailed comparison is also reported in [32].

DNS Visualization Method

A new visualization method named "Ω criterion" proposed by Liu [56] is used in this book chapter to identify the vortices. The "Ω" is defined as the proportion of vorticity and deformation in fluid element motion:

$$\nabla V = \frac{1}{2}(\nabla V + \nabla V^T) + \frac{1}{2}(\nabla V - \nabla V^T) = S + W \qquad (3.22)$$

where S is the symmetric while W is the anti-symmetric part of the velocity gradient tensor. S represents deformation and W is related to the whole vorticity.

The square of Frobenius norms of S and W are $a = \text{trace}(SS^T)$, $b = \text{trace}(WW^T)$, then:

$$\Omega = \frac{b + \varepsilon}{(a + \varepsilon) + (b + \varepsilon)} \tag{3.23}$$

$\Omega = 0.52$ is set as the threshold to identify the region where rotation plays a dominant role rather than deformation as a vortex.

Compared with traditional Q [57] or λ_2 [58] criteria, Ω criteria doesn't need to tune the threshold and have clear physical meaning. In addition, it successes to capture both strong and weak vortices simultaneously while both Q and λ_2 criteria fails. When $\Omega = 1.0$, the flow has pure rotation and has no deformation which means the fluid is stiff like solid, but vortex itself is also very stiff like solid. Although Ω is a measurement of fluid stiffness, but it is a good measurement of flow rotation, *i.e.*, vortex. Therefore, Ω criteria is the best visualization method for us to detect and track the generation of vortices.

INSTABILITY OF TWO-DIMENSIONAL SHEAR FLOW

Linear Stability Equation

Consider the Non-dimensional Navier-Stokes equations for incompressible flow:

$$\begin{cases} \dfrac{\partial V}{\partial t} + V \cdot \nabla V = -\nabla p + \dfrac{1}{Re} \nabla^2 V \\ \nabla \cdot V = 0 \end{cases} \tag{4.1}$$

where $V = (u, v, w)$ is the velocity vector with u denoting the streamwise component, v denoting the normal component, and w denoting the spanwise component.

For the linear stability theory, we can write:

$$q(x, y, z, t) = q_0(y) + q'(x, y, z, t) \tag{4.2}$$

where q can be specified as (u,v,w,p) and $q_0 = (u_0, v_0, w_0, p_0)$ represents the value of base flow. Q' denotes the corresponding linear perturbation.

By eliminating the second order perturbation terms, the linearized governing equation for small perturbations can be written as,

$$\begin{cases} \dfrac{\partial V'}{\partial t} + (V_0 \cdot \nabla)V' + (V' \cdot \nabla)V_0 + \nabla p' = \dfrac{1}{Re} \nabla^2 V' \\ \\ \nabla \cdot V' = 0 \end{cases} \quad (4.3)$$

As a first step, a localized 2-D incompressible temporal stability for shear layer is studied. It relates to the distance among two neighboring vortices in the central streamwise plane. Assume the normal mode is

$$V' = \widehat{V}(y)e^{i(\alpha x + \beta z - \omega t)} + c.c. = \widehat{V}(y)e^{i\alpha\left(x + \frac{\beta}{\alpha}z - ct\right)} + c.c. \quad (4.4)$$

$$p' = \widehat{p}(y)e^{i(\alpha x + \beta z - \omega t)} + c.c. = \widehat{p}(y)e^{i\alpha\left(x + \frac{\beta}{\alpha}z - ct\right)} + c.c. \quad (4.5)$$

with $c = \frac{\omega}{\alpha}$. Here $\widehat{V} = (\widehat{u}, \widehat{v}, \widehat{w})$, the wavenumber α and β are given real numbers. The parameter c should be a complex number. Plugging Eq. (4.4) and Eq. (4.5) into Eq. (4.3) yields:

$$L\widehat{u} = Re(Du_0)\widehat{v} + i\alpha Re\widehat{p}$$
$$L\widehat{v} = Re(D\widehat{p})$$
$$L\widehat{w} = i\beta Re\widehat{p}$$
$$i(\alpha\widehat{u} + \beta\widehat{w}) + D\widehat{v} = 0 \quad (4.6)$$

where $L = [D^2 - (\alpha^2 + \beta^2) - iRe(\alpha u_0 - \omega)]$, and $D = \dfrac{d}{dy}$.

By eliminating $\widehat{u}, \widehat{w}, \widehat{p}$, we can obtain the Orr-Sommerfeld equation on $\widehat{v}$,

$$\left(-Uk^2 - U'' - \dfrac{k^4}{i\alpha Re}\right)\widehat{v} + \left(U + \dfrac{2k^2}{i\alpha Re}\right)\widehat{v}'' - \dfrac{1}{i\alpha Re}\widehat{v}'''' = c(\widehat{v}'' - k^2\widehat{v}) \quad (4.7)$$

where $U = u_0$ and $k^2 = \alpha^2 + \beta^2$.

With boundary conditions

$$\widehat{v}(\pm 1) = \widehat{v}'(\pm 1) = 0,$$

Eq. (4.7) is an eigenvalue problem about $\widehat{v}$ with eigenvalue c. The eigenvalue c determines the property of stability of the equation. Let $c = c_r + ic_i$, if $c_i > 0$, then the disturbance will continuously grow, and the flow will be instable. While if c_r is greater, the disturbance will grow faster, and the flow will be more unstable. But if $c_i < 0$, the flow will be stable.

Orr-Sommerfeld equation is named after William McFadden Orr and Arnold Sommerfled, who derived is at the beginning of the 20th century. It describes the perturbation of two-dimensional parallel flow and it is widely used in boundary-layer linear stability theory.

Chebyshev Discretization of the Orr-Sommerfeld Equation

In this section, a spectral collocation method based on Chebyshev polynomials is applied to the Orr-Sommerfeld equation. This method has advantages to compute the stability characteristics of shear flows.

Recall Eq. (4.7), the eigenfunction $\hat{v}$ could be approximated by Chebyshev expansion,

$$\hat{v}(y) = \sum_{n=0}^{\infty} a_n T_n(y) \approx \sum_{n=0}^{N} a_n T_n(y) \tag{4.8}$$

The derivatives of the eigenfunctions are obtained by differentiating the expansion above. For example, the approximation of the second derivative is,

$$D^2 \hat{v}(y) = \sum_{n=0}^{N} a_n T_n''(y) \tag{4.9}$$

And similarly, for the fourth derivative.

Upon substitution into the Orr-Sommerfeld equation we get

$$\sum_{n=0}^{N} \left[\left(-Uk^2 - U'' - \frac{k^4}{i\alpha Re} \right) T_n + \left(U + \frac{2k^2}{i\alpha Re} \right) T_n'' - \frac{1}{i\alpha Re} T_n'''' \right] a_n =$$

$$c \sum_{n=0}^{N} a_n (T_n'' - k^2 T_n) \tag{4.10}$$

Chebyshev collocation method is then used, require this equation to be satisfied at the Gauss point (Chebyshev nodes):

$$y_j = \cos\left(\frac{2j+1}{2N}\pi\right), \qquad j = 0,1,\cdots,N-1 \qquad (4.11)$$

Recurrence relations are used to evaluate the derivatives of the Chebyshev polynomials.

The discretized boundary conditions read

$$\sum_{n=0}^{N} a_n T_n(1) = 0 \qquad\qquad \sum_{n=0}^{N} a_n T_n(-1) = 0$$

$$\sum_{n=0}^{N} a_n T_n'(1) = 0 \qquad\qquad \sum_{n=0}^{N} a_n T_n'(-1) = 0 \qquad (4.12)$$

Applying Eq. (4.10) on the collocation points grid with boundary conditions above, a matrix form of generalized eigenvalue problem is given by the form and similar for the left-hand side Aa. We have chosen to use the first, second, last and next-to-last row of B to implement the four boundary conditions. The same rows in the matrix A can be chosen as a complex multiple of the corresponding rows in B. By carefully selecting this complex multiple, the spurious modes associated with the boundary conditions can be mapped to an arbitrary location in the complex plane.

$$Aa = cBa \qquad (4.13)$$

With the right-hand side

$$cBa =$$

$$c\begin{pmatrix} T_0(1) & T_1(1) & \cdots \\ T_0'(1) & T_1'(1) & \cdots \\ T_0''(y_1) - k^2 T_0(y_1) & T_1''(y_1) - k^2 T_1(y_1) & \cdots \\ \vdots & \vdots & \vdots \\ T_0''(y_N) - k^2 T_0(y_M) & T_1''(y_N) - k^2 T_1(y_M) & \cdots \\ T_0'(-1) & T_1'(-1) & \cdots \\ T_0(-1) & T_1(-1) & \cdots \end{pmatrix}\begin{pmatrix} a_0 \\ a_1 \\ a_2 \\ \vdots \\ a_{N-3} \\ a_{N-2} \\ a_{N-1} \end{pmatrix} \qquad (4.14)$$

Numerical Results for Typical Shear Flows

Typical shear flow (Fig. (4-1)):

$$U(y) = tanh(by), \; y \in [-1,1]$$

Computational conditions:

The number of O-S nodes is $N = 100$, streamwise wave number is $\alpha = 1$ and spanwise wave number is $\beta = 0$.

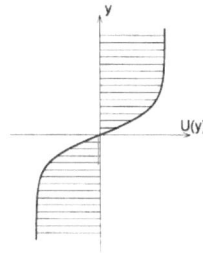

Fig. (4-1). Sketch of typical shear flow velocity profile.

First, set $b = 2$, Fig. **(4-2)** provides the comparison of spectrums upon base flow $U(y) = \tanh(2y)$ with Reynolds number $Re = 10, 100, 1000$.

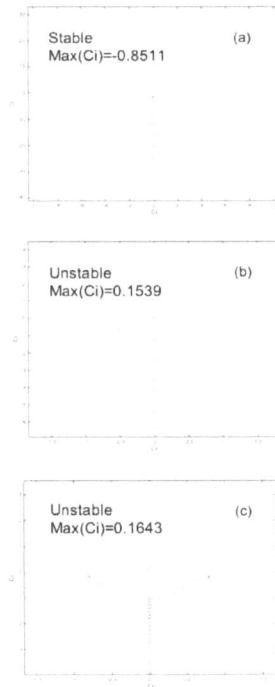

Stable   (a)
Max(Ci)=-0.8511

Unstable   (b)
Max(Ci)=0.1539

Unstable   (c)
Max(Ci)=0.1643

Fig. (4-2). Graphs of spectrum on shear flow $U(y) = \tanh(2y)$. (**a**) $Re = 10$; (**b**) $Re = 100$; (**c**) $Re = 1000$.

Unstable mode (c_i =0.1643) only obtained at $Re = 1000$, the associated eigenfunction $\hat{v}$ is as Fig. **(4-3)**:

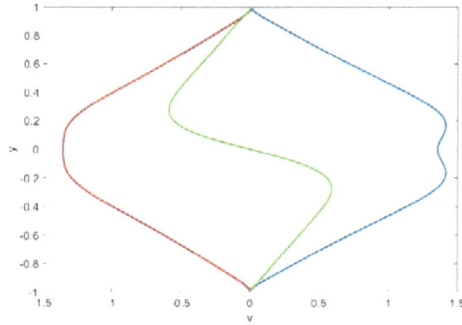

Fig. (4-3). Eigenfunction $\hat{v}$ on [-1, 1]: red-imag($\hat{v}$); green-real($\hat{v}$); blue-$|\hat{v}|$.

Next, set $b = 8$, with base flow $U(y) = \tanh(8y)$, Fig. **(4-4)** provides the comparison of spectrums with Reynolds number $Re = 10, 100, 1000$.

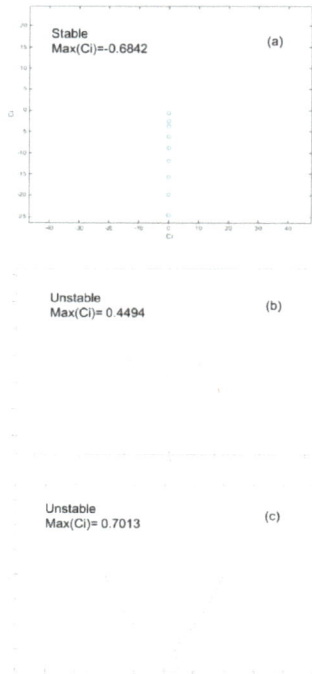

Fig. (4-4). Graphs of spectrum on shear flow $U(y) = \tanh(8y)$. **(a)** $Re = 10$; **(b)**$Re = 100$; **(c)**$Re = 1000$.

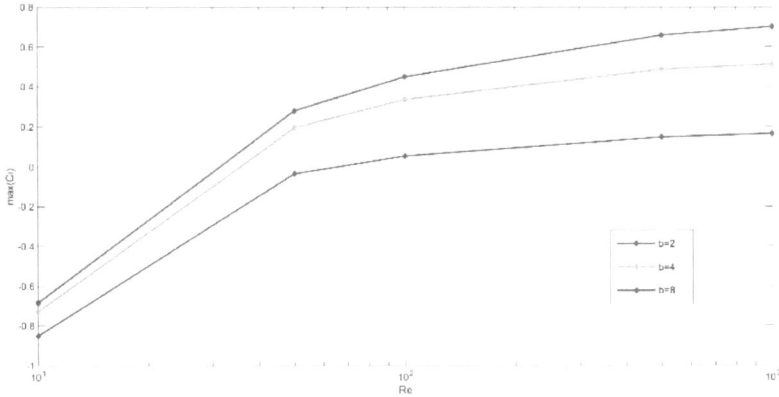

Fig. (4-5). Graph of Re and least stable c_i on $b = 2,4,8$.

Consider Fig. (**4-5**), the graph shows that: for 1-D typical shear flow, it is unstable when Reynolds number is large enough. The flow is more unstable with the larger Reynolds number and stronger shear stress.

LINEAR STABILITY EQUATION FOR QUASI-ROTATION FLOW IN CYLINDRICAL COORDINATES

Derivation of Linear Perturbation System

The dimensionless NS equations of 2D incompressible flow can be written in cylindrical coordinates as:

Continuity:

$$\frac{1}{r}\frac{\partial(ru_r)}{\partial r} + \frac{1}{r}\frac{\partial u_\theta}{\partial \theta} = 0 \tag{5.1}$$

Radial conservation of momentum:

$$\frac{Du_r}{Dt} - \frac{u_\theta^2}{r} = -\frac{\partial p}{\partial r} + \frac{1}{Re}\left[\nabla^2 u_r - \frac{u_r}{r^2} - \frac{2}{r^2}\frac{\partial u_\theta}{\partial \theta}\right] \tag{5.2}$$

Azimuthal conservation of momentum:

$$\frac{Du_\theta}{Dt} + \frac{u_r u_\theta}{r} = -\frac{\partial p}{\partial \theta} + \frac{1}{Re}\left[\nabla^2 u_\theta - \frac{u_\theta}{r^2} + \frac{2}{r^2}\frac{\partial u_r}{\partial \theta}\right] \tag{5.3}$$

Here,

$$\frac{D}{Dt} \equiv \frac{\partial}{\partial t} + u_r \frac{\partial}{\partial r} + \frac{u_\theta}{r} \frac{\partial}{\partial \theta} \tag{5.5}$$

and

$$\nabla^2 \equiv \frac{\partial^2}{\partial r^2} + \frac{1}{r} \frac{\partial}{\partial r} + \frac{1}{r^2} \frac{\partial^2}{\partial \theta^2} \tag{5.5}$$

with u_r, u_θ representing the radial, azimuthal velocity components, p is the pressure, Re is the Reynolds number.

Given the class of steady-state solutions to this system of equations where $u_r = U_0, u_\theta = V_0$, with $p = P_0$, these velocities and pressure must then be solutions to:

$$\frac{1}{r} \frac{\partial(rU_0)}{r} + \frac{1}{r} \frac{\partial V_0}{\partial \theta} = 0 \tag{5.6}$$

$$\frac{DU_0}{Dt} - \frac{V_0^2}{r} = -\frac{\partial P_0}{\partial r} + \frac{1}{Re} \left[\nabla^2 U_0 - \frac{U_0}{r^2} - \frac{2}{r^2} \frac{\partial V_0}{\partial \theta} \right] \tag{5.7}$$

$$\frac{DU_0}{Dt} + \frac{U_0 V_0}{r} = -\frac{\partial P_0}{\partial \theta} + \frac{1}{Re} \left[\nabla^2 V_0 - \frac{V_0}{r^2} + \frac{2}{r^2} \frac{\partial U_0}{\partial \theta} \right] \tag{5.8}$$

while satisfying appropriate boundary conditions.

Suppose that the steady-state solution (U_0, V_0, P_0) is subjected to a set of small fluctuations, where

$$u_r = U_0 + u'(r, \theta, t) \tag{5.9}$$

$$u_\theta = V_0 + v'(r, \theta, t) \tag{5.10}$$

$$p = P_0 + p'(r, \theta, t) \tag{5.11}$$

Those fluctuation variables incorporated with their steady-state components are solutions to the governing equation. According to Eq. (5.6) - (5.8), the original steady-state part can be removed, leaving:

$$\frac{1}{r}\frac{\partial(ru')}{r} + \frac{1}{r}\frac{\partial v'}{\partial \theta} = 0 \tag{5.12}$$

$$\frac{Du'}{Dt} + u'\frac{\partial U_0}{\partial r} - 2\frac{v'V_0}{r} + \frac{\partial u'^2}{\partial r} + \frac{1}{r}\frac{\partial u'v'}{\partial \theta} + \frac{u'^2 - v'^2}{r}$$
$$= -\frac{\partial p'}{\partial r} + \frac{1}{Re}\left[\nabla^2 u' - \frac{u'}{r^2} - \frac{2}{r^2}\frac{\partial v'}{\partial \theta}\right] \tag{5.13}$$

$$\frac{Dv'}{Dt} + u'\frac{\partial V_0}{\partial r} + \frac{v'U_0 + u'V_0}{r} + \frac{\partial u'v'}{\partial r} + \frac{1}{r}\frac{\partial v'^2}{\partial \theta} + 2\frac{u'v'}{r}$$
$$= -\frac{1}{r}\frac{\partial p'}{\partial \theta} + \frac{1}{Re}\left[\nabla^2 v' - \frac{v'}{r^2} + \frac{2}{r^2}\frac{\partial u'}{\partial \theta}\right] \tag{5.14}$$

where

$$\frac{D}{Dt} \equiv \frac{\partial}{\partial t} + U_0\frac{\partial}{\partial r} + \frac{V_0}{r}\frac{\partial}{\partial \theta} \tag{5.15}$$

The linear stability equations for rotational flow evolve from these fluctuation equations by assuming that quadratic fluctuation terms are negligibly small. Furthermore, we would like to work these equations on cylindrical base flow, hence the general base flow for a vortex in 2D coordinates can be approximated by

$$U_0 \approx 0, V_0 = V_0(r) \tag{5.16}$$

Therefore, the dimensionless, linearized perturbation equations in 2D cylindrical coordinates can be written as:

Continuity:

$$r\frac{\partial u'}{\partial r} + u' + \frac{\partial v'}{\partial \theta} = 0 \tag{5.17}$$

r-Momentum:

$$\frac{\partial u'}{\partial t} + \frac{V_0}{r}\frac{\partial u'}{\partial \theta} - 2\frac{v'V_0}{r} + \frac{\partial u'^2}{\partial r} = -\frac{\partial p'}{\partial r} + \frac{1}{Re}\left[\nabla^2 u' - \frac{u'}{r^2} - \frac{2}{r^2}\frac{\partial v'}{\partial \theta}\right] \tag{5.18}$$

θ-Momentum:

$$\frac{\partial v'}{\partial t} + u'\frac{\partial V_0}{\partial r} + \frac{V_0}{r}\frac{\partial v'}{\partial \theta} + \frac{u'V_0}{r} + = -\frac{1}{r}\frac{\partial p'}{\partial \theta} + \frac{1}{Re}\left[\nabla^2 v' - \frac{v'}{r^2} + \frac{2}{r^2}\frac{\partial u'}{\partial \theta}\right] \quad (5.19)$$

and

$$\nabla^2 \equiv \frac{\partial^2}{\partial r^2} + \frac{1}{r}\frac{\partial}{\partial r} + \frac{1}{r^2}\frac{\partial^2}{\partial \theta^2} \quad (5.20)$$

Eigenvalue Function

To solve the Eq. (5.17) – (5.20), we first subject that system to a normal mode representation of the disturbance field, where

$$\{u', v', p'\} = \{\hat{u}(r), \hat{v}(r), \hat{p}(r)\}e^{i(\alpha\theta - \omega t)} \quad (5.21)$$

Here, u, v and p are amplitude functions which are dependent on the radial coordinate only. For the temporal solution α is the given real and constant wavenumbers in the tangential direction. The angular frequency, ω, is complex and must be computed. The disturbances will grow if the imaginary part of the frequency, ω_i, is positive, and if it is negative the disturbances decay with time. Substituting Eq. (5.21) into the governing Eq. (5.17) – (5.19), we obtain:

Continuity:

$$\hat{u} + rD\hat{u} + i\alpha\hat{v} = 0 \quad (5.22)$$

r-Momentum:

$$(-i\omega)\hat{u} + \frac{i\alpha V_0}{r}\hat{u} - \frac{2V_0}{r}\hat{v} =$$

$$-D\hat{p} + \frac{1}{Re}\left[\frac{1}{r}D\hat{u} + D^2\hat{u} - \frac{\alpha^2}{r^2}\hat{u} - \frac{2i\alpha}{r^2}\hat{v} - \frac{\hat{u}}{r^2}\right] \quad (5.23)$$

θ-Momentum:

$$(-i\omega)\hat{v} + DV_0\hat{u} + \frac{i\alpha V_0}{r}\hat{v} + \frac{V_0}{r}\hat{u} =$$

$$-\frac{i\alpha}{r}\hat{p} + \frac{1}{Re}\left[\frac{1}{r}D\hat{v} + D^2\hat{v} - \frac{\alpha^2}{r^2}\hat{v} + \frac{2i\alpha}{r^2}\hat{u} - \frac{\hat{v}}{r^2}\right] \tag{5.24}$$

where $D = \dfrac{\partial}{\partial r}$.

Eq. (5.22) – (5.24) represent the most general form for this normal model solution to the vortex stability equations for the 2D quasi-cylindrical approximation.

According to the three linear equations system with three variables, it is possible to eliminate two variables $\hat{v}, \hat{p}$ and get a fourth-order ordinary differential equation with respect to $\hat{u}$.

From Eq. (5.24), the pressure $\hat{p}$ can be expressed in terms of $\hat{u}$ and $\hat{v}$ as follows:

$$\left(\frac{i\alpha}{r}\right)\hat{p} = -\left[(-ic)\hat{v} + DV_0\hat{u} + \frac{i\alpha V_0}{r}\hat{v} + \frac{V_0}{r}\hat{u}\right] +$$
$$\frac{1}{Re}\left[\frac{1}{r}D\hat{v} + D^2\hat{v} - \frac{\alpha^2}{r^2}\hat{v} + \frac{2i\alpha}{r^2}\hat{u} - \frac{\hat{v}}{r^2}\right] \tag{5.25}$$

$$\hat{p} = \frac{cr}{\alpha}\hat{v} + \frac{ir}{\alpha}DV_0 \cdot \hat{u} - V_0\hat{v} + \frac{i}{\alpha}V_0\hat{u} -$$
$$\frac{1}{\alpha Re}\left(iD\hat{v} + irD^2\hat{v} - \frac{i\alpha^2}{r}\hat{v} - \frac{2\alpha}{r}\hat{u} - \frac{i}{r}\hat{v}\right) \tag{5.26}$$

By taking the derivatives of p, we can get the expression of Dp:

$$Dp = \frac{\partial p}{\partial r} = \frac{c}{\alpha}(v + rDv) + \frac{i}{\alpha}DV_0 \cdot u +$$
$$\frac{ir}{\alpha}(D^2V_0 \cdot u + DV_0 \cdot Du) - DV_0 \cdot v - V_0Dv + \frac{i}{\alpha}(DV_0 \cdot u + V_0 \cdot Du) -$$

$$\frac{i}{\alpha Re}\left[2D^2v + rD^3v - \alpha^2\left(-\frac{1}{r^2}v + \frac{Dv}{r}\right) - \left(-\frac{1}{r^2}v + \frac{Dv}{r}\right)\right] +$$

$$\frac{2}{Re}\left(-\frac{1}{r^2}u + \frac{Du}{r}\right) \tag{5.27}$$

By rearranging the equation, get:

$$Dp =$$

$$\left[\left(\frac{c}{\alpha} - DV_0\right) + \left(-\frac{i}{\alpha Re}\right)\left(\frac{\alpha^2 + 1}{r^2}\right)\right]v + \left[\left(\frac{cr}{\alpha} - V_0\right) + \left(\frac{i}{\alpha Re}\right)\left(\frac{\alpha^2 + 1}{r}\right)\right]Dv$$

$$+ \left[\left(\frac{cr}{\alpha} - V_0\right) + \left(\frac{i}{\alpha Re}\right)\left(\frac{\alpha^2 + 1}{r}\right)\right]Dv + \left(-\frac{2i}{\alpha Re}\right)D^2v$$

$$+ \left(-\frac{ir}{\alpha Re}\right)D^3v + \left[\frac{2}{Re}\left(-\frac{1}{r^2}\right) + \frac{i}{\alpha}(2DV_0 + rD^2V_0)\right]u$$

$$+ \left[\frac{2}{rRe} + \frac{i}{\alpha(r \cdot DV_0 + V_0)}\right]Du \tag{5.28}$$

By combining Eq. (5.23) and Eq. (5.28), we get an equation LS=0. Terms with respect to each velocity component on LS is as following individually:

u: $\quad -ic + \dfrac{i\alpha V_0}{r} + \dfrac{1}{Re}\left(\dfrac{\alpha^2 + 1}{r^2}\right) + \dfrac{2}{Re}\left(-\dfrac{1}{r^2}\right) + \dfrac{i}{\alpha}(2DV_0 + rD^2V_0)$

Du: $\quad +\dfrac{1}{rRe} + \dfrac{i}{\alpha}(r \cdot DV_0 + V_0)$

D^2u: $-\dfrac{1}{Re}$

v: $\quad -\dfrac{2V_0}{r} + \dfrac{1}{Re}\left(\dfrac{2i\alpha}{r^2}\right) + \left(\dfrac{c}{\alpha} - DV_0\right) + \left(-\dfrac{i}{\alpha Re}\right)\left(\dfrac{\alpha^2 + 1}{r^2}\right)$

Dv: $\quad +\dfrac{cr}{\alpha} - V_0 + \dfrac{i}{\alpha Re}\left(\dfrac{\alpha^2 + 1}{r}\right)$

D^2v: $-\dfrac{2i}{\alpha Re}$

D^3v: $-\dfrac{i}{\alpha r Re}$

$$(5.29)$$

By considering Eq. (5.22), all v-velocity components can be replaced by u-velocity components since they have the relations as below:

$$v = \frac{i}{\alpha}(u + r \cdot Du)$$

$$Dv = \frac{i}{\alpha}(2Du + r \cdot D^2u)$$

$$D^2v = \frac{i}{\alpha}(3D^2u + r \cdot D^3u)$$

$$D^3v = \frac{i}{\alpha}(4D^3u + r \cdot D^4u)$$

$$(5.30)$$

By considering Eq. (5.29) and Eq. (5.30), a new equation $LS^* = 0$ can be obtained with only u-velocity components on LS^*:

u: $\dfrac{1}{r^2 Re}\left(\alpha - \dfrac{1}{\alpha}\right)^2 + i\left[\dfrac{V_0}{r}\left(\alpha - \dfrac{2}{\alpha}\right) + \dfrac{1}{\alpha}(DV_0 + r \cdot D^2 V_0)\right] + ic\left(\dfrac{1}{\alpha} - \alpha\right)$

Du: $-\dfrac{1}{r Re}\left(2 + \dfrac{1}{\alpha^2}\right) - i\left(\dfrac{3V_0}{\alpha}\right) + ic\left(\dfrac{3r}{\alpha}\right)$

D^2u: $\dfrac{1}{Re}\left(\dfrac{5}{\alpha^2} - 2\right) + i\left(-\dfrac{V_0 r}{\alpha}\right) + ic(\dfrac{r^2}{\alpha})$

D^3u: $\dfrac{1}{Re}(\dfrac{6r}{\alpha^2})$

$$(5.31)$$

D^4u: $\dfrac{1}{Re}(\dfrac{r^2}{\alpha^2})$

By multiplying LS^* by $\alpha^2 Re$, we get:

u: $\dfrac{1}{r^2}(\alpha^2 - 1)^2 + iRe\left[\dfrac{\alpha V_0}{r}(\alpha^2 - 2) + \alpha(DV_0 + r \cdot D^2 V_0)\right] + icRe(1 - \alpha^2)$

$$Du: \quad -\frac{1}{r}(2\alpha^2 + 1) - iRe(3V_0\alpha) + icRe(3r)$$

$$D^2u: \quad (5 - 2\alpha^2) - iRe(2V_0r) + icRer^2$$

$$D^3u: \quad 6r \tag{5.32}$$

$$D^4u: \quad r^2$$

By rearranging Eq. (5.32), we obtain an ordinary differential equation with respect to the u-velocity components, it is also an eigenvalue function describing the linear modes of disturbance to a quasi-rotation flow. Similar with Orr-Sommerfeld equation, this equation determines what the conditions for flow stability are

$$D^4u \cdot r^4 + D^3u \cdot 6r^3 + D^2u \cdot r^2[(5 - 2\alpha^2) - i\alpha ReV_0r]$$
$$+Du \cdot rw[-(2\alpha^2 + 1) - 3i\alpha ReV_0r]$$
$$+u\{(\alpha^2 - 1)^2 + i\alpha Rer[V_0(\alpha^2 - 2) + rDV_0 + r^2D^2V_0]\}$$
$$= -c \cdot iRer[D^2u \cdot r^3 + Du \cdot 3r^2 + u \cdot (1 - \alpha^2)r] \tag{5.33}$$

with the boundary conditions:

$$u(0) = 0, \quad u(1) = 0$$

$$u'(0) = 0, \quad u'(1) = 0 \tag{5.34}$$

c represents the angular frequency. Let $c = c_r + c_i$. The imaginary part, c_i, determines the stability of the perturbation. It is stable if c_i is negative and unstable if c_i is positive.

Shifted Chebyshev Polynomials and Discretization

Since Eq. (5.25) is based on cylindrical coordinates on a finite domain in radius, the domain is always being normalized to $r \in [0,1]$, Chebyshev polynomials are no longer available for this problem. A common way to solve it is Shifted Chebyshev polynomials by a linear change-of-coordinate.

Shifted Chebyshev Polynomials with Linear Argument

Define n orthogonal polynomials satisfying

$$T_n^*(r) = T_n(2r - 1) = cos(n\,cos^{-1}(2r - 1)), \qquad n = 0,1,2,\cdots \qquad (5.35)$$

for $r \in [0,1]$ with n is the order of polynomials.

Therefore, $|T_n^*(r)|$ also bounded by 1.

By setting $2r - 1 = cos\theta$, we have:

$$T_n^* = cosn\theta \qquad (5.36)$$

Shifted Chebyshev polynomials keep most characteristics of Chebyshev polynomials. First few polynomials of Shifted Chebyshev of linear argument are:

$$T_0^* = 1$$
$$T_1^* = 2r - 1$$
$$T_2^* = 8r^2 - 8r + 1$$
$$T_3^* = 32r^3 - 48r^2 + 18r - 1$$
$$T_4^* = 128r^4 - 256r^3 + 160r^2 - 32r + 1 \qquad (5.37)$$

Similar with Chebyshev polynomials, Shifted Chebyshev polynomials also have the recurrence relationship for polynomials:

$$T_{n+1}^*(r) = (4r - 2)T_n^*(r) - T_{n-1}^*(r) \qquad (5.38)$$

A recurrence relation on the derivative also can easily be obtained. First, the differentiation of $T_n^*(r)$ gives:

$$
\begin{aligned}
T_n^{*\prime(r)} &= \frac{d}{d\theta}(cosn\theta)\frac{d\theta}{dr} \\
&= \frac{d}{d\theta}(cosn\theta)\frac{1}{\dfrac{dr}{d\theta}} \\
&= -nsin(n\theta)\frac{1}{\dfrac{(-1)sin\theta}{2}} \\
&= 2n\frac{sin(n\theta)}{sin\theta}.
\end{aligned}
\qquad (5.39)
$$

Then, by the application of trigonometrical formulas, we get the relation:

$$\frac{T^{*\prime}_{n+1}(r)}{n+1} - \frac{T^{*\prime}_{n-1}(r)}{n-1} = 4T^*_n(r) \qquad (5.40)$$

with $T^{*\prime}_0(r) = 0$ and $T^{*\prime}_1(r) = 2$.

Shifted Chebyshev Polynomials with Quadratic Argument

In cylindrical or polar coordinates, the Shifted Chebyshev polynomials with linear argument is usually a bad option. The reason is that the Shifted Chebyshev grid has points clustered near both $r = 0$ and $r = 1$. However, the disk bounded by $r = \rho$ has an area which is only the fraction ρ^2 of the area of the unit disk. Near the origin, points are separated by $O(1/N^2)$. It follows that the high density of points near the origin is giving high resolution of only a tiny, $O(1/N^4)$ in the area portion of the disk.

The Shifted Chebyshev polynomials of quadratic argument is a good alternate of the linear argument. They are defined as:

$$T^{**}_n(r) = T_n(2r^2 - 1) = cos(n\,cos^{-1}(2r^2 - 1)), \qquad n = 0,1,2,\cdots \qquad (5.41)$$

for $r \in [0,1]$ with n is the order of polynomials.

Fig. (5-1) shows the points of two different Shifted Chebyshev series on $[0,1]$.

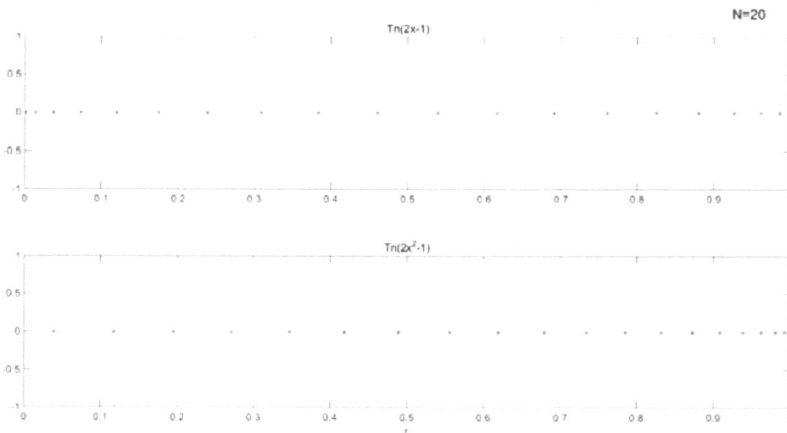

Fig. (5-1). Chebyshev nodes distribution of two Shifted Chebyshev series on [0,1] with N=20.

By setting $2r^2 - 1 = cos\theta$, we have:

$$T_n^{**} = cosn\theta \tag{5.42}$$

First few polynomials of Shifted Chebyshev of quadratic argument are:

$$T_0^{**} = 1$$
$$T_1^{**} = 2r^2 - 1$$
$$T_2^{**} = 8r^4 - 8r^2 + 1$$
$$T_3^{**} = 32r^6 - 48r^4 + 24r^2 - 3r - 4$$

$$\tag{5.43}$$

The recurrence relationship for Shifted Chebyshev polynomials of quadratic is:

$$T_{n+1}^{**}(r) = (4r^2 - 2)T_n^{**}(r) - T_{n-1}^{**}(r) \tag{5.44}$$

A recurrence relation on the derivative also can easily be obtained. First, the differentiation of $T_n^*(r)$ gives:

$$
\begin{aligned}
T_n^{*\prime(r)} &= \frac{d}{d\theta}(cosn\theta)\frac{d\theta}{dr} \\
&= \frac{d}{d\theta}(cosn\theta)\frac{1}{\dfrac{dr}{d\theta}} \\
&= -nsin(n\theta)\frac{1}{\dfrac{(-1)sin\theta}{4r}} \\
&= 4nr\frac{sin(n\theta)}{sin\theta}.
\end{aligned}
\tag{5.45}
$$

Then, by the application of trigonometrical formulas, we get the relation:

$$\frac{T_{n+1}^{**\prime}(r)}{n+1} - \frac{T_{n-1}^{**\prime}(r)}{n-1} = 4rT_n^{**}(r) \tag{5.46}$$

with $T_0^{*\prime}(r) = 0$ and $T_1^{*\prime}(r) = 4r$.

Equation Discretization

Recall Eq. (5.25), the independent variable of this ordinary differential equation is $u(r)$. In linear stability analysis, the function $u(r)$ could be approximated by

$$u(r) = \sum_{n=0}^{\infty} a_n T_n^*(r) \approx \sum_{n=0}^{N-1} a_n T_n^*(r), \tag{5.47}$$

or

$$u(r) = \sum_{n=0}^{\infty} a_n T_n^{**}(r) \approx \sum_{n=0}^{N-1} a_n T_n^{**}(r), \tag{5.48}$$

where N is the number of Chebyshev polynomials used to approximate the velocity profile, T_n^* is Shifted Cheybshev polynomials of linear argument and T_n^{**} is of quadratic argument. a_n are the coefficients.

Eq. (5.25) and Eq. (5.33) give

$$\sum_{n=0}^{N-1} \{ r^4 \cdot T_n'''' + 6r^3 \cdot T_n''' + r^2[(5 - 2\alpha^2) - i\alpha Re V_0 r] \cdot T_n'' + r[-(2\alpha^2 + 1) - 3i\alpha Re V_0 r] \cdot T_n'$$
$$+ \{(\alpha^2 - 1)^2 + i\alpha Re r[V_0(\alpha^2 - 2) + rDV_0 + r^2 D^2 V_0]\} \cdot T_n \} a_n$$
$$= c \sum_{n=0}^{N-1} \{ i Re r[r^3 \cdot T_n'' + 3r^2 \cdot T_n' + (1 - \alpha^2) r \cdot T_n]\} a_n$$

$$\tag{5.49}$$

For convenience, here T_n represents T_n^* or T_n^{**}.

Use Chebyshev nodes in the interval $(0,1)$ to determine r_j:

$$r_j = \frac{1}{2}\left[\cos\left(\frac{2j+1}{2M}\pi\right) + 1\right], \qquad j = 0,1,\dots,M-1 \tag{5.50}$$

$$r_j = \frac{1}{2}\left[\cos\left(\frac{2j+1}{2M}\pi\right) + 1\right]^{\frac{1}{2}}, \qquad j = 0,1,\dots,M-1 \tag{5.51}$$

The boundary conditions are $u(0) = u'(0) = 0$ and $u(1) = u'(1) = 0$:

$$\sum_{n=0}^{N-1} a_n T_n(0) = 0, \qquad \sum_{n=0}^{N-1} a_n T_n'(0) = 0,$$

$$\sum_{n=0}^{N-1} a_n T_n(1) = 0, \qquad \sum_{n=0}^{N-1} a_n T_n'(1) = 0. \tag{5.52}$$

The derivatives have the following recurrence relation for linear argument:

$$T_n^{(k)}(r_j) = 4n T_{n-1}^{(k-1)}(r_j) + \frac{n}{n-2} T_{n-2}^{(k)}(r_j), \qquad n = 3, 4, \cdots \tag{5.53}$$

The first three terms from the first derivative to the forth derivative are:

$$T_0'(r_j) = 0, \qquad T_1'(r_j) = 2, \qquad T_2'(r_j) = 8r - 16,$$
$$T_0''(r_j) = 0, \qquad T_1''(r_j) = 0, \qquad T_2''(r_j) = 16,$$
$$T_0'''(r_j) = 0, \qquad T_1'''(r_j) = 0, \qquad T_2'''(r_j) = 0,$$
$$T_0''''(r_j) = 0, \qquad T_1''''(r_j) = 0, \qquad T_2''''(r_j) = 0. \tag{5.54}$$

Apply Eq. (5.41) on the whole Chebyshev nodes with boundary conditions Eq. (5.52), a matrix form of generalized eigenvalue problem is given by the form

$$\boldsymbol{Aa} = c\boldsymbol{Ba}. \tag{5.55}$$

Matrix $\boldsymbol{A}$ and $\boldsymbol{B}$ have $(M + 4) * N$ dimension. The size of row is $M + 4$ not M, the reason is that the Chebyshev Nodes do not include two boundary points, four rows of boundary conditions are thus added to Matrix A.

The right-hand side $c\boldsymbol{Ba} =$

$$c \begin{pmatrix} T_0(1) & T_1(1) & \cdots \\ T_0'(1) & T_1'(1) & \cdots \\ T_0''(y_1) - k^2 T_0(y_1) & T_1''(y_1) - k^2 T_1(y_1) & \cdots \\ \vdots & \vdots & \vdots \\ T_0''(y_M) - k^2 T_0(y_M) & T_1''(y_M) - k^2 T_1(y_M) & \cdots \\ T_0'(0) & T_1'(0) & \cdots \\ T_0(0) & T_1(0) & \cdots \end{pmatrix} \begin{pmatrix} a_0 \\ a_1 \\ a_2 \\ \vdots \\ a_{N-3} \\ a_{N-2} \\ a_{N-1} \end{pmatrix} \tag{5.56}$$

The matrix form of this system can be written as

$$B^g A a = c a. \tag{5.57}$$

B^g represents the generalized inverse matrix of B with $(M + 4) * N$, it can be obtained by using least squares. c appears as eigenvalue of matrix $B^g A$ with the associated eigenfunction $\hat{u}(r) = \sum_{n=0}^{N-1} a_n T_n(r)$.

Then we have a set of flow modes, denoted as $\{u_n, c_n\}_{n=0}^{N-1}$. Note that if c_n has greatest imaginary part, then associated u_n is most unstable.

For quadratic argument, the derivatives have the following recurrence relation:

$$T_n^{(k)}(r_j) = 4nr T_{n-1}^{(k-1)}(r_j) + \frac{n}{n-2} T_{n-2}^{(k)}(r_j), \qquad n = 3, 4, \cdots \tag{5.58}$$

The first three terms from the first derivative to the forth derivative are:

$$
\begin{aligned}
T_0'(r_j) &= 0, & T_1'(r_j) &= 4r, & T_2'(r_j) &= 16r(2r^2 - 1), \\
T_0''(r_j) &= 0, & T_1''(r_j) &= 4, & T_2''(r_j) &= 96r^2 - 16, \\
T_0'''(r_j) &= 0, & T_1'''(r_j) &= 0, & T_2'''(r_j) &= 192r, \\
T_0''''(r_j) &= 0, & T_1''''(r_j) &= 0, & T_2''''(r_j) &= 192.
\end{aligned}
\tag{5.59}
$$

Recall Eq. (5.53) and Eq. (5.54), for quadratic case, the dimension of A, B and B^g is $(M + 3) * N$, the reason is that the derivative of $T_n(0)$ is always be zero since every term contains r:

$$T_n'(0) = 0 \ for \ all \ n = 0, 1, \cdots, N - 1.$$

The right-hand side $c B a =$

$$c \begin{pmatrix} T_0(1) & T_1(1) & \cdots \\ T_0'(1) & T_1'(1) & \cdots \\ T_0''(y_1) - k^2 T_0(y_1) & T_1''(y_1) - k^2 T_1(y_1) & \cdots \\ \vdots & \vdots & \vdots \\ T_0''(y_M) - k^2 T_0(y_M) & T_1''(y_M) - k^2 T_1(y_M) & \cdots \\ T_0(0) & T_1(0) & \cdots \\ T_0(0) & T_1(0) & \cdots \end{pmatrix} \begin{pmatrix} a_0 \\ a_1 \\ a_2 \\ \vdots \\ a_{N-3} \\ a_{N-2} \\ a_{N-1} \end{pmatrix} \qquad (5.60)$$

DNS OBSERVATIONS AND NUMERICAL RESULTS

Comparison of Two Shifted Chebyshev Polynomials in a Hyperbolic Case

To test the convergence property of two Shifted Chebyshev polynomials, a hyperbolic function $V_0 = tanh(5r)$ is used as the velocity of base flow to solve the linear system, Eq. (5.55).

Fig. (6-1) shows the velocity profile of the base flow, the first derivative term DV_0 and the second derivative term DDV_0 can be easily obtained by the hyperbolic function. For this base flow, the gradient of velocity is large near $r = 0$ and is almost zero near $r = 1$.

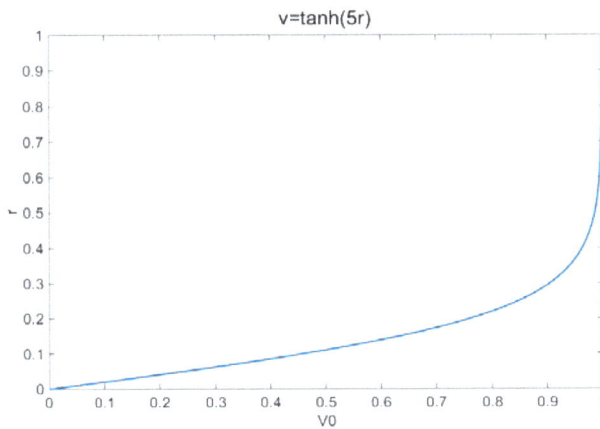

Fig. (6-1). Illustration of hyperbolic function $v = tanh(5r)$ on $r \in [0,1]$.

Recall Eq. (5.49). The problem is solved by using MATLAB, given $Re = 1000$, $\alpha = 1$, and $N = M = 100$. Fig. (**6-2**) gives the distribution of imaginary part of eigenvalues c (*i.e.* c_i) with Shifted Chebyshev of linear transformation. In Fig. (**6-2b**), one positive c_i is presented, it represents one unstable mode with $c_i = 0.2004$. The graph of this unstable mode is shown in Fig. (**6-3**). The eigenfunction $v = \sum_0^{N-1} a_n T_n$ is complex, thus Fig. (**6-3**) provides the real part, imaginary part and modulus of it.

The distribution of imaginary eigenvalues c_i and graph of unstable mode ($c_i = 0.2213$) under Shifted Chebyshev polynomials of quadratic transformation are presented in Figs. (**6-4** and **6-5**).

Recall the previous section, the different basis functions present different collocation points. Upon Linear Shifted Chebyshev polynomials, points clustered near both $r = 0$ and $r = 1$; while upon Quadratic Shifted Chebyshev polynomials, they only clustered near $r = 1$.

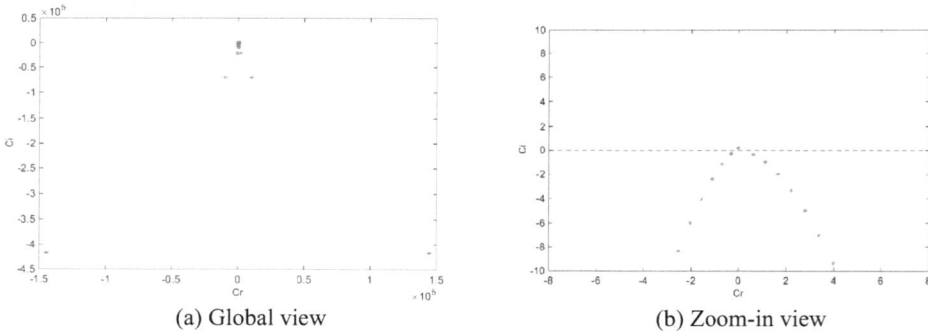

(a) Global view   (b) Zoom-in view

Fig. (6-2). Spectrum distribution of using $T_n(2r - 1)$.

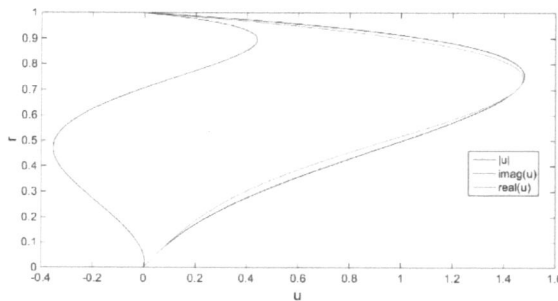

Fig. (6-3). Eigenfunction $\hat{u}$ associated with $c = -0.0102 + 0.2004i$ on $T_n(2r - 1)$.

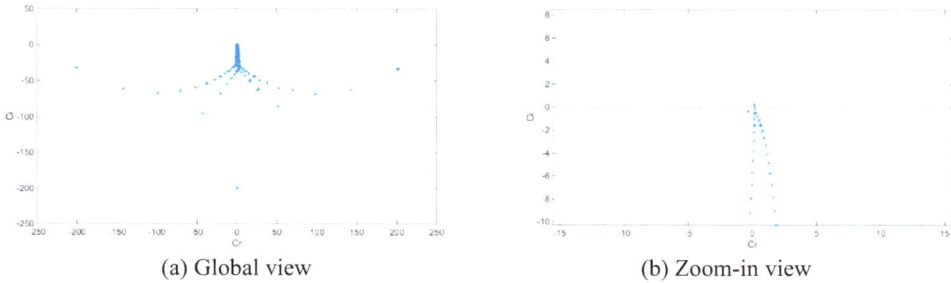

(a) Global view   (b) Zoom-in view

Fig. (6-4). Spectrum distribution of using $T_n(2r^2 - 1)$.

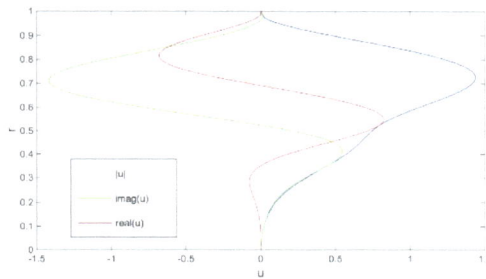

Fig. (6-5). Eigenfunction û associated with c = $0.1709 + 0.22146i$ on $T_n(2r^2 - 1)$.

The convergence property of the Chebyshev Collocation Method upon two Shifted Chebyshev polynomials is listed in Table **6-1**. M is the number of grid points (Chebyshev nodes). In Table **6-1**, with M increased gradually, the convergence of the eigenvalues on quadratic Shifted Chebyshev polynomials is observed, while eigenvalues on Linear Shifted Chebyshev polynomials are not convergent. The results by using linear transformation are not reliable.

In quadratic case, four-digit accuracy is obtained.

Table 6-1. The Convergence property of c_i on two Shifted Chebyshev Polynomials.

M	$T_n(2r - 1)$	$T_n(2r^2 - 1)$
100	0.20044245	0.22127542
120	0.20780676	0.22145990
140	0.21363761	0.22146397
160	0.21835979	0.22145546
180	0.22225536	0.22147179
200	0.22552000	0.22146576
400	0.24190921	0.22145805
800	0.25114780	0.22146244

DNS Leg-Like Vortices Cases

Section 6.1 shows the convergence property of Chebyshev collocation method with different basis functions (linear and quadratic transformation) with base flow as a mathematical hyperbolic function. In this and next section, quasi-rotation flow in our DNS data are carefully discussed in this and next section. Quasi-rotation flow locates on the core of vortex structure, it can be well captured with Ω vortex identification method.

Hairpin vortices are widely recognized as a fundamental coherent structure since their appearance in every significant process during transition. The hairpin vortex usually consists of three parts: (1) Two counter-rotating quasi-streamwise vortices, known as two legs; (2) A ring-like vortex named as the vortex head, where the spanwise vorticity is dominant, sitting on top farther from the wall, (3) Necks connect the head and legs.

In this section, the DNS observation and numerical results of stability based on leg-like vortex are discussed. The other typical vortex, ring-like vortex, is discussed in next section.

DNS Observations

Fig. (**6-6**) provides the structure of transition flow at $t = 6.87T$, T is the period. The spanwise tubes and hairpin vortices are captured at $\Omega = 0.5$, the spanwise tubes cannot represent rotation. With stretch and distortion, the tubes become Hairpin vortices gradually. Leg-like vortices are generated within this process.

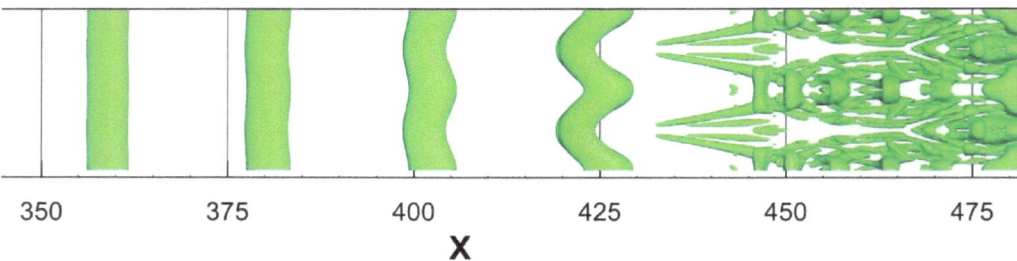

Fig. (6-6). X-Z view of the transition flow at $\Omega = 0.5$ on $t = 6.87T$.

Fig. (**6-7**) gives the distribution of shear on three directions on certain sections of spanwise tubes. Fig. (**6-8**) shows the generation of leg-like vortices with $\Omega = 0.52$, the isosurface describes rotation with the reason $\Omega > 0.5$ based on the Ω vortex identification method.

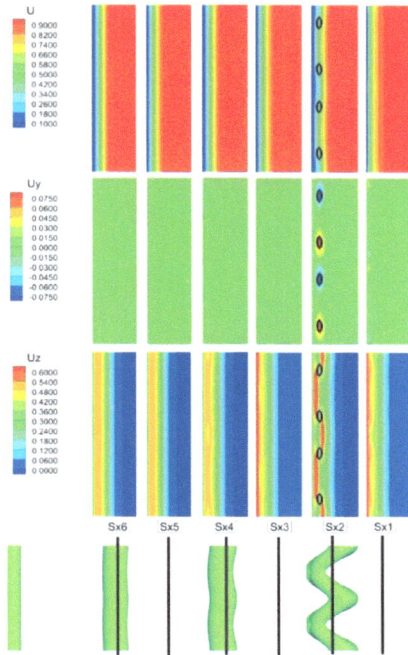

Fig. (6-7). Distribution of u-velocity and correlated shear stress on certain six slices.

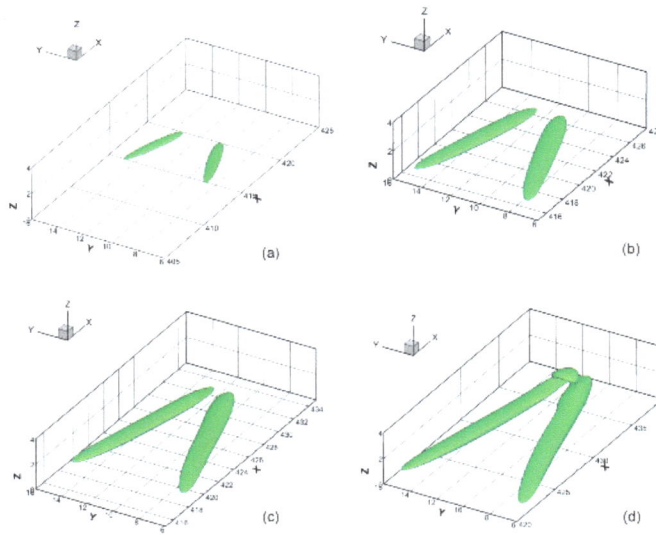

Fig. (6-8). Generation of one leg-like vortex at $\Omega = 0.52$ on (**a**) t=5.87T; (**b**) t=6.01T; (**c**) t=6.16T; (**d**) t=6.30T.

To investigate the quasi-rotation flow with leg-like vortex, a fully developed leg is shown in Fig. (**6-9**). Fig. (**6-10**) depicts three y-section located in $x = 430.6$, 434.7 and 439.8.

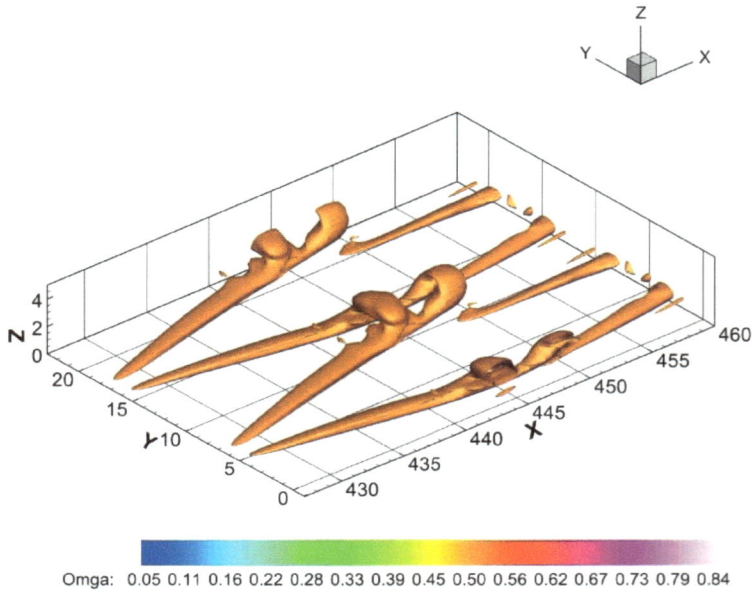

Omga: 0.05 0.11 0.16 0.22 0.28 0.33 0.39 0.45 0.50 0.56 0.62 0.67 0.73 0.79 0.84

Fig. (6-9). Fully developed leg-like vortices ($x = 427$ to 440) at $\Omega = 0.52$ on $t = 6.60T$.

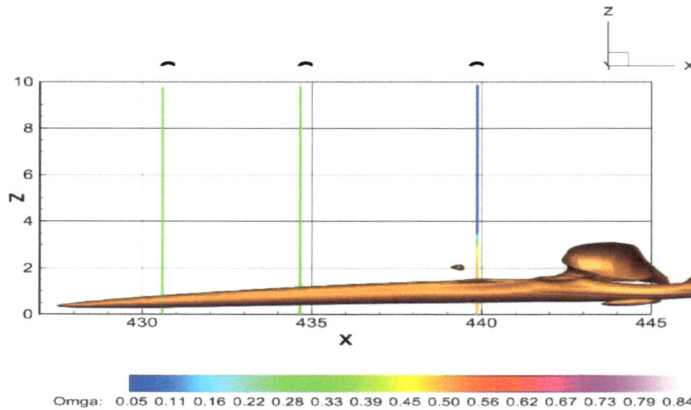

Omga: 0.05 0.11 0.16 0.22 0.28 0.33 0.39 0.45 0.50 0.56 0.62 0.67 0.73 0.79 0.84

Fig. (6-10). x-z view of leg-like vortex with three y-sections.

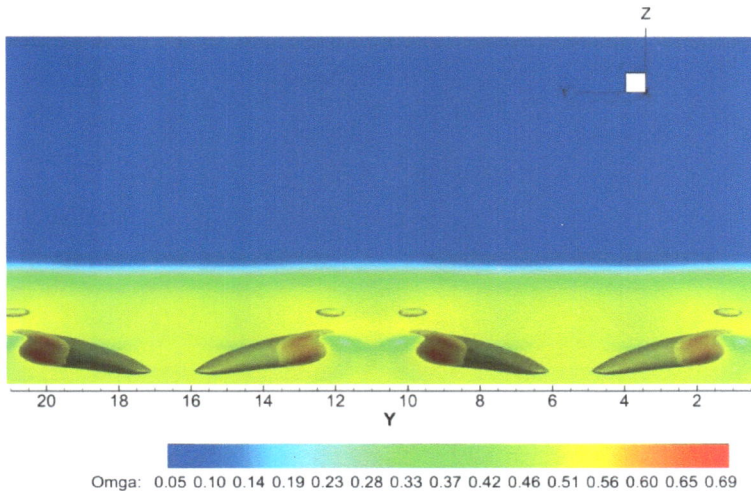

Fig. (6-11). y-section slice at $x = 439.8$ with Ω value contour.

A certain y-section slice located at $x = 439.8$ with Ω value and spanwise velocity v contours are shown in Fig. (**6-11, 12** and **13**). Fig. (**6-11**) gives the distribution of Ω value on this slice, highest Ω values at about 0.65 are obtained inside the leg-like vortices. Fig. (**6-12**) verifies Liu's theory that high Ω value region presents the core of rotation. The reason is the existence of the rotational streamline on high Ω region. In Fig. (**6-13**), distribution of v velocity provides a typical rotation flow profile.

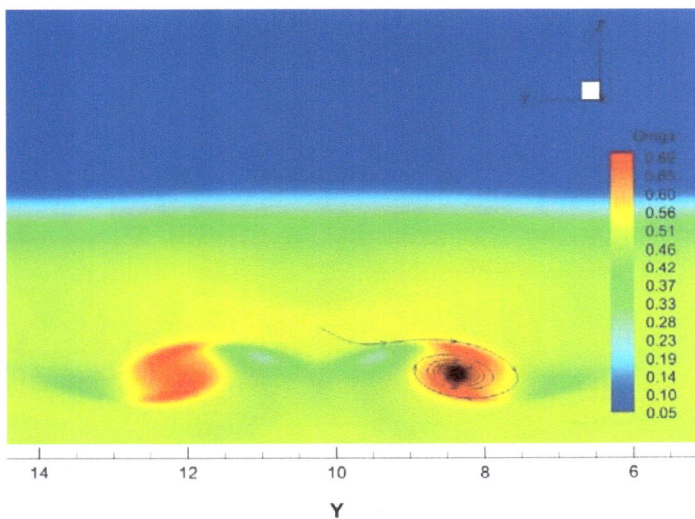

Fig. (6-12). y-section slice at $x = 439.8$, Ω value contour with a streamline.

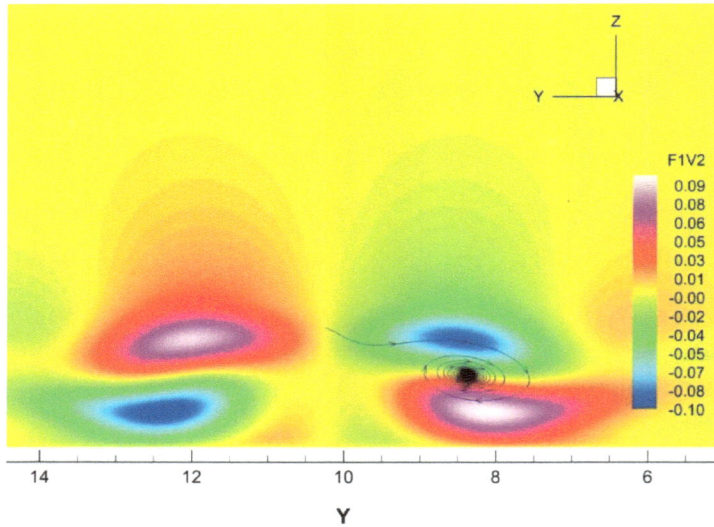

Fig. (6-13). y-section slice at $x = 439.8$, v value contour with a streamline.

Numerical Results

Extracted spanwise v velocity profiles from slices S1, S2 and S3 (see Fig. **6-10**), then applied them as the quasi-rotational base flow to solve Eq. (5.49) by using Quadratic Shifted Chebyshev polynomials associated with collocation points (Chebyshev nodes).

The Reynolds number $Re=1000$ in DNS case, given $\alpha = 1$, $N = 100$ and $M = 200$. N is the order of Shifted Chebyshev polynomials, M is the number of collocation points.

Three base flow velocity profiles related to S1, S2 and S3 are shown in Fig. (**6-14**). The least stable imaginary eigenvalues c_i for three cases are given in Table **6-2**. Spectrum distributions are depicted in Fig. (**6-15**). Fig. (**6-16**) shows the eigenfunction $\hat{u}$ of these unstable modes.

Table **6-2** also provides the Ω values of each core of vortex, largest Ω value connected with smallest positive imaginary eigenvalue, and smallest Ω with the largest positive. However, three cases are not enough to show the relationship between Ω value and stability. To study the stability of leg-like vortex deeply, Fig. (**6-17**) is presented, the Ω values and least stable c_i are obtained by tracking a vortex over time. The result shows that with the Ω increasing over time, the vortex tends to be more stable with the decrement of the positive imaginary part of eigenvalues.

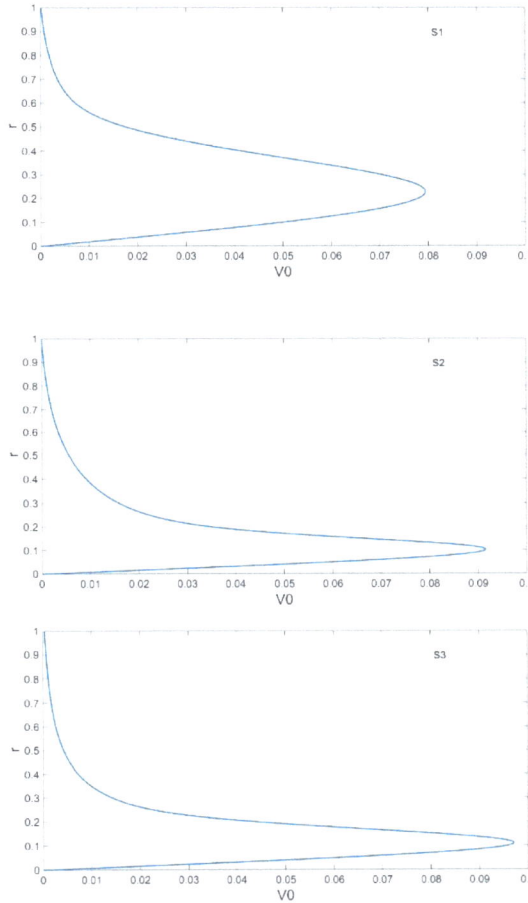

Fig. (6-14). Velocity profiles from (**a**) s1; (**b**) s2; (**c**) s3.

Table 6-2. Most unstable eigenvalues related with Omega value.

Slices	c_i	Ω
1	0.0303338	0.632872
2	0.0161457	0.662620
3	0.0148870	0.668555

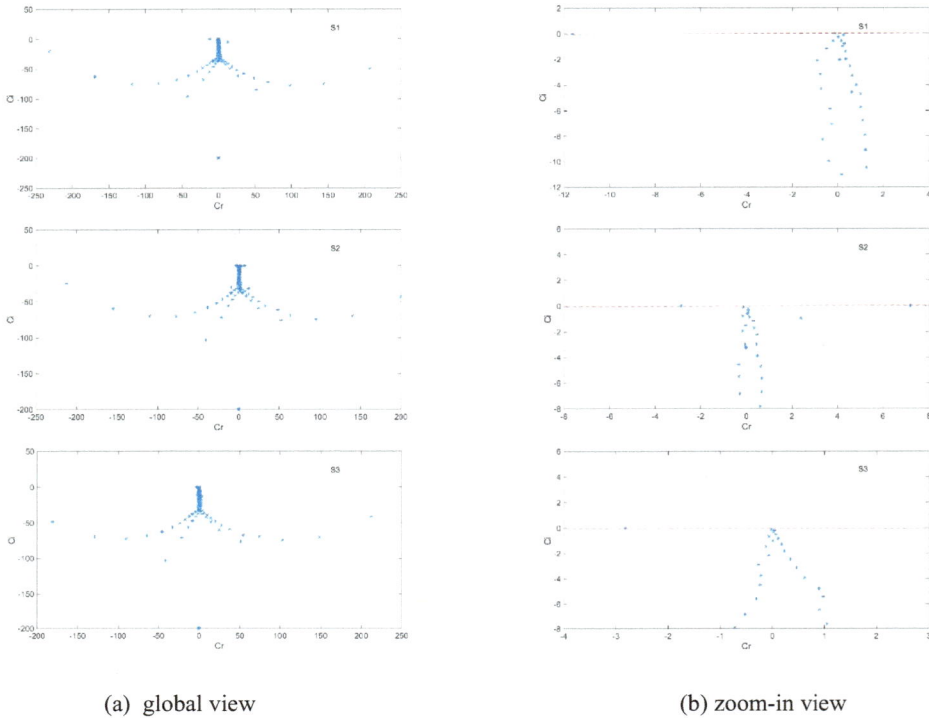

(a) global view   (b) zoom-in view

Fig. (6-15). Distribution of spectrums on three base flows.

DNS Ring-Like Vortices Cases

DNS Observations

Ring-like vortices are generated at the head of the leg-like vortices (see Fig. **6-18**). Similar with leg-like vortices, high shear stress locates the same region before ring-like vortices appear. Figs. (**6-19** and **6-20**) provide some illustrations of u velocity and shear layers in certain regions.

Numerical Results

The technique used to solve ring-like vortices problems is same with leg-like vortices. Fig. (**6-21**) gives the section of ring-like vortices along with streamwise direction at $t = 10.914$. Fig. (**6-22**) shows the Ω distribution on the certain slice. The velocity profile of the given slice is shown in Fig. (**6-23**). The numerical results under this base flow, is provided in Figs. (**6-24** and **6-25**), showing the eigenvalues and eigenfunctions respectively.

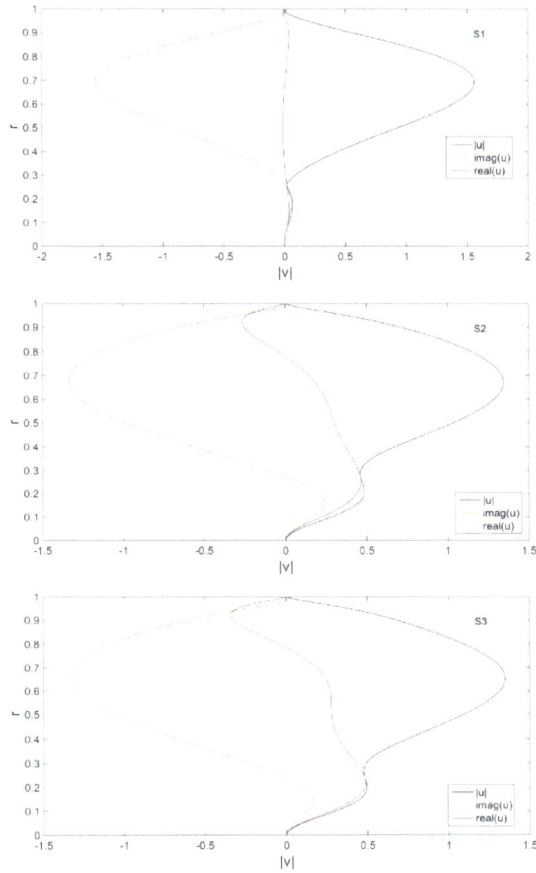

Fig. (6-16). Eigenfunction $\hat{u}$ on three base flows.

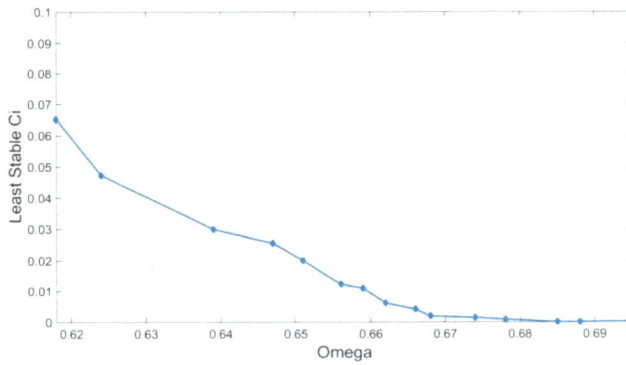

Fig. (6-17). Least stable c_i *vs* Ω value on leg-like vortex.

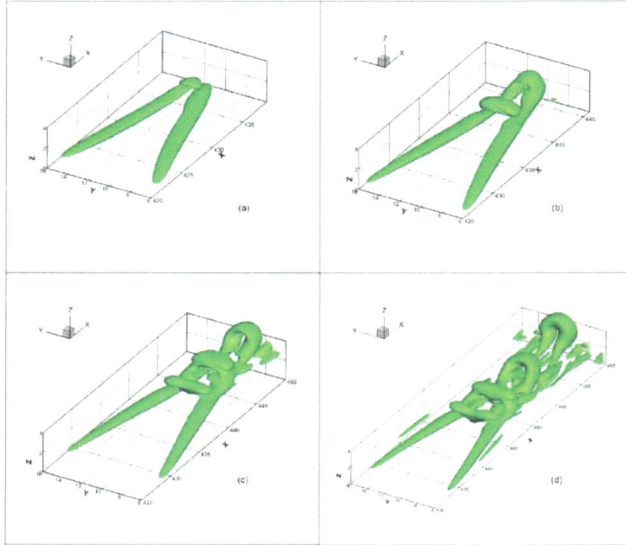

Fig. (6-18). Generation of ring-like vortex at (**a**) t=6.30T; (**b**) t= 6.52T; (**c**) t=6.66T; (**d**) t=6.87T.

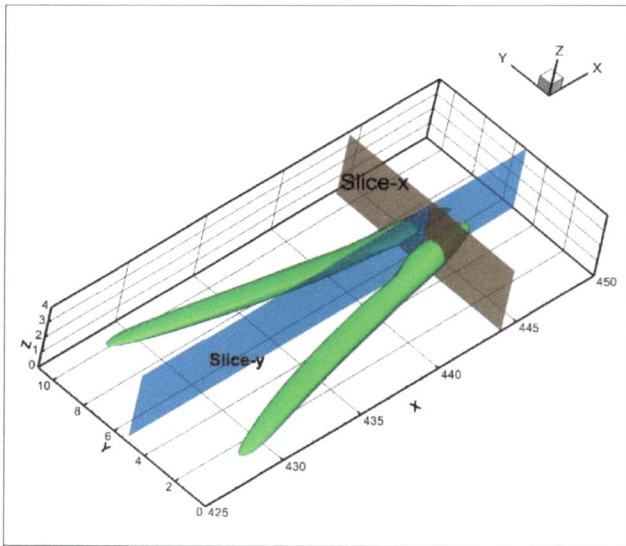

Fig. (6-19). Illustrations of Slice-x and y locations on DNS flow field.

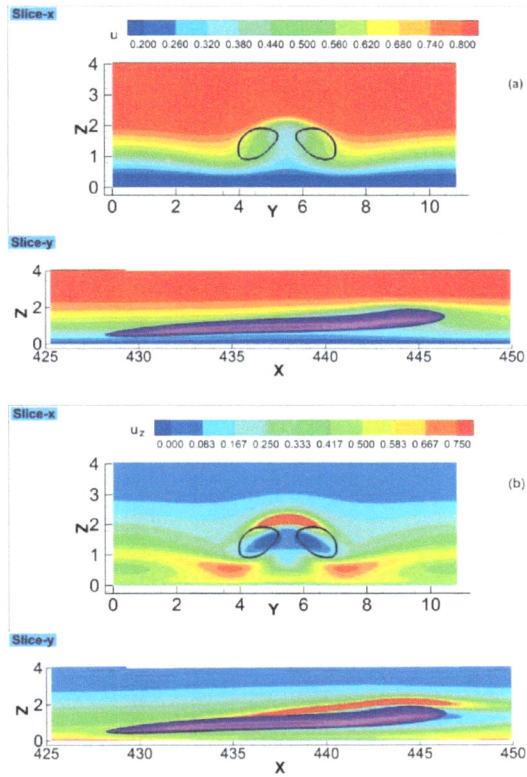

Fig. (6-20). Distribution of u velocity and gradient in z direction.

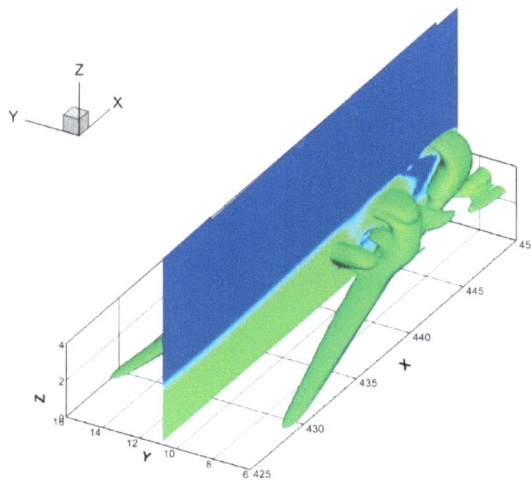

Fig. (6-21). Section of ring-like vortices at $y = 10.914$ on $t = 6.66T$.

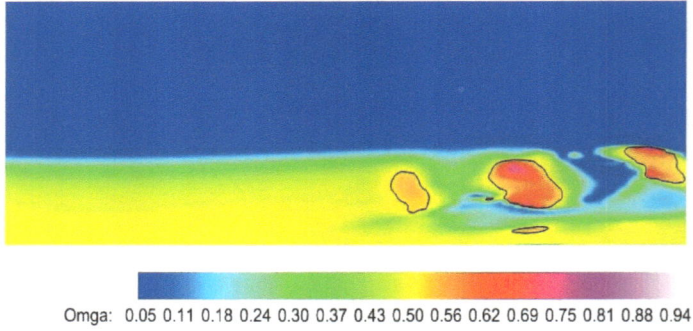

Omga: 0.05 0.11 0.18 0.24 0.30 0.37 0.43 0.50 0.56 0.62 0.69 0.75 0.81 0.88 0.94

Fig. (6-22). Distribution of Ω value with three rings.

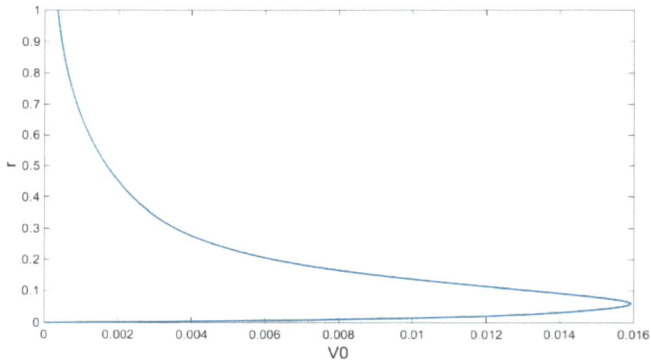

Fig. (6-23). Velocity Profile on the certain slice at $t = 6.66T$.

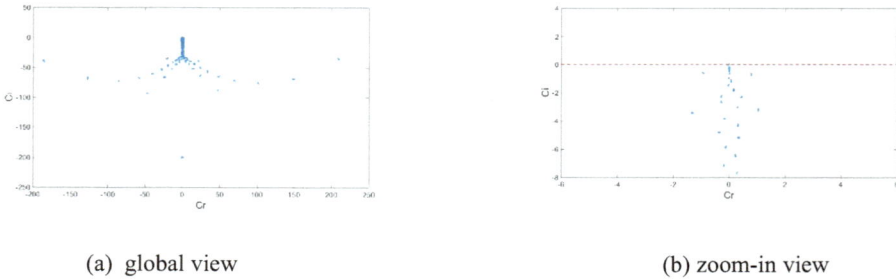

(a) global view   (b) zoom-in view

Fig. (6-24). Eigenvalues distribution with unstable $c_i = 0.0095$.

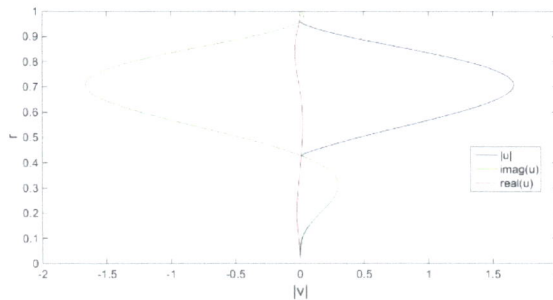

Fig. (6-25). Eigenfunction $\hat{u}$ with $c_i = 0.0095$.

By tracking a fixed streamwise ring-like vortex from t = 6.59T to 6.87T, the relationship between imaginary part of the eigenvalue with respect to the least stable mode and the Ω value on the core of vortex over time is shown on Fig. (**6-26**). It clearly describes that the vortex trends to be more stable as the Ω increased over time.

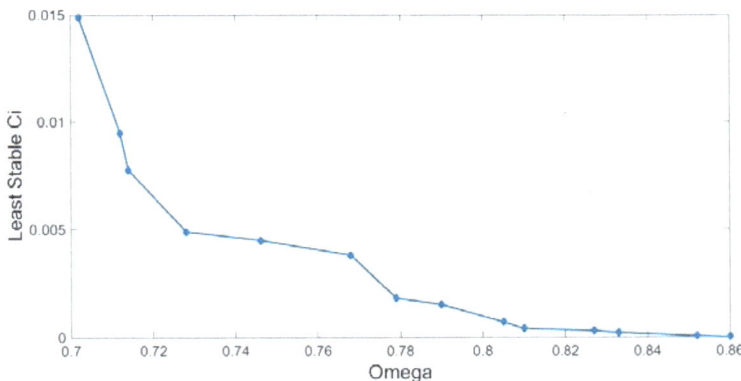

Fig. (6-26). Least stable c_i *vs* Ω value on ring-like vortex.

CONCLUSION

1. Chebyshev collocation method associated with Quadratic Shifted Chebyshev polynomials performs well in solving the fourth order ODE, the linear stability equation for symmetric flow.

2. Linear Shifted Chebyshev polynomials are invalid in interpolating certain equation. The clustering points near the singular point r = 0 in cylindrical coordinates might be dangerous in convergence.

3. The linear stability equation for symmetric flow can be used to analyze the stability of real-case quasi-rotation flow obtained from DNS data. The result shows that, in the flow transition process, new vortices are generated with vortex legs and vortex rings. Omega becomes larger and the vortex become less unstable in the fact of decreasing positive eigenvalues. When the Omega of vortex becomes larger, the flow becomes more stable.

4. Liu's theory that turbulent flow which is dominated by rotations is more stable than laminar flow which is dominated by shears has been numerically checked and verified.

CONSENT FOR PUBLICATION

Not applicable.

CONFLICT OF INTEREST

The authors confirm that this chapter contents have no conflict of interest.

ACKNOWLEDGEMENTS

The author is grateful to Dr. Chaoqun Liu for his guidance and countless helpful discussions. The author also thanks Texas Advanced Computation Center (TACC) for providing computation hours.

REFERENCES

[1]   G.K. Batchelor, *An Introduction to Fluid Dynamics*. Book, 1967, p. 631.
[2]   C.J. Geankoplis, *"Transport Processes and Separation Process Principles: Includes Unit Operations,"* in *Transport Processes and Separation Process Principles*. Includes Unit Operations, 2003, p. 1026.
[3]   R. Nave, *Laminar Flow.* .http://hyperphysics.phy-astr.gsu.edu/hbase/ pfric.html
[4]   I. Tani, "Boundary-Layer Transition", *Annu. Rev. Fluid Mech.,* vol. 1, pp. 169-196, 1969. [http://dx.doi.org/10.1146/annurev.fl.01.010169.001125]
[5]   Y.S. Kachanov, "Physical mechanisms of laminar-boundary-layer transition", *Annu. Rev. Fluid Mech.,* vol. 26, pp. 411-482, 1994.
        [http://dx.doi.org/10.1146/annurev.fl.26.010194.002211]
[6]   V. M. Epifanov, *Boundary Layer*. [http://dx.doi.org/10.1615/AtoZ.b.boundary_layer]
[7]   L. Prandtl, *Motion of Fluids with Very Little Viscosity*. Int. Math, 1904, pp. 1-8.
[8]   W.S. Saric, H.L. Reed, and E.J. Kerschen, "Boundary-Layer Receptivity to Freestream Disturbances", *Annu. Rev. Fluid Mech.,* vol. 34, no. 1, pp. 291-319, 2002. [http://dx.doi.org/10.1146/annurev.fluid.34.082701.161921]
[9]   L. M. Mack, "Boundary-Layer Linear Stability Theory", *AGARD Special Course on Stability and Transition of Laminar Flow,* pp. 1-81, 1984.
[10] M. V Morkovin, and W. Patterson, *Critical evaluation of transition from laminar to turbulent shear layer with emphasis on hypersonically traveling bodies,* 1969.
[11] M.V. Morkovin, *"Bypass transition to turbulence and research desiderata,"* NASA CP-2386 Transit. Turbines, 1985, pp. 161-204.
[12] M.V. Morkovin, and E. Reshotko, "Dialogue on progress and issues in stability and transition research", *IUTAM*

Symposium, pp. 3-29 Toulouse/France [http://dx.doi.org/10.1007/978-3-642-84103-3_1]

[13]   R.I. Loehrke, M.V. Morkovin, and A.A. Fejer, "Review — Transition in nonreversing oscillating boundary layers", *J. Fluids Eng.*, vol. 97, no. 4, pp. 534-549, 1975. [http://dx.doi.org/10.1115/1.3448111]

[14]   E. Reshotko, "Boundary-Layer Stability and Transition", *Annu. Rev. Fluid Mech.*, vol. 8, no. 1, pp. 311-349, 1976. [http://dx.doi.org/10.1146/annurev.fl.08.010176.001523]

[15]   V. P. Maksimov, "Origin of Tollmien–Schlichting waves in oscillating boundary layers", *Development of Disturbances in Boundary Layer*, pp. 68-75, 1979.

[16]   Y.S. Kachanov, "The transformation of external disturbances into the boundary layer waves", *Sixth International Conference on Numerical Methods in Fluid Dynamics*, 1979pp. 299-307 [http://dx.doi.org/10.1007/3540091157_181]

[17]   A.M. Zhigulyov, and V.N. Tumin, *Onset of turbulence. Dynamical theory of excitation and development of instabilities in boundary layers*, 1987.

[18]   M. Nishioka, and M. Morkovin, "Boundary-layer receptivity to unsteady pressure gradients: experiments and overview", *J. Fluid Mech.*, vol. 171, pp. 219-261, 1986. [http://dx.doi.org/10.1017/S002211208600143X]

[19]   M.E. Goldstein, and L.S. Hultgren, "Boundary-Layer Receptivity to Long-Wave Free-Stream Disturbances", *Annu. Rev. Fluid Mech.*, vol. 21, no. 1, pp. 137-166, 1989. [http://dx.doi.org/10.1146/annurev.fl.21.010189.001033]

[20]   R.A.E. Heinrich, and E.J. Kerschen, "Leading-edge boundary layer receptivity to various free-stream disturbance structures", *Z. Angew. Math. Mech.*, vol. 69, no. 6, pp. T596-T598, 1989.

[21]   E.J. Kerschen, "Boundary Layer Receptivity Theory", *Appl. Mech. Rev.*, vol. 43, no. 5S, p. S152, 1990. [http://dx.doi.org/10.1115/1.3120795]

[22]   W. Tollmien, *Ueber die Entstehung der Turbulenz - The production of turbulence*, 1929.

[23]   D.H. Schlichting, and K. Gersten, "Boundary-layer theory", *Eur. J. Mech. BFluids*, vol. 20, p. 817, 1979.

[24]   S.C. Lin, E.L. Resler, and A. Kantrowitz, "Electrical conductivity of highly ionized argon produced by shock waves", *J. Appl. Phys.*, vol. 26, no. 1, pp. 95-109, 1955. [http://dx.doi.org/10.1063/1.1721870]

[25]   T. Herbert, "On finite amplitudes of periodic disturbances of the boundary layer along a flat plate", *Proceedings of the Fourth International Conference on Numerical Methods in Fluid Dynamics*, 1975. [http://dx.doi.org/10.1007/BFb0019753]

[26]   L. Kleiser, and T.A. Zang, "Numerical simulation of transition in wall-bounded shear flows", *Annu. Rev. Fluid Mech.*, vol. 23, no. 1, pp. 495-537, 1991. [http://dx.doi.org/10.1146/annurev.fl.23.010191.002431]

[27]   N.D. Sandham, and L. Kleiser, "The late stages of transition to turbulence in channel flow", *J. Fluid Mech.*, vol. 245, pp. 319-348, 1992. [http://dx.doi.org/10.1017/S002211209200048X]

[28]   U. Rist, and Y.S. Kachanov, "Numerical and experimental investigation of the K-regime of boundary- layer transition", *IUTAM Symposium*, pp. 405-412 Sendai/Japan/. [http://dx.doi.org/10.1007/978-3-642-79765-1_48]

[29]   S. Bake, D.G.W. Meyer, and U. Rist, "Turbulence mechanism in Klebanoff transition: a quantitative comparison of experiment and direct numerical simulation", *J. Fluid Mech.*, vol. 459, pp. 217-243, 2002. [http://dx.doi.org/10.1017/S0022112002007954]

[30]   V.I. Borodulin, V.R. Gaponenko, Y.S. Kachanov, D.G.W. Meyer, U. Rist, Q.X. Lian, and C.B. Lee, "Late-stage transitional boundary-layer structures. Direct numerical simulation and experiment", *Theor. Comput. Fluid Dyn.*, vol. 15, no. 5, pp. 317-337, 2002. [http://dx.doi.org/10.1007/s001620100054]

[31]   Y. S. Kachanov, "On a Universal Mechanism of Turbulence Production in Wall Shear Flows", *Recent Results in Laminar-Turbulent Transition*, pp. 4-5, 2004. [http://dx.doi.org/10.1007/978-3-540-45060-3_1]

[32]   C. Liu, Y. Yan, and P. Lu, "Physics of turbulence generation and sustenance in a boundary layer", *Comput. Fluids*, vol. 102, pp. 353-384, 2014. [http://dx.doi.org/10.1016/j.compfluid.2014.06.032]

[33]   L. Chen, and C. Liu, ""Vortical structure, sweep and ejection events in transitional boundary layer," Sci. China, Ser. G, Physics", *Mech. Astron.*, vol. 39, no. 10, pp. 1520-1526, 2009.

[34]   C. Liu, and L. Chen, "DNS for Ring - Like Vortices Formation and Roles in", *48th AIAA Aerospace Sciences Meeting*, 2010p. 1471

[35]   L. Chen, and D. Lin, "Tang, P. Lu and C. Liu, "Evolution of the vortex structures and turbulent spots at the late-stage of transitional boundary layers", *Sci. China Phys. Mech. Astron.*, vol. 54, no. 5, pp. 986-990, 2011. [http://dx.doi.org/10.1007/s11433-011-4266-4]

[36]   L. Chen, and C. Liu, "Numerical study on mechanisms of second sweep and positive spikes in transitional flow on a flat plate", *Comput. Fluids*, vol. 40, no. 1, pp. 28-41, 2011. [http://dx.doi.org/10.1016/j.compfluid.2010.07.016]

[37]   C. Liu, and L. Chen, "Study of Mechanism of Ring-Like Vortex Formation in Late Flow Transition", *AIAA Aerospace Sciences Meeting*, vol. vol. 1456, 2010pp. 1-21 [http://dx.doi.org/10.2514/6.2010-1456]

[38]   L. Chen, X. Liu, M. Oliveira, and C. Liu, *DNS for Late Stage Structure of Flow Transition on a Flat- Plate Boundary Layer*, 2010.
       [http://dx.doi.org/10.2514/6.2010-1470]

[39]   C. Liu, L. Chen, and P. Lu, *New findings by high-order DNS for late flow transition in a boundary layer.* vol. Vol. 2011. Model. Simul. Eng., 2011.

[40]   C. Liu, and L. Chen, "Parallel DNS for vortex structure of late stages of flow transition", *Comput. Fluids,* vol. 45, no. 1, pp. 129-137, 2011.
       [http://dx.doi.org/10.1016/j.compfluid.2010.11.006]

[41]   C. Liu, "Numerical and theoretical study on 'vortex breakdown,'", *Int. J. Comput. Math.,* vol. 88, no. 17, pp. 3702-3708, 2011.
       [http://dx.doi.org/10.1080/00207160.2011.617438]

[42]   C. Liu, L. Chen, P. Lu, and X. Liu, "Study on multiple ring-like vortex formation and small vortex generation in late flow transition on a flat plate", *Theor. Comput. Fluid Dyn.,* vol. 27, no. 1–2, pp. 41- 70, 2013.
       [http://dx.doi.org/10.1007/s00162-011-0247-5]

[43]   P. Lu, Q. Li, and C. Liu, "Numerical Study of Mechanism of U-shaped Vortex Formation", *49th AIAA Aerospace Sciences Meeting,* 2011p. 286.
       [http://dx.doi.org/10.2514/6.2011-286]

[44]   C. Liu, and Z. Liu, "Multigrid mapping and box relaxation for simulation of the whole process of flow transition in 3D boundary layers", *J. Comput. Phys.,* vol. 119, no. 2, pp. 325-341, 1995. [http://dx.doi.org/10.1006/jcph.1995.1138]

[45]   Z. Liu, G. Xiong, and C. Liu, "Direct numerical simulation for the whole process of transition on 3-D airfoils", *27th AIAA Fluid Dynamics Conference,* 1996p. 2081 [http://dx.doi.org/10.2514/6.1996-2081]

[46]   Z. Liu, and C. Liu, *Direct Numerical Simulation for Flow Transition Around Airfoils,* 1997.

[47]   P. Lu, Z. Wang, L. Chen, and C. Liu, "Numerical study on U-shaped vortex formation in late boundary layer transition", *Comput. Fluids,* vol. 55, no. January, pp. 36-47, 2012. [http://dx.doi.org/10.1016/j.compfluid.2011.10.014]

[48]   Y. Yan, and C. Liu, "Shear layer stability analysis in later boundary layer transition and MVG controlled flow", *51st AIAA Aerospace Sciences Meeting,* 2013p. 531 [http://dx.doi.org/10.2514/6.2013-531]

[49]   Y. Yan, C. Chen, H. Fu, and C. Liu, "DNS study on Δ-vortex and vortex ring formation in flow transition at Mach number 0.5", *J. Turbul.,* vol. 15, no. 1, pp. 1-21, 2014. [http://dx.doi.org/10.1080/14685248.2013.871023]

[50]   Y. Wang, Y. Yang, G. Yang, and C. Liu, "DNS Study on Vortex and Vorticity in Late Boundary Layer Transition", *Commun. Comput. Phys.,* vol. 22, no. 2, pp. 441-459, •••. [http://dx.doi.org/10.4208/cicp.OA-2016-0183]

[51]   C. Liu, and X. Cai, "New theory on turbulence generation and structure—DNS and experiment", *Sci. China Phys. Mech. Astron.,* vol. 60, no. 8, 2017.
       [http://dx.doi.org/10.1007/s11433-017-9047-2]

[52]   M. By Claudio Canuto, *Yousuff Hussaini, Alfio Quarteroni, Thomas A. and Jr. Zang, Spectral Methods in Fluid Dynamics.* Springer Science & Business Media, 1988.
       [http://dx.doi.org/10.1007/978-3-642-84108-8]

[53]   R. Peyret, *"Chebyshev method," in Spectral methods for incompressible viscous flow.* Springer Science & Business Media, 2013, p. 39.

[54]   S.A. Orszag, "Accurate solution of the Orr-Sommerfeld stability equation", *J. Fluid Mech.,* vol. 50, no. 4, pp. 659-703, 1971.
       [http://dx.doi.org/10.1017/S0022112071002842]

[55]   L. Jiang, "Chang, Chau-Lyan, Choudhari, Meelan and Chaoqun Liu, "Cross-Validation of DNS and PSE Results for Instability-Wave Propagation in Compressible Boundary Layers past Curvilinear Surfaces", *16th AIAA Computational Fluid Dynamics Conference,* 2003 [http://dx.doi.org/10.2514/6.2003-3555]

[56]   C. Liu, Y. Wang, and Y. Yang, "Yong and Z. Duan, "New omega vortex identification method", *Sci. China Phys. Mech. Astron.,* vol. 59, no. 684711, 2016.

[57]   J.C.R. Hunt, A.A. Wray, and P. Moin, "Eddies, streams, and convergence zones in turbulent flows", *Proceedings of the Summer Program,* 1988pp. 193-208

[58]   J. Jeong, and F. Hussain, "On the identification of a vortex", *J. Fluid Mech.,* vol. 285, no. February, pp. 69-94, 1995.
       [http://dx.doi.org/10.1017/S0022112095000462]

CHAPTER 5

POD and DMD Analysis in Late Flow Transition with Omega Method

Sita Charkrit and Chaoqun Liu*

Department of Mathematics, University of Texas at Arlington, Arlington, Texas 76019, USA

Abstract: In this paper, the proper orthogonal decomposition (POD) and dynamic mode decomposition (DMD) are applied to analyze the 3D late transitional flow on the flat plate obtained from direct numerical simulation (DNS). POD is used to find the most persistent spatial structures while DMD is used to find single frequency modes. The Omesga method is applied as a vortex identification to visualize vortices with isosurfaces $\Omega = 0.52$. The results in POD and DMD are discussed and compared to show the same and different features such as shapes, amplitudes and time evolutions.

Keywords: Dynamic mode decomposition, Identification method, Modal decomposition, Late flow transition, Omega method, Proper orthogonal decomposition.

INTRODUCTION

In the study of fluid flow structure in computation fluid dynamics (CFD), the modal decomposition is a useful tool to extract the whole structure into coherent structures in different features such as energy content, mode shape, amplitude and frequency since a complex flow structure often consists of a combination of coherent structure. The two popular modal decomposition methods, *i.e.*, the proper orthogonal decomposition (POD) and dynamic mode decomposition (DMD), are presented in this paper. The POD and DMD have been widely applied to explore the complex flow fields since they can be used to decouple the spatial and temporal applications. These two methods help in further understanding of fundamental fluid processes since they can examine the dominant and coherent structures in fluid flows.

The POD is one of the most widely used techniques to analyze fluid flows. There are two versions of POD technique. The original POD, proposed by Lumley [1] in

*Corresponding author Chaoqun Liu:** Department of Mathematics, University of Texas at Arlington, Arlington, Texas 76019, USA; Tel: +1-8172725151; Fax: +1-8172725802; E-mail: cliu@uta.edu;

1967, is used to investigate the turbulent flow. The other version is called snapshot POD, which was introduced by Sirovich [2] in 1987. The snapshot POD is applied in order to optimize the computation. In POD, the flow structure is decomposed into orthogonal mode ranking by their kinetic energy content. In recent years, there have been many applications about POD in many fields of fluid dynamics. For examples, POD was used to study the turbulent pipe flow [3, 4]. In some studies [5, 6], POD was applied to identify turbulent discontinuous and nonlinear flows. A mixing layer downstream on a thick splitter plate obtain form DNS was analyzed by POD in a study [7]. The flow structure in transition stage has been analyzed by POD. For example, POD was applied to analyze coherent structures in pipe flow [8, 9] and a transitional boundary layer with and without control [10]. POD was also used to investigate asymmetric structures of flow on the flat plate [11]. The vortex structure in MVG wake was examined [12]. The entropy generation in a laminar separation boundary layer was analyzed by POD [13].

The other popular decomposition method used in this paper is DMD, which was first introduced by Schmid [14] in 2010 to extract the dynamic features by finding the relationship between each time step. Each coherent structure has a single feature in temporal mode. This method relates to Koopman operator explained in the paper of Rowley [15]. Many researchers applied DMD and compared both POD and DMD to analyze the flows in CFD. For instance, Premaratne and Hu [16] studied turbine wake characteristics by DMD. Alina and Navon [17] used DMD in shallow water and a swirling flow problem. Mohan and Gaitondey [18] worked on analysis of deep dynamic stall on a plunging airfoil by DMD. Both methods were compared in terms of identification of multi-dominant coherent structures and high-order harmonics buried in fluid flow in a study [19]. The POD and DMD on LES of subsonic jets were presented in a study [20].

The purpose of this work is to apply POD and DMD to investigate the complex flow on the flat plate in late transition stage obtained from direct numerical simulation (DNS) and to better understand its three-dimensional structure. Moreover, the vortex identification method called the omega method is introduced to visualize the vortex structures of spatial modes. This paper is organized as follows. In section 2, the detail of DNS data used in this paper is introduced. In section 3, the concept of Omega method is described. In section 4, descriptions of POD and DMD are explained. In addition, the analyses in both POD and DMD are discussed and compared in section 4.

CASE SETUP AND CODE VALIDATION

The case and code validation are presented in studies [21, 22]. The computational domain is demonstrated in Figs. (**1** and **2**). The grid level is 1920 x 128 x 241, representing the number of grids in streamwise (x), spanwise (y), and wall normal (z) directions. The grid is stretched in the normal direction and uniform in the streamwise and spanwise directions. The length of the first grid interval in the normal direction at the entrance is found to be 0.43 in wall units (Z+ = 0.43). The flow parameters, including Mach number, Reynolds number, etc. are listed in Table **1**. The DNS code, DNSUTA, has been validated by NASA Langley and UTA researchers carefully to make sure the DNS results are correct. For more detail about case setup and code validation, see [21, 22].

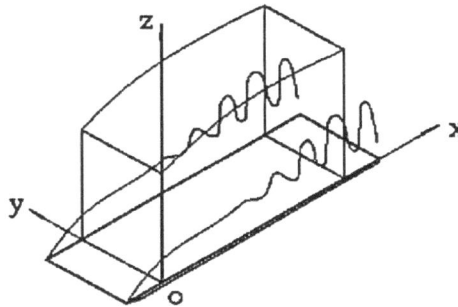

Fig. (1). Computation domain.

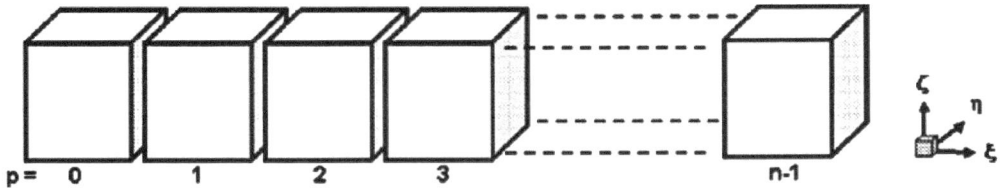

Fig. (2). Domain decomposition along the streamwise direction in the computational space.

Table 1. DNS parameters.

M_∞	Re	x_{in}	Lx	Ly	Lz_{in}
0.5	1000	$300.79\delta_{in}$	$798.03\delta_{in}$	$22\delta_{in}$	$40\delta_{in}$

The parameters in Table **1** are defined as:

M_∞ =   Mach number
Re =   Reynolds number
x_{in} =   distance between leading edge of flat plate and upstream boundary of computational domain
δ_{in} =   inflow displacement thickness
Lz_{in} =   height at inflow boundary
Lx =   length of computational domain along x direction
Ly =   length of computational domain along y direction

THE VORTEX IDENTIFICATION AND THE OMEGA METHOD

Vortex is one of typical structures of turbulent flows. In order to reveal the complexities of vortex structures in the flows, vortex identification methods have been developed to visualize vortices. In 1988, Hunt *et al.* [23] introduced Q-criterion, which is defined as a positive second invariant Q of the velocity gradient tensor. Q can be derived by

$$Q = \frac{1}{2}(\|B\|^2 - \|A\|^2),$$

which represents the balance between shear strain rate and vorticity magnitude. A and B are the symmetric and antisymmetric components of ∇V by the expressions

$$A = \frac{1}{2}(\nabla V + \nabla V^t)$$

and

$$B = \frac{1}{2}(\nabla V - \nabla V^t).$$

Another well-known method called the λ_2 method was introduced by Jeong and Hussain [24] in 1995. λ_2 is defined by the second eigenvalue of the symmetric tensor $A^2 + B^2$. It can capture the pressure minimum in a plane normal to the vortex axis.

However, there are some limitations of using Q and λ_2 as follows. Firstly, the physical meaning of Q and λ_2 is unclear. Secondly, a proper threshold is required for each related case. Moreover, Q and λ_2 are not able to capture both strong and weak vortices simultaneously. Therefore, it is difficult to obtain the accurate surface of vortex structures by applying Q and λ_2 or other methods since they have widely varying thresholds.

Recently, the Ω-method (Omega method) was developed in 2016 by Liu *et al.* [25] as a new identification method to overcome the weaknesses of such identification schemes. By giving a clear physical meaning of the Omega$-\Omega$, it is defined by a ratio of vorticity squared over the sum of vorticity squared and deformation squared:

$$\Omega = \frac{\|B\|_F^2}{\|A\|_F^2 + \|B\|_F^2} = \frac{b}{a+b+\varepsilon,} \tag{1}$$

where $a = Trace(A^t A)$, $b = Trace(B^t B)$, and ε is a small positive number used to avoid division by zero.

According to Liu *et al.* [25], the Omega method was compared with the other identification methods like Q-criterion and λ_2-method. It was confirmed that this method presents some advantages as follows:

    1) the method is able to capture vortex well and very easy to perform

    2) the physical meaning of Ω is clear while the interpretations of iso-surface values of Q and λ_2 chosen to visualize vortices are obscure

    3) being different from Q and λ_2 iso-surface visualization which requires wildly various thresholds to capture the vortex structure properly, Ω is pretty universal and does not need much adjustment in different cases and the iso-surfaces of $\Omega=0.52$ can always capture the vortices properly in all the cases at different time steps, which we investigated

    4) both strong and weak vortices can be captured well simultaneously while improper Q and λ_2 threshold may lead to strong vortex capture while weak vortices are lost or weak vortices are captured but strong vortices are smeared

    5) $\Omega =0.52$ is a quantity to approximately define the vortex boundary."

This method has been evaluated by some researchers [25-30] to compare with the existing vortex identification methods and shown that it is very effective to capture vortices. Nevertheless, the epsilon$-\varepsilon$ in eq. (1) is case-related and required an appropriate adjustment.

Then, Dong *et al.* [31] gave a new determination of epsilon$-\varepsilon$ in explicit form as

$$\varepsilon = 0.001(b - a)_{max} = 0.002Q_{max}. \tag{2}$$

The term $(b - a)_{max}$ is easy to obtain as a fixed number at each time step in a certain case. A study [31], showed that the epsilon in eq. (2) is a proper number for many cases. After the epsilon is determined, the adjustment of ε in any case in vortex visualization is not necessary. Therefore, by using the Omega method with this ε determination, vortex visualization can be easily obtained without much adjustment.

Vortex Visualization

In this paper, the vortices are visualized by applying the Omega method with isosurfaces of $\Omega = 0.52$ to capture vortex structures in 3D with $\varepsilon = 0.001(b - a)_{max}$. The vortex structure of flow on flat plate at $t = 17.60$ where T is a period of T-S wave is shown in Fig. (3) along the streamwise direction, $x = 630$ to $x = 950$.

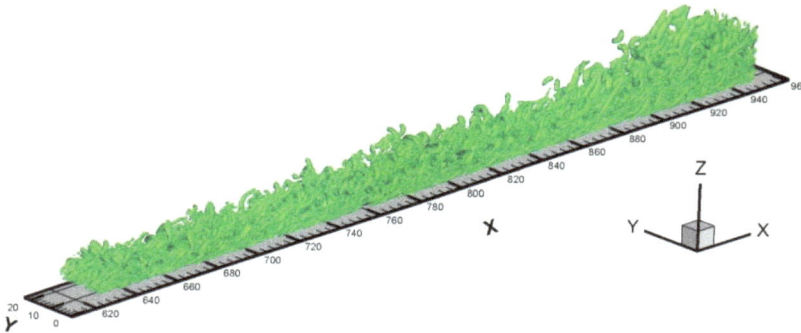

Fig. (3). Vortex structures (isosurfaces of $\Omega = 0.52$) along $x = 630$ to $x = 950$ at $t = 17.60T$.

MODAL DECOMPOSITION

The main propose of this work is to extract coherent structures from DNS data by using modal decomposition schemes. Two different types of modal analyses are applied, *i.e.,* POD and DMD. Both methods are used to analyze the data, extract the coherent structure and reduce the dimension to keep the most important mode as a basis. The area of interest by using both methods is the transition area of the flow on flat plate.

The data in this study are discrete and in the matrix form. Hence, the concept of modal decomposition will be given in terms of the matrix decomposition. The original matrix P is given by a product of three matrices as follows:

$$P = M_1 M_2 M_3.$$

Each component matrix is defined as different physical meanings. The first matrix, M_1, is a matrix of spatial structures, which represents a shape of each mode. The second matrix, M_2, is a matrix of amplitudes, which indicates how important each mode is. The third matrix, M_3, is a matrix of temporal structures, which represents the dynamics of modes over time.

The three matrices in POD and DMD are obtained in different ways and concepts. Firstly, an overview of POD is given. Then, it is applied to the DNS data. Secondly, a concept of DMD is presented. Next, it is also applied to the DNS data. Some illustrations of vortex structures are demonstrated by the Omega method with isosurfaces of $\Omega = 0.52$ and $\varepsilon = 0.001(b - a)_{max}$.

The Proper Orthogonal Decomposition (POD)

POD is applied to discrete data as a matrix in our case. The studies [32, 33] showed that POD coincides with the singular value decomposition (SVD), which is a method of matrix decomposition. The POD algorithm is processed in the same way as SVD. Thus, the method of SVD is shown in the POD algorithm to extract the dataset. The main advantages of POD in our case are for dimensional reduction to keep the most important mode as a basis and to extract the coherent structure.

POD Algorithm

Step 1: Collect the data

Collect the data from numerical simulation or experiment into the matrix $P \in \mathbb{R}^{m \times n}$ and assume that $m \gg n$. Let:

$$P = \begin{bmatrix} | & | & & | \\ x_1 & x_2 & \cdots & x_n \\ | & | & & | \end{bmatrix}_{m \times n,}$$

where n is the number of snapshot (time step) and m the number of spatial point. Each column in the matrix P is a snapshot in time and it can be the components such as velocity, vorticity or pressure.

Step 2: Compute the SVD of matrix P

The SVD of matrix P is represented by

$$P = USV^T, \qquad (3)$$

where
  (a) T stands for the transpose of the matrix.
  (b) U is an $m \times m$ orthogonal matrix
      V is an $n \times n$ orthogonal matrix.
      The matrix U is called the left matrix. It is the matrix of spatial structure.
      The matrix V is called the right matrix. It is the matrix of temporal structure.
  (c) S is an $m \times n$ matrix with all elements zero except along the diagonal.
      The diagonal elements of S consist of $S_{ii} = \sigma_i \geq 0$. σ_i is called a singular
      values of P with $\sigma_1 \geq \sigma_2 \geq \cdots \geq \sigma_n$. In CFD, σ_i stands for the kinetic energy
      of fluid flow.

$$U = \begin{bmatrix} | & | & & | \\ u_1 & u_2 & \cdots & u_m \\ | & | & & | \end{bmatrix}, \quad S = \begin{bmatrix} \sigma_1 & 0 & \cdots & 0 \\ 0 & \sigma_2 & \cdots & 0 \\ \vdots & \vdots & \ddots & \vdots \\ 0 & 0 & 0 & \sigma_n \\ 0 & 0 & 0 & 0 \\ \vdots & \vdots & \vdots & \vdots \\ 0 & 0 & 0 & 0 \end{bmatrix}, \quad V = \begin{bmatrix} | & | & & | \\ v_1 & v_2 & \cdots & v_n \\ | & | & & | \end{bmatrix}. \qquad (4)$$

Thus, we can write eq. (3) as

$$P = \begin{bmatrix} | & | & & | \\ u_1 & u_2 & \cdots & u_m \\ | & | & & | \end{bmatrix}_{m \times m} \begin{bmatrix} \sigma_1 & 0 & \cdots & 0 \\ 0 & \sigma_2 & \cdots & 0 \\ \vdots & \vdots & \ddots & \vdots \\ 0 & 0 & 0 & \sigma_n \\ 0 & 0 & 0 & 0 \\ \vdots & \vdots & \vdots & \vdots \\ 0 & 0 & 0 & 0 \end{bmatrix}_{m \times n} \begin{bmatrix} - & v_1 & - \\ - & v_2 & - \\ & \vdots & \\ - & v_n & - \end{bmatrix}_{n \times n}. \qquad (5)$$

Next, we compute U, S, V from eq. (3). The procedure to obtain U, S, V is as follows:

  1. Solve for the right matrix V and the singular matrix S by computing $P^T P$.
Since $P^T P = V S^T S V^T = V S^2 V^T$,   then $P^T P V = V S^2$,   which   is   the
eigendecomposition of matrix $P^T P$. Then we obtain

$$V = \begin{bmatrix} | & | & & | \\ v_1 & v_2 & \cdots & v_n \\ | & | & & | \end{bmatrix},$$

where v_i are eigenvectors of $P^T P$ corresponding to eigenvalues, and

$$S^2 = \begin{bmatrix} \lambda_1 & 0 & \cdots & 0 \\ 0 & \lambda_2 & \cdots & 0 \\ \vdots & \vdots & \ddots & \vdots \\ 0 & 0 & 0 & \lambda_n \\ 0 & 0 & 0 & 0 \\ \vdots & \vdots & \vdots & \vdots \\ 0 & 0 & 0 & 0 \end{bmatrix},$$

where λ_i are eigenvalues of $P^T P$.

Then, we can get the singular matrix S by

$$S = \begin{bmatrix} \sigma_1 & 0 & \cdots & 0 \\ 0 & \sigma_2 & \cdots & 0 \\ \vdots & \vdots & \ddots & \vdots \\ 0 & 0 & 0 & \sigma_n \\ 0 & 0 & 0 & 0 \\ \vdots & \vdots & \vdots & \vdots \\ 0 & 0 & 0 & 0 \end{bmatrix},$$

where $\sigma_i = \sqrt{\lambda_i}$.

2. Solve for the left matrix U by plugging S, V into eq. (3). Thus, we obtain $U = PVS^{-1}$.

Dimension Reduction and the Matrix Reconstruction

If we use S, V in eq. (4) to solve for U, the matrix U will be a huge $m x m$ matrix. This is not a good idea to obtain U with the high dimension matrix. To avoid this, we can obtain U with the lower dimension. Since the rank of S is n, we can write S in terms of $n x n$ matrix,

$$S = \begin{bmatrix} \sigma_1 & 0 & \cdots & 0 \\ 0 & \sigma_2 & \cdots & 0 \\ \vdots & \vdots & \ddots & \vdots \\ 0 & 0 & 0 & \sigma_n \end{bmatrix}.$$

Then, eq. (5) becomes

$$P = \begin{bmatrix} | & | & & | \\ u_1 & u_2 & \cdots & u_n \\ | & | & & | \end{bmatrix}_{mxn} \begin{bmatrix} \sigma_1 & 0 & \cdots & 0 \\ 0 & \sigma_2 & \cdots & 0 \\ \vdots & \vdots & \ddots & \vdots \\ 0 & 0 & 0 & \sigma_n \end{bmatrix}_{nxn} \begin{bmatrix} - & v_1 & - \\ - & v_2 & - \\ & \vdots & \\ - & v_n & - \end{bmatrix}_{nxn}. \qquad (6)$$

Similarly, we can solve for U, S, V. The matrix U is reduced to a lower dimension mxn, which is much more convenience than a higher dimension mxm. Moreover, the matrix U can be reduced to a lower dimension mxr with $r \le n$ by truncating the last $m - r$ columns. Then by truncating the last $n - r$ rows and columns of S, the dimension of S becomes rxr. For the matrix V, to truncate the last $n - r$ rows of V, the dimension of matrix V becomes rxn. Thus, the eq. (7) will be an approximated form of eq. (6) as follows:

$$P \approx \begin{bmatrix} | & | & & | \\ u_1 & u_2 & \cdots & u_r \\ | & | & & | \end{bmatrix}_{mxr} \begin{bmatrix} \sigma_1 & 0 & \cdots & 0 \\ 0 & \sigma_2 & \cdots & 0 \\ \vdots & \vdots & \ddots & \vdots \\ 0 & 0 & 0 & \sigma_r \end{bmatrix}_{rxr} \begin{bmatrix} - & v_1 & - \\ - & v_2 & - \\ & \vdots & \\ - & v_r & - \end{bmatrix}_{rxn}. \qquad (7)$$

Note that the approximated matrix P in eq. (7) is called the matrix reconstruction.

Reduction criterion: To choose the size r of the reduced-dimension matrix, it can be computed from the relative energy of the snapshots by the first r POD basis vectors as the following formula:

$$\epsilon(r) = \frac{\sum_{i=1}^{r} \sigma_i^2}{\sum_{i=1}^{n} \sigma_i^2}, \qquad (8)$$

r is usually chosen as the minimum integer such that

$$1 - \epsilon(r) \le tol,$$

where tol means a given tolerance with $0 < tol < 1$. For example, $tol = 0.11\%$

Linear Combination of POD Mode

By equations (3) and (6), we can write P in the form

$$P = \sum_{k=1}^{n} \sigma_k u_k v_k^T,$$

or

$$P = \sigma_1 u_1 v_1^T + \sigma_2 u_2 v_2^T + \cdots + \sigma_n u_n v_n^T.$$

The matrix can be written into the superposition of basis. We call u_i a POD mode. The modes are the bases. The whole structure can be extracted into n coherent structures. Dimensional reduction can keep the most important mode as a basis. The first mode will be the most dominant energy structure and the last mode will be the least dominant energy structure.

If P is reduced to the rank $r < n$ as in eq. (7), the approximated P is

$$P \approx \sum_{k=1}^{r} \sigma_k u_k v_k^T$$

or

$$P \approx \underbrace{\sigma_1 u_1 v_1^T}_{\text{mode1}} + \underbrace{\sigma_2 u_2 v_2^T}_{\text{mode 2}} + \cdots + \underbrace{\sigma_r u_r v_r^T}_{\text{mode r}}.$$

POD Analysis for Late Transition Flow

In our case, the POD is applied over 120 snapshots in time between $t = 17.11T$ to $t = 18.30T$ to investigate the principal components of the coherent structures. The area of POD analysis is from $x = 880$ to $x = 920$ as shown in Fig. (**4**). A subzone is extracted to reduce the computation complexity. The parameters of the subzone are given in Table **2**.

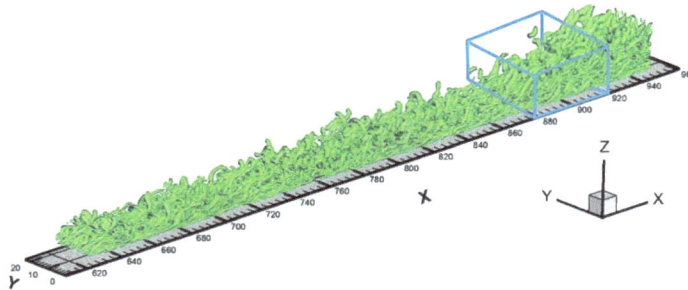

Fig. (4). The area of POD analysis (in the blue box).

Table 2. Parameters of subzone.

	Start Index	End Index
I (in x direction)	511	590
J (in y direction)	1	128
K (in z direction)	1	200

The snapshot x_j is defined by

$$x_j = \begin{pmatrix} u^{(j)}_{511,1,1} \\ \vdots \\ u^{(j)}_{590,1,1} \\ u^{(j)}_{511,2,1} \\ \vdots \\ u^{(j)}_{590,2,1} \\ \vdots \\ u^{(j)}_{I,J,K} \\ \vdots \\ u^{(j)}_{590,128,200} \\ \vdots \\ v^{(j)}_{I,J,K} \\ \vdots \\ w^{(j)}_{I,J,K} \\ \vdots \\ w^{(j)}_{590,128,200} \end{pmatrix} \quad \text{for } j = 1, \dots, 120,$$

where $u^{(j)}$, $v^{(j)}$ and $w^{(j)}$ are velocity fields at $t = (17.10 + 0.01j)T$. In the POD method, the singular values are computed and ordered. The results of singular values are shown in Figs (**5a-b**). These show that the singular values converge to zero in this case. The first mode has the highest amplitude compared with other modes. By dimensional reduction, the number of mode r is chosen by eq. (8) and shown in Fig. (**5c**). The appropriate number of modes to reconstruct the vortex structure is 30 modes since it can keep the most cumulative energy as 100%.

Fig. (**6**) shows that the reconstruction performs very well with the first 30 modes in different time steps comparing to the original flows data. The spatial shapes of mode 1 to mode 30 are illustrated as in Fig. (**7**).

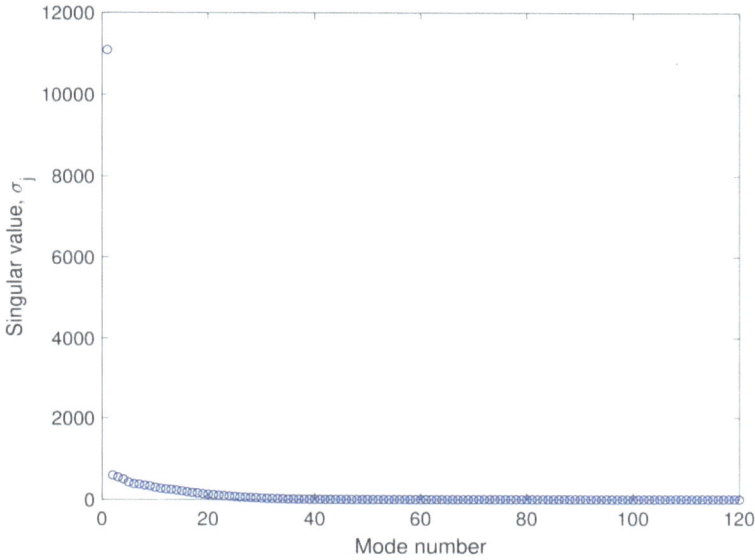

(a) Singular values σ_i of the matrix

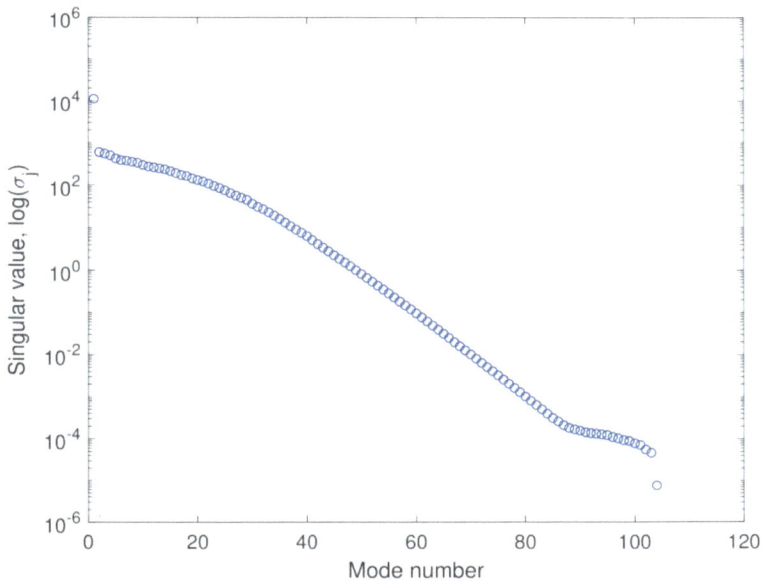

(b) Singular values σ_i of the matrix S in log scale

Fig. 5 cont.....

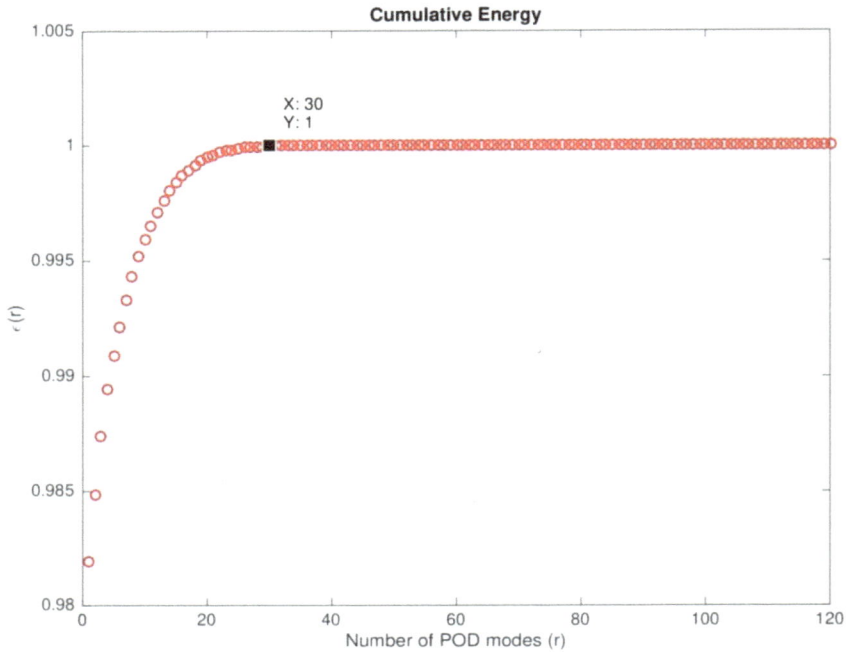

(c) Distribution of $\epsilon(r)$

Fig. (5). Singular values and the relative energy.

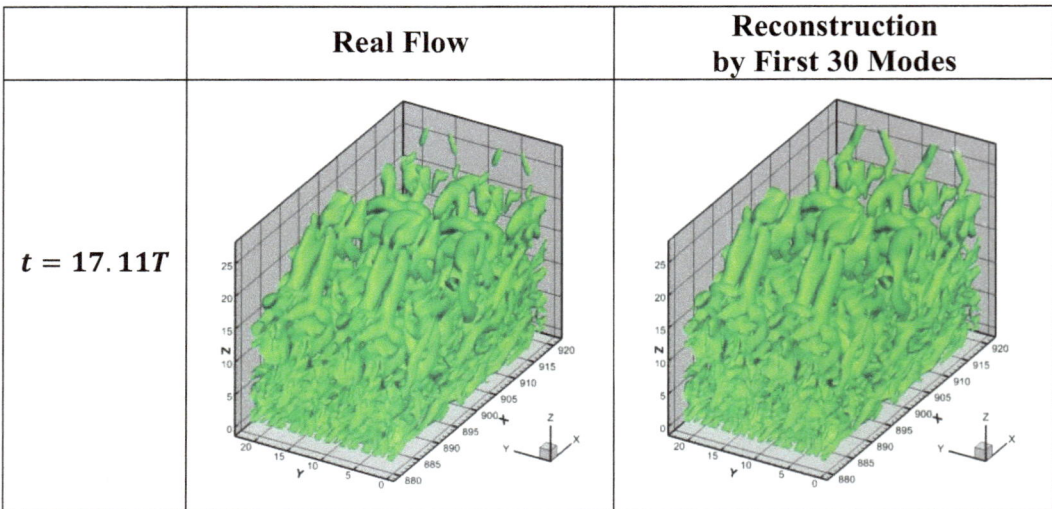

	Real Flow	**Reconstruction by First 30 Modes**
$t = 17.11T$		

t = 17.70T

t = 18.00T

t = 18.25T

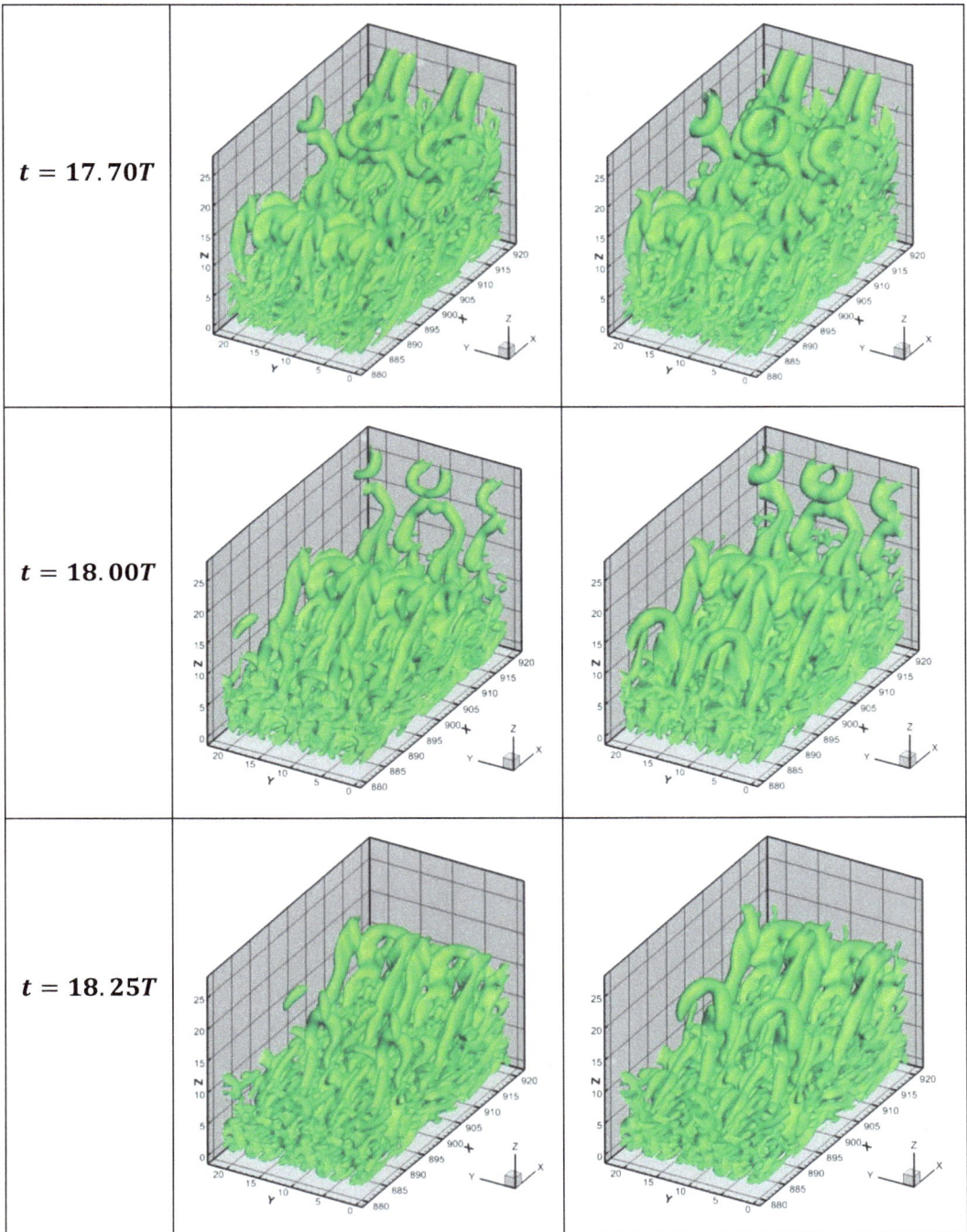

Fig. (6). Original data and reconstruction by first 30 modes with isosurfaces of $\Omega = 0.52$.

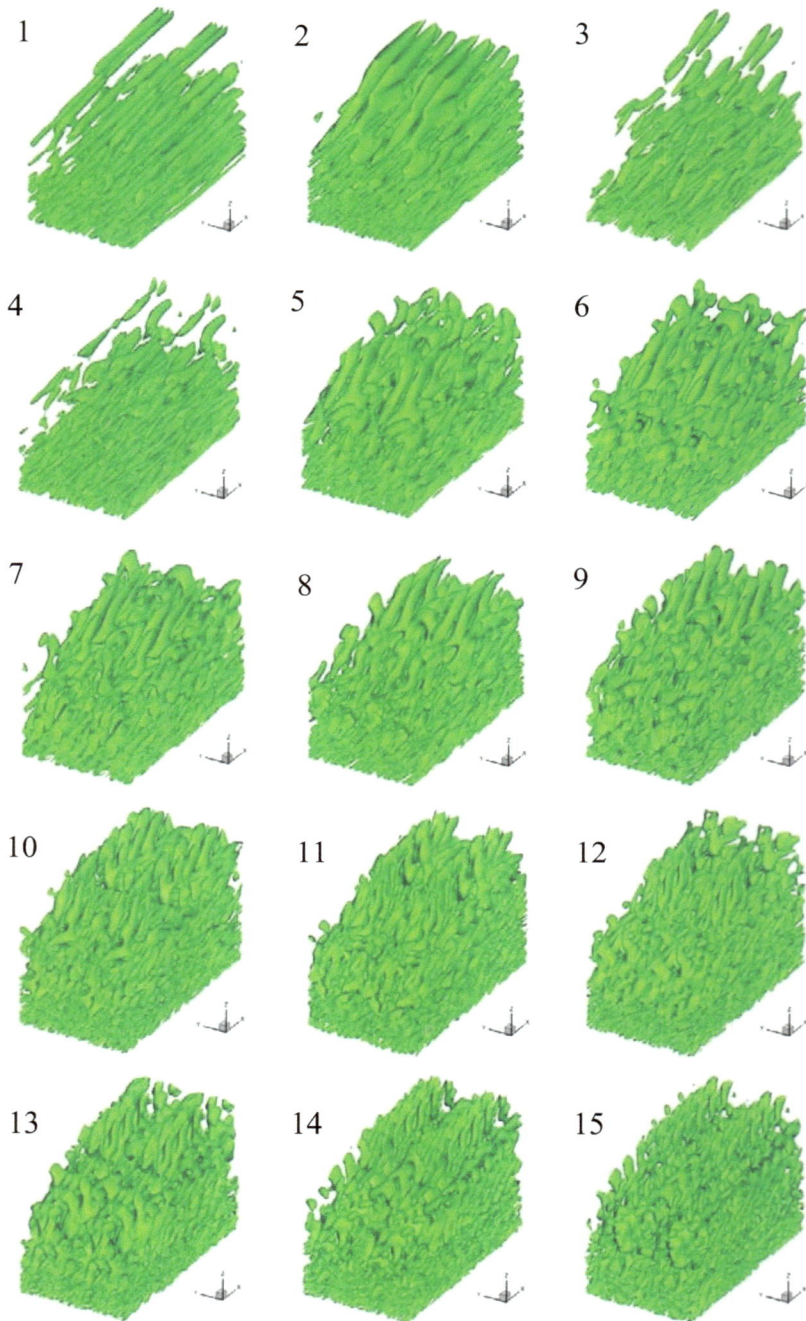

Fig. 7 cont.....

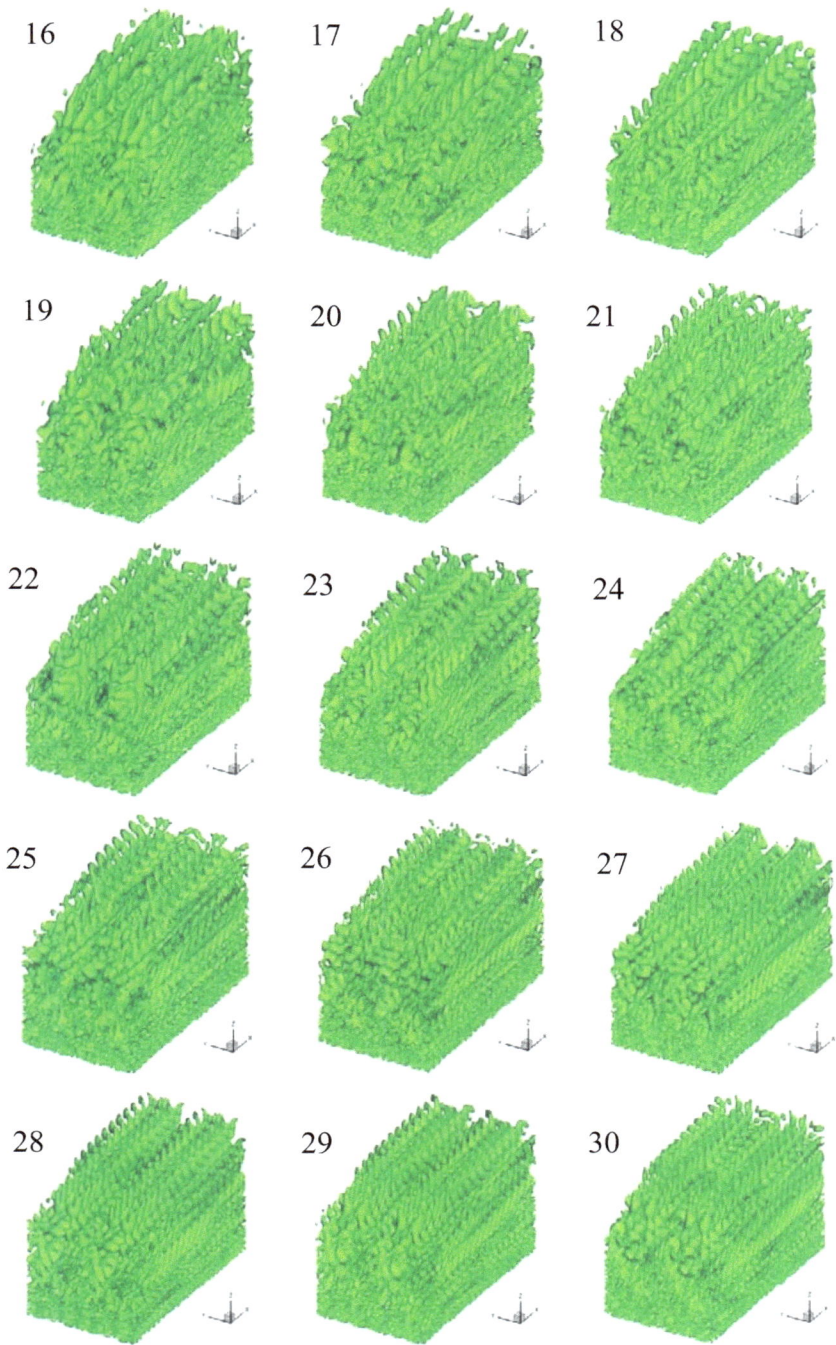

Fig. (7). Vortex structures of typical modes (mode 1 – mode 30) with isosurfaces of $\Omega = 0.52$.

In the original flow, the vortex structures are in different shapes in different time steps as shown in Fig (**6**). However, the shapes of vortex structures in each mode are the same in every time step from $t = 17.11T$ to $t = 18.30T$ as shown in Fig. (**7**). All modes still keep the same shapes and structures over time in the time domain since the POD mode represents a spatial structure of the flow with the time average.

In terms of kinetic energy content, the first mode is the most dominant since it is the most principal component with $\epsilon(r) = 98.2\%$. As we can see in Fig. (**7**), the first POD mode is in a streamwise characteristic. We can imply that the mean flow is presented by the streamwise structure. We can also see a streamwise structure in mode 2. In mode 3 and mode 4, they are also in streamwise structures but the structures are smaller than the first two modes. Hairpin typed structures are found in modes 5,6,7. The spanwise characteristics are presented from mode 8 to mode 30. Moreover, we can see that the shapes of vortex structures in higher-order modes get smaller than the lower-order modes. The first modes will usually be associated with the largest scale flow structures. This is because the modes are ordered by amplitudes.

Since the streamwise structures in modes 1,2,3 have significantly higher energy content than the other modes. We can also conclude that the vortex structures in the late transitional flow on the flat plate can be represented by the reconstruction of the first three modes as in Fig. (**8**).

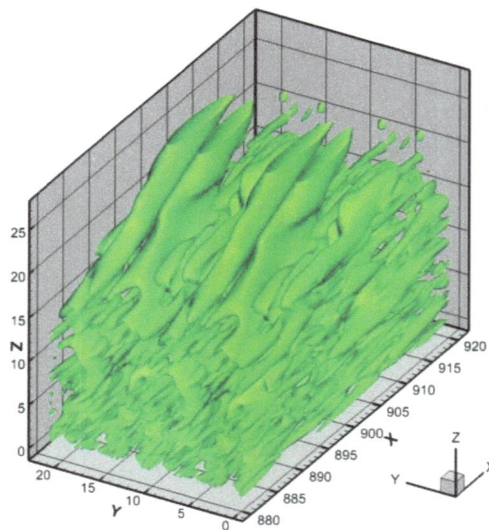

Fig. (8). The reconstruction by first three modes with isosurfaces of $\Omega = 0.52$.

To investigate the interiors of vortex structures, the X-Z cross sections and X-Y cross sections are considered. Some examples of cross sections in some modes are demonstrated in Fig. (**9**).

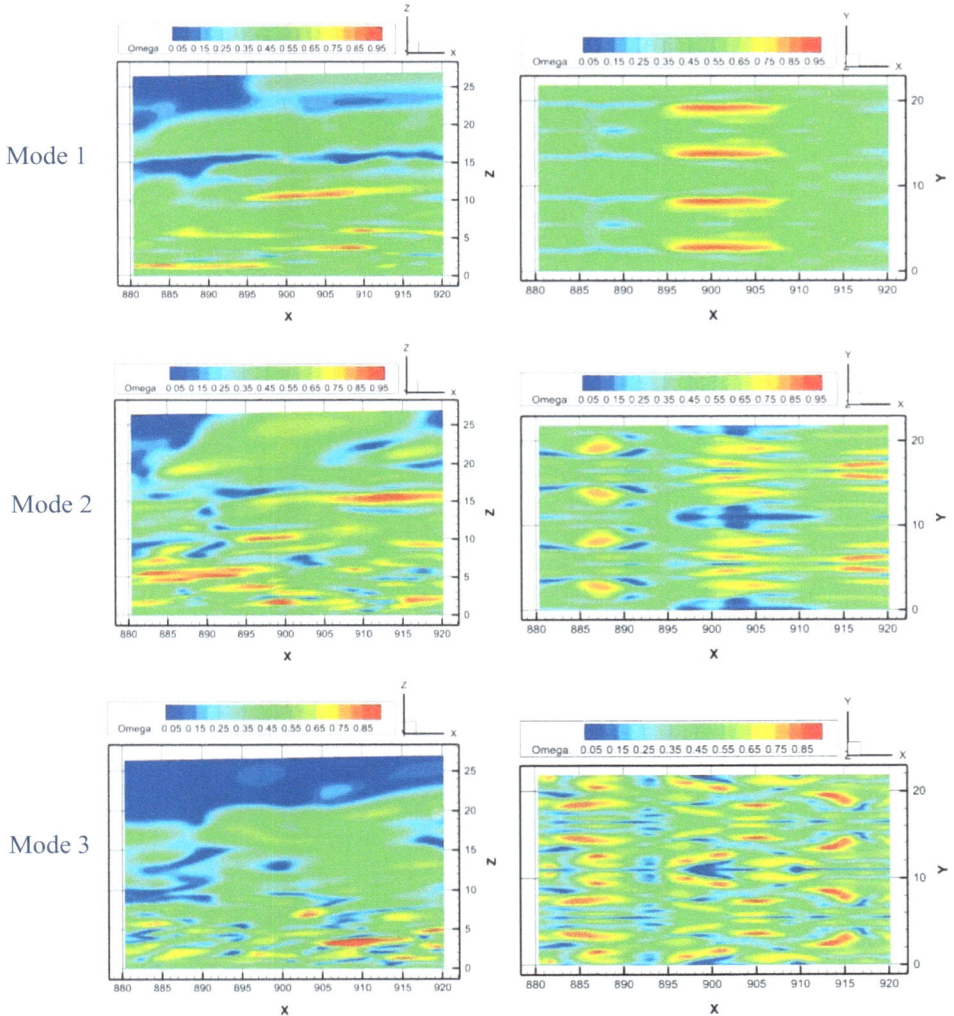

Fig. 9 cont.....

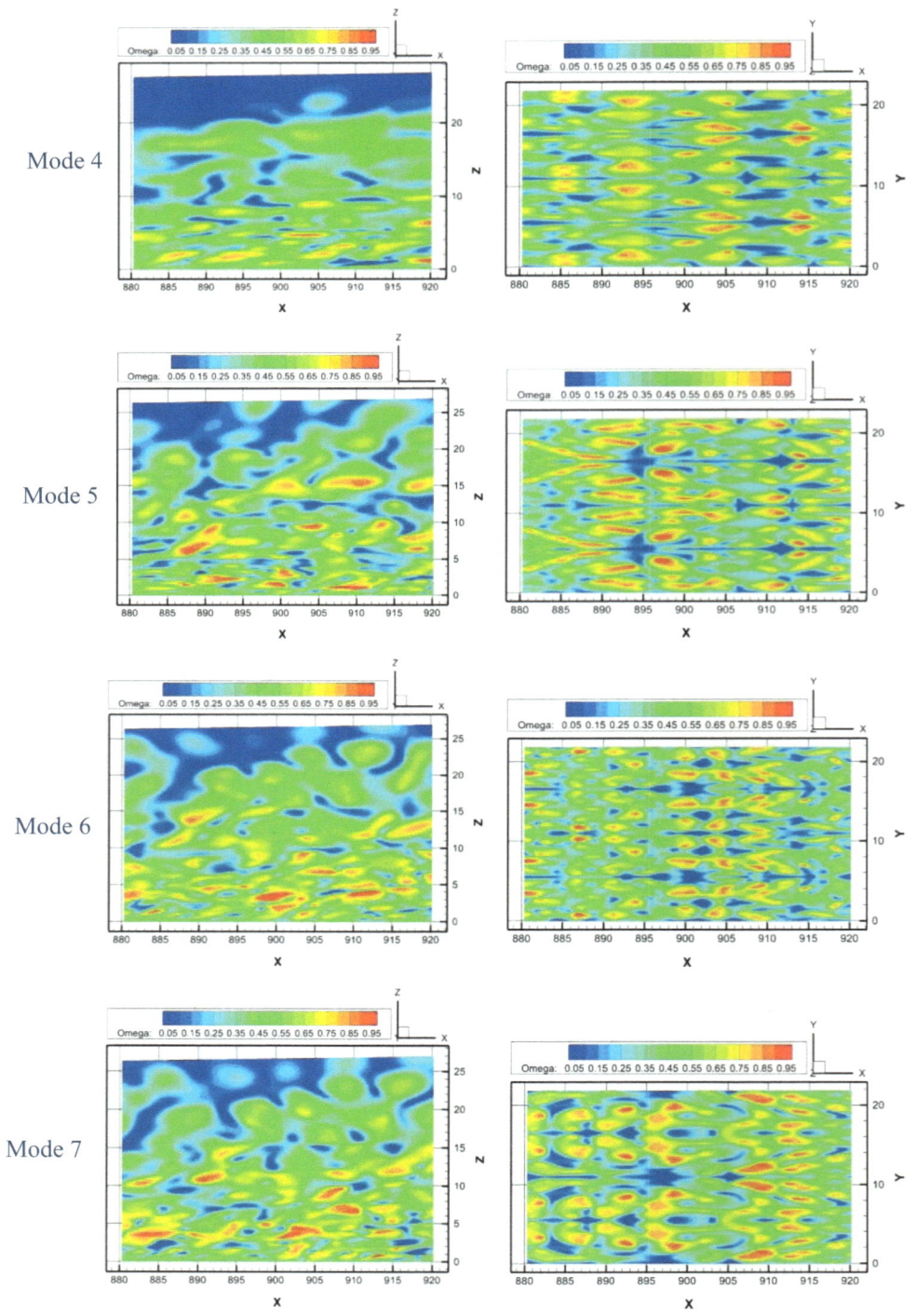

Mode 4

Mode 5

Mode 6

Mode 7

Fig. 9 cont.....

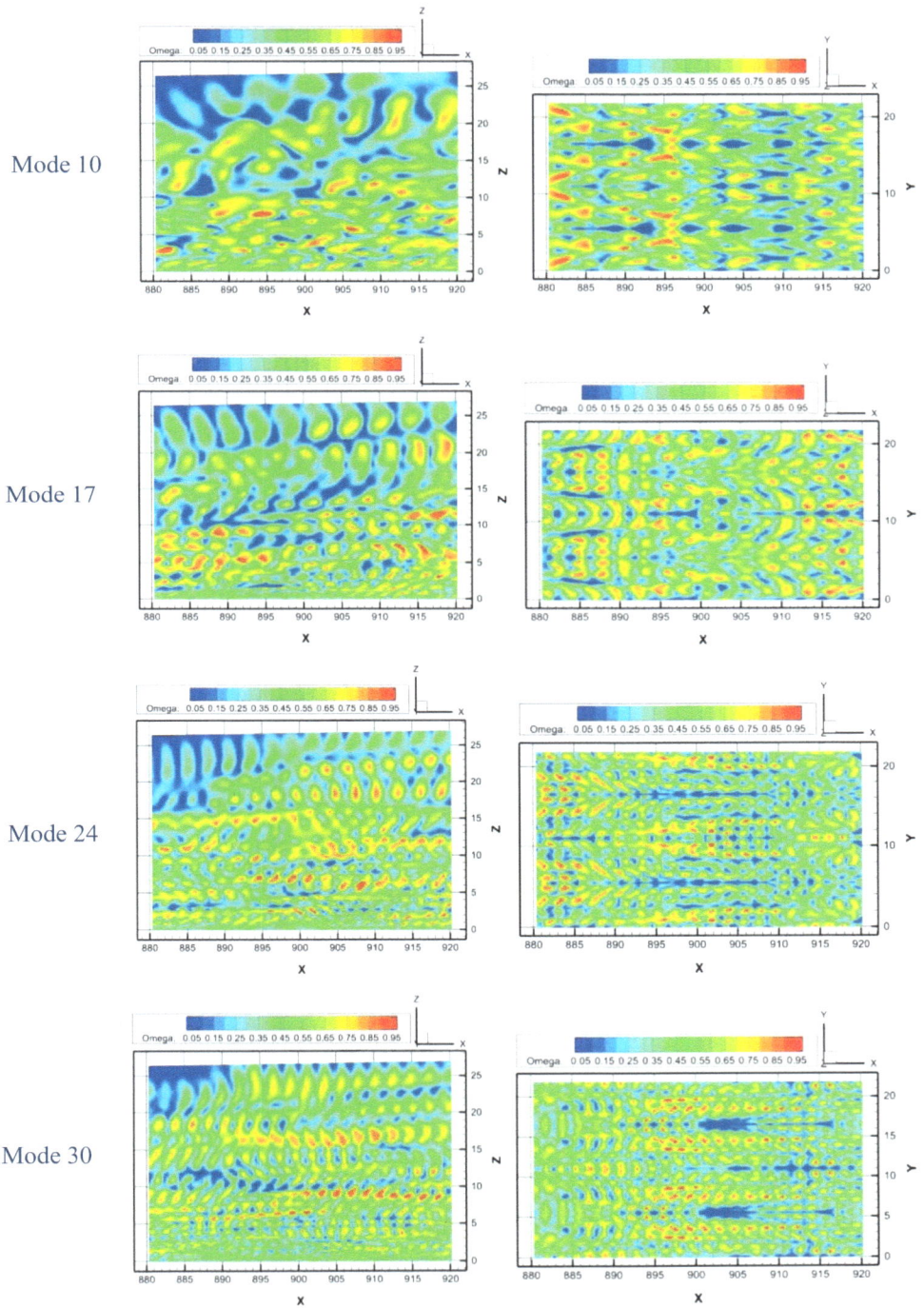

Fig. (9). Interior views of some modes in X-Z cross sections (left) and X-Y cross sections (right).

We can see in Fig. (**9**) that the lower-order modes have larger vortex structures corresponding to the angle views in 3D as shown in Fig. (**7**). As in Fig. (**9**), mode 1 and mode 2 are obviously in the streamwise vortex structures in interior views while the higher-order modes show fluctuation distributions of vortex structures, which are resulted by the generation of spanwise characteristics.

POD time coefficients are shown in Fig. (**10**). They can be obtained from the matrix of temporal structure V in eqs (3)-(4). Each column of matrix V represents the time evolution. The POD time coefficient of mode 1 is from the column 1 of V. Time coefficient of mode 2 is from column 2 of V. Similarly, POD time coefficient of mode r is from the column r of matrix V. These graphs show fluctuations of fluid flows. We can see that the higher-order modes have higher fluctuations. We can also see some similar fluctuations of some modes such as modes 11-12, modes 17-18, and modes 25-30, *etc*. It is also clear that mode 1 represents the mean flow separating from the other modes, which represent fluctuating flows.

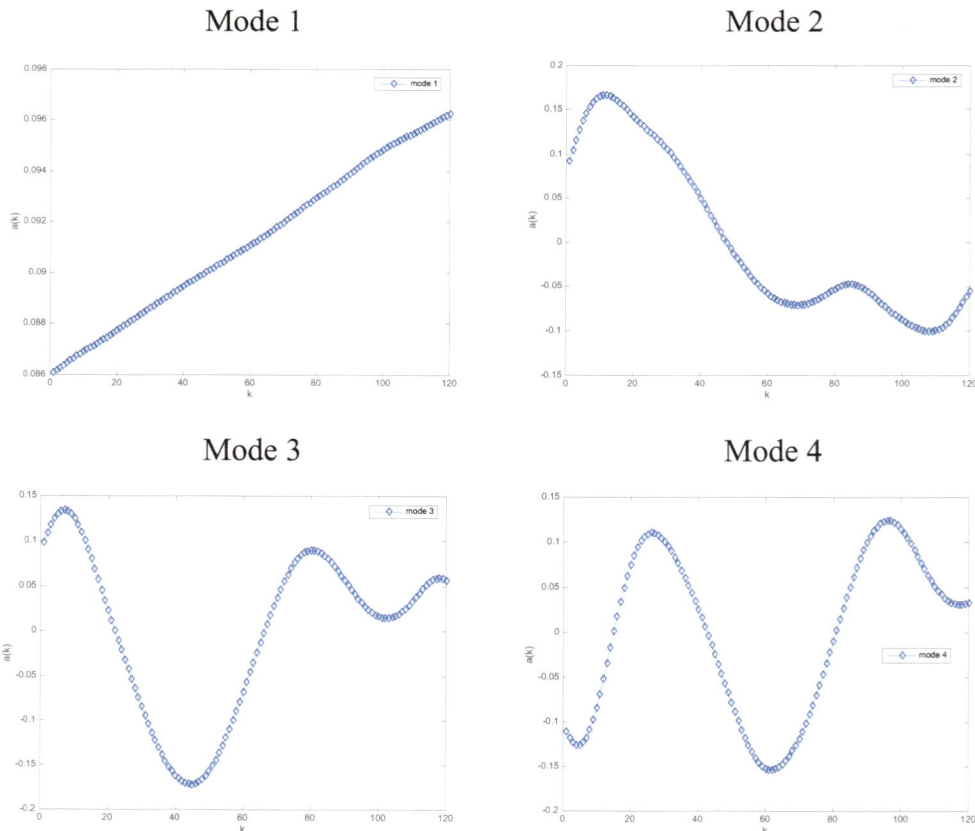

Mode 1

Mode 2

Mode 3

Mode 4

Fig. 10 cont.....

Mode 5

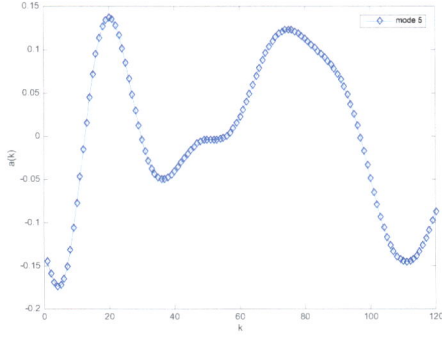

Mode 6

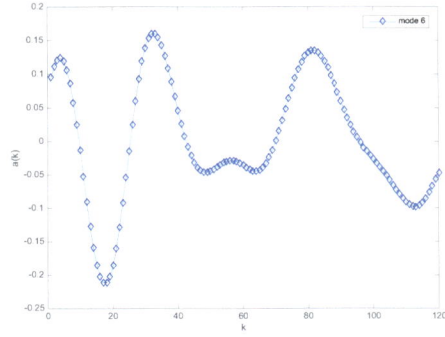

Mode 7

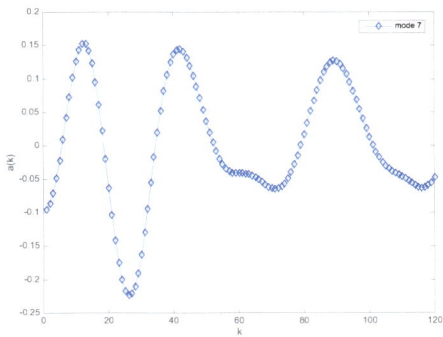

Mode 10

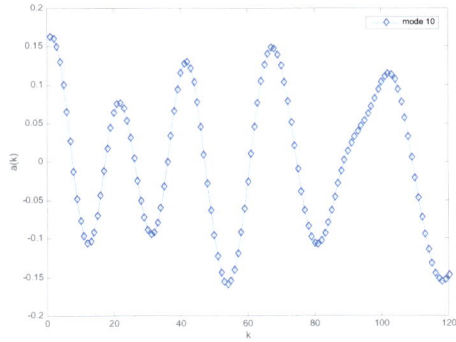

Mode 11

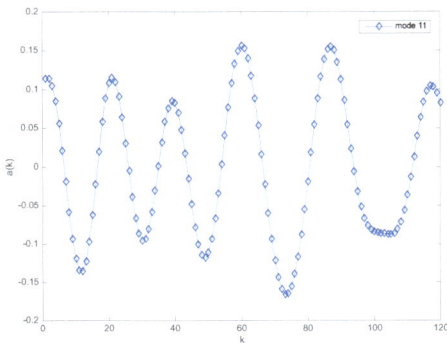

Mode 12

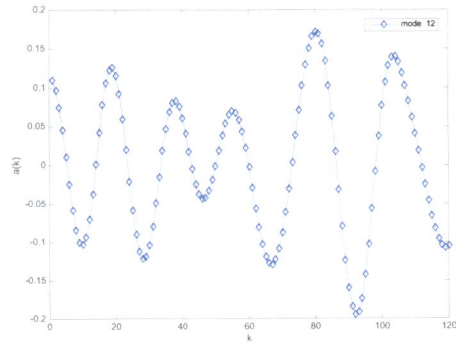

Fig. 10 cont.....

Mode 17

Mode 18

Mode 24

Mode 30

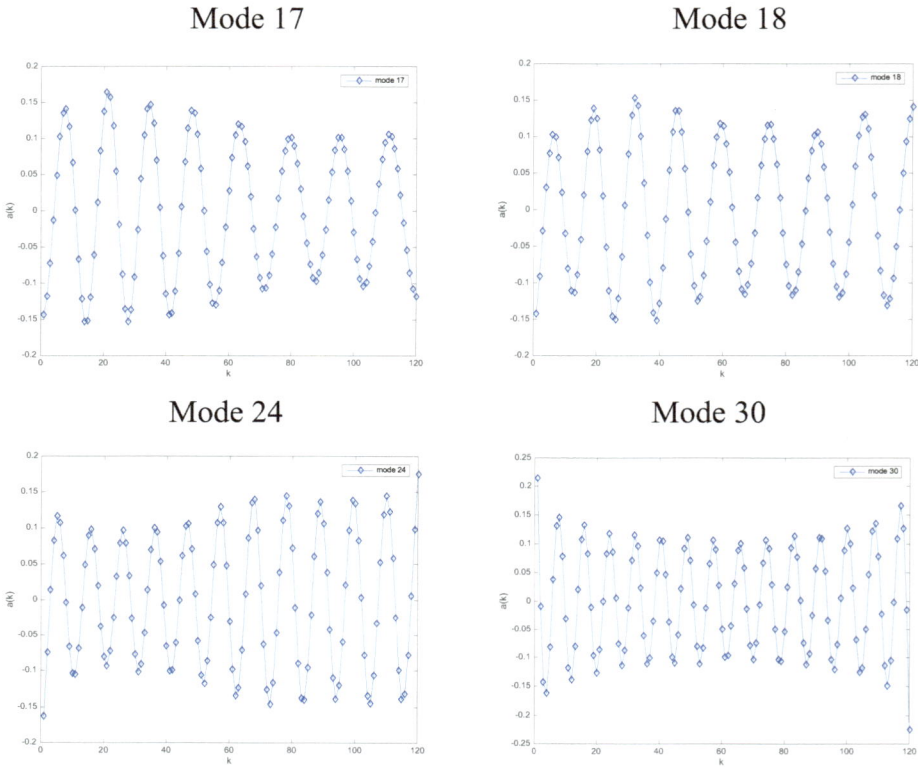

Fig. (10). POD time coefficients of some modes.

Although POD is a useful tool to identify the coherent structures of flows based on the energy rank, there are still some weaknesses. In POD, the mode shapes correspond to multiple frequencies in the flow. Thus, POD cannot be used to extract the single frequency per mode. Moreover, POD mode is a time averaged structure and it does not present the dynamic features.

The other modal decomposition method called DMD is introduced in the next section. This method is efficient for a particular phenomenon occurring at a certain frequency.

Dynamic Mode Decomposition (DMD)

The DMD is applied to extract the dynamic features by finding the relationship between each time step. The purpose of DMD is to identify the coherent structure from the data set. Moreover, each coherent structure has a single feature in temporal mode, unlike the POD. This method is the basis of Koopman analysis of nonlinear dynamical systems [15, 34]. The description of DMD technique is as follows. The data matrix is defined by

$$P = \begin{bmatrix} | & | & & | \\ x_1 & x_2 & \cdots & x_n \\ | & | & & | \end{bmatrix}_{m \times n}$$

which is the same data matrix used in POD.

To find the relationship between each time step, the matrix P is split into two matrices, X and Y, as follows

$$X = \begin{bmatrix} | & | & & | \\ x_1 & x_2 & \cdots & x_{n-1} \\ | & | & & | \end{bmatrix}_{m \times (n-1)}$$

and

$$Y = \begin{bmatrix} | & | & & | \\ x_2 & x_3 & \cdots & x_n \\ | & | & & | \end{bmatrix}_{m \times (n-1)}$$

with

$$Y = AX, \tag{9}$$

where A is a linear operator relating to Koopman analysis, which presents that a nonlinear dynamic system can be represented by a linear system with a linear operator A. Then by eq. (9), every snapshot depends on the previous one and has a relation by a linear operator A as follows.

$$x_1 \xrightarrow{A} x_2 \xrightarrow{A} \cdots \xrightarrow{A} x_{n-1} \xrightarrow{A} x_n.$$

In general, for $k = 1, 2, \dots, n-1$, the relationship in each snapshot is expressed by

$$x_{k+1} = Ax_k. \tag{10}$$

DMD Algorithm

The goal of DMD is to determine two quantities of the unknown matrix A, *i.e.*, eigenvalues and eigenvectors of A. The procedure to obtain eigenvalues and eigenvectors of A is as follows.

Step 1: Find SVD for X

Let $X = USV^T$, then solve for U, S, V as shown in POD method.

Step 2: Find eigenvalues of A

Since A is an mxm matrix, which is a huge matrix. In general, it is impossible to find A. To avoid computing from a high dimension matrix, we will use a lower dimension matrix instead. From $Y = AX$ and from step 1, we get

$$Y = AUSV^T.$$

We can have

$$U^T Y V S^{-1} = U^T A U.$$

Let $\tilde{A} = U^T Y V S^{-1}$, which is an rxr matrix where $r \leq n - 1$ (by dimension reduction from SVD of X). Then,

$$\tilde{A} = U^T A U.$$

We see that $\tilde{A}$ is similar to A. Thus $\tilde{A}$ has the same eigenvalues as A.

Thus, we can find the eigenvalues of the lower dimensional of rank r matrix $\tilde{A}$ instead of the high dimensional matrix A as follows

$$\tilde{A}W = W\Lambda$$

where Λ is an rxr matrix with all elements zero except along the diagonal.

The diagonal elements of Λ consist of λ_i, eigenvectors of $\tilde{A}$ and A.

$$\Lambda = \begin{bmatrix} \lambda_1 & 0 & \cdots & 0 \\ 0 & \lambda_2 & \cdots & 0 \\ \vdots & \vdots & \ddots & \vdots \\ 0 & 0 & 0 & \lambda_r \end{bmatrix}$$

and W is a matrix of eigenvectors of $\tilde{A}$ with w_i are eigenvectors corresponding to eigenvalues λ_i,

$$W = \begin{bmatrix} | & | & & | \\ w_1 & w_2 & \cdots & w_r \\ | & | & & | \end{bmatrix}_{r \times r}.$$

Step 3: Find eigenvectors of A

The eigenvectors of A can be obtained from projecting the vector w_i in W onto the basis made from orthogonal columns in U. Then, we obtain the matrix Φ of eigenvectors of A by

$$\Phi = UW$$

where

$$\Phi = \begin{bmatrix} | & | & & | \\ \varphi_1 & \varphi_2 & \cdots & \varphi_r \\ | & | & & | \end{bmatrix}_{m \times r}$$

and φ_i are eigenvectors of A called DMD modes.

Linear Combination of DMD Mode

From the eigendecomposition of matrix A, we have

$$A\Phi = \Phi\Lambda.$$

Then

$$A = \Phi\Lambda\Phi^{-1}. \tag{11}$$

where Φ^{-1} is a pseudoinverse of Φ.

By eqs (10)-(11) for $k = 1,2,\ldots,n-1$, then

$$x_{k+1} = \Phi\Lambda\Phi^{-1}x_k$$

and

$$x_{k+1} = (\Phi\Lambda\Phi^{-1})^k x_1,$$

$$x_{k+1} = \Phi\Lambda^k\Phi^{-1}x_1.$$

Thus, we can predict the dynamic by the equation

$$x_{k+1} = \Phi\Lambda^k b, \tag{12}$$

where $b = \Phi^{-1}x_1$ is called the initial DMD amplitude and x_1 is the first column in matrix X.

Then, eq. (12) can be written as

$$x_{k+1} = \sum_{j=1}^{r} \varphi_j \lambda_j^k b_j,$$

$$x_{k+1} = \underset{\text{mode}1}{\varphi_1\lambda_1^k b_1} + \underset{\text{mode 2}}{\varphi_2\lambda_2^k b_2} + \cdots + \underset{\text{mode r}}{\varphi_r\lambda_r^k b_r} \tag{13}$$

where $r \leq n - 1$, k is the kth time step and $\lambda_i = e^{\omega_i}$ with ω_i is a complex frequency in terms of pulsation. We can see from eq. (13) that the structure for each time step can be extracted into r modes. φ_i are eigenvectors, which represent spatial modes, λ_i^k are eigenvalues, which represent temporal modes at time k and b_i are coefficients at the initial time, which represent amplitudes. Each mode has a unique characteristic such as structure, frequency, or growth rate.

The Matrix Reconstruction

The reconstructed flow involving the DMD modes is given by

$$P_{DMD} \approx \Phi BC, \tag{14}$$

which approximates the original data matrix P. The matrices in eq. (14) are as follows.

Φ is the matrix of eigenvectors of A with $r \leq n - 1$,

$$\Phi = \begin{bmatrix} | & | & & | \\ \varphi_1 & \varphi_2 & \cdots & \varphi_r \\ | & | & & | \end{bmatrix}_{m \times r}.$$

B is the matrix obtained from the initial DMD amplitude $b = \Phi^{-1}x_1$ such that

$$b = (b_1 \; b_2 \; \cdots \; b_r)^T$$

and

$$B = \begin{bmatrix} b_1 & 0 & \cdots & 0 \\ 0 & b_2 & \cdots & 0 \\ \vdots & \vdots & \ddots & \vdots \\ 0 & 0 & 0 & b_r \end{bmatrix}_{rxr}.$$

C is the Vandermonde matrix in the form

$$C = \begin{bmatrix} 1 & \lambda_1 & \lambda_1^2 & \cdots & \lambda_1^{n-1} \\ 1 & \lambda_2 & \lambda_2^2 & \cdots & \lambda_2^{n-1} \\ \vdots & \vdots & \vdots & \ddots & \vdots \\ 1 & \lambda_r & \lambda_r^2 & \cdots & \lambda_r^{n-1} \end{bmatrix}_{rxn}.$$

Diagnostic from Eigenvalues and Eigenvectors of A

Since the eigenvectors are not orthogonal in general, the eigenvalues and eigenvectors are complex numbers. We can analyze some behaviors of dynamical features from Floquet values, which are represented by the logarithm of eigenvalues as the following expression.

$$\alpha_i = \frac{\log(\lambda_i)}{\Delta t}. \tag{15}$$

By using the relation in eq. (15), the growth or decay rates and frequencies of DMD modes can be explained by the real and imaginary parts of α_i, respectively.

    (a) Stability characteristics:
        The growth or decaying can be examined by the real components of α_i. We can also predict the oscillation by plotting the real parts and imaginary parts of eigenvalues of A with the unit circle. The x-axis is for the real parts and the y-axis is for imaginary parts of eigenvalues λ_i. There are three types of stability characteristics.
            • $\text{Re}(\alpha_i) < 0$: the component is inside the unit circle, which implies decaying
            • $\text{Re}(\alpha_i) = 0$: the component is on the unit circle, which implies

neutrally stable

• $\text{Re}(\alpha_i) > 0$: the component is outside the unit circle, which implies growing or unstable

(b) Frequency:

The unique frequency of each DMD mode can be described by the imaginary component of α_i,

$$\omega_i = \frac{\text{Im}(\alpha_i)}{2\pi}$$

It can also be obtained by the eigenvalues λ_i as the following expression

$$\omega_i = \frac{\text{arc}(\lambda_i)}{2\pi}$$

In addition, the eigenvecters φ_i are used to interpret the DMD mode amplitudes. It can be obtained from b_i in eq. (13) as the following expression

$$\text{b} = \Phi^{-1}x_1 \quad \text{or} \quad \text{b} = \frac{X(:,1)}{\Phi}.$$

Since DMD provides the single feature for each mode, we may select DMD modes based on their amplitude, frequency, or growth rate depending on the objective. According to Rowley *et al.* [15], they suggest that the DMD modes can be ranked based on the modal energy (amplitude) of each mode, unlike the POD ranking by the kinetic energy content.

DMD Analysis for Late Transition Flow

By using the same data with POD snapshots, the DNS data is evaluated by DMD method. The real and imaginary component of eigenvalues are plotted in the unit circle as shown in Fig. (**11**). Note that the same specific DMD modes as indicated in POD modes are used in this paper.

The eigenvalues of DMD usually arise as complex conjugate pairs. Thus, this leads to complex conjugate eigenvectors, which represent spatial modes. In this case, there are two types of eigenvalues computed by DMD algorithm. First, eigenvalues with only real parts are found in modes 1,4,27. Second, eigenvalues with pairs of complex conjugate are introduced in the other modes.

DMD modes are considered corresponding to the mode amplitudes, which represent the modal energy. The six dominant modes are selected ranging by their amplitudes, *i.e.*, modes 1,4,13,14,30,31, respectively. DMD modes 19,20,48,49 are also considered as they have intermediate amplitudes. The least dominant mode as modes 32,33 are selected to compare.

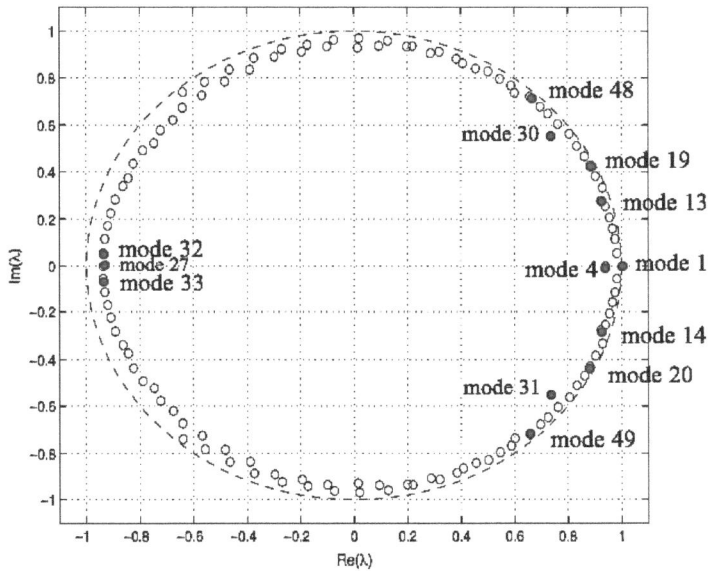

Fig. (11). DMD eigenvalues for all temporal modes in unit circle.

The eigenvalues are used to calculate dynamic features. Some selected modes are presented as follows:

DMD mode 1: $\lambda = 1.0008 + 0i$
DMD mode 4: $\lambda = 0.9399 + 0i$
DMD mode 13: $\lambda = 0.9223 + 0.2783i$
DMD mode 14: $\lambda = 0.9223 - 0.2783i$
DMD mode 19: $\lambda = 0.8817 + 0.4277i$
DMD mode 20: $\lambda = 0.8817 - 0.4277i$
DMD mode 27: $\lambda = -0.9372 + 0i$
DMD mode 30: $\lambda = 0.7349 + 0.5536i$
DMD mode 31: $\lambda = 0.7349 - 0.5536i$
DMD mode 32: $\lambda = -0.9371 + 0.0555i$
DMD mode 33: $\lambda = -0.9371 - 0.0555i$
DMD mode 48: $\lambda = 0.6573 + 0.7808i$
DMD mode 49: $\lambda = 0.6573 - 0.7808i$

The real and imaginary parts of eigenvalues are plotted as points $\big(\mathrm{Re}(\lambda_i),$ $\mathrm{Im}(\lambda_i)\big)$ in the unit circle shown in Fig. (**11**). Only one point of mode 1 lies on the unit circle but we notice that this point has positive growth rate close to zero. This indicates that only DMD mode 1 is likely neutrally stable since we consider the phenomena in a short period of time not the far future. The other points in the circle and close to the boundary represent slow decay rates. The closer point to the boundary means the slower decay approximation. In our case, all modes except mode 1 show the decay oscillation.

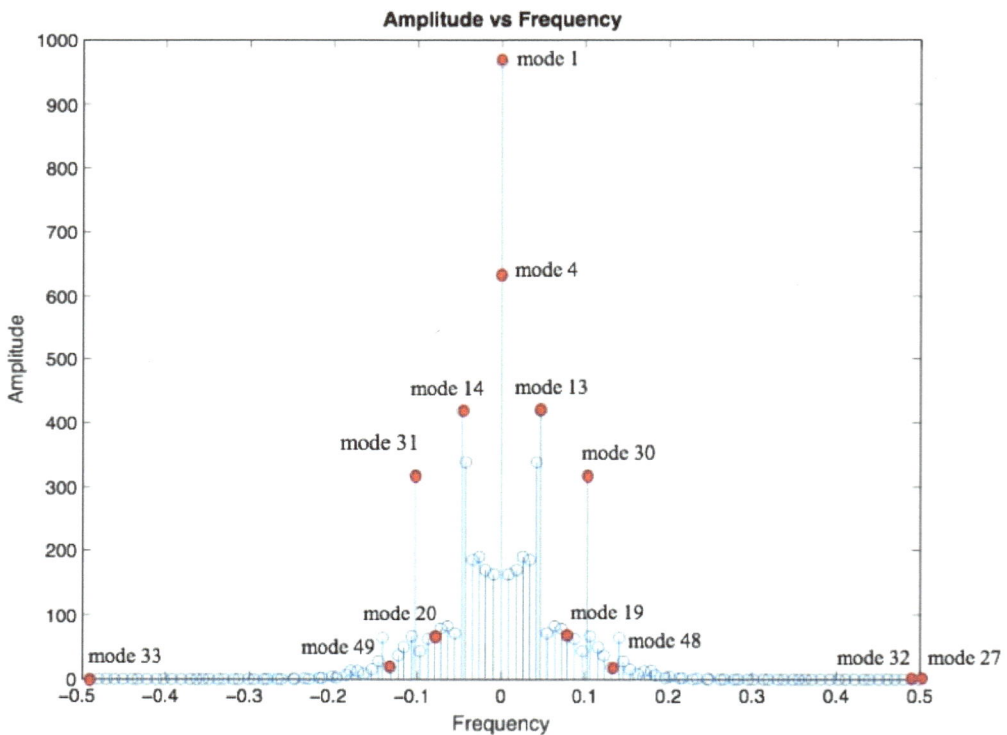

Fig. (12). Amplitude and frequency in each DMD mode.

The frequencies shown in Fig. (**12**) are presented in both positive and negative. The negative frequencies are resulted from the computation with the complex conjugate. These conjugate modes have the same stability characteristics and frequencies but in different signs as shown in Figs. (**12 & 13**). The largest-energy mode 1 has the lowest frequency since the eigenvalue of mode 1 is a positive real number. The lowest frequency is also shown in mode 4. However, mode 1 has a larger-scaled structure than mode 4 because of the larger modal energy. The lowest energy mode as mode 32 introduces the highest frequency with a faster decaying than the other selected modes. Mode 1 is the most important mode since it has the largest amplitude, lowest frequency and stability.

Fig. (**13**) shows the amplitudes and growth rates of DMD modes. The lower number of $|\alpha_i|$ gets the slower decay oscillation. The highest modal energy content in mode 1 corresponds the stability of mode. The second rank of energy content mode as shown in mode 4 decays slowly.

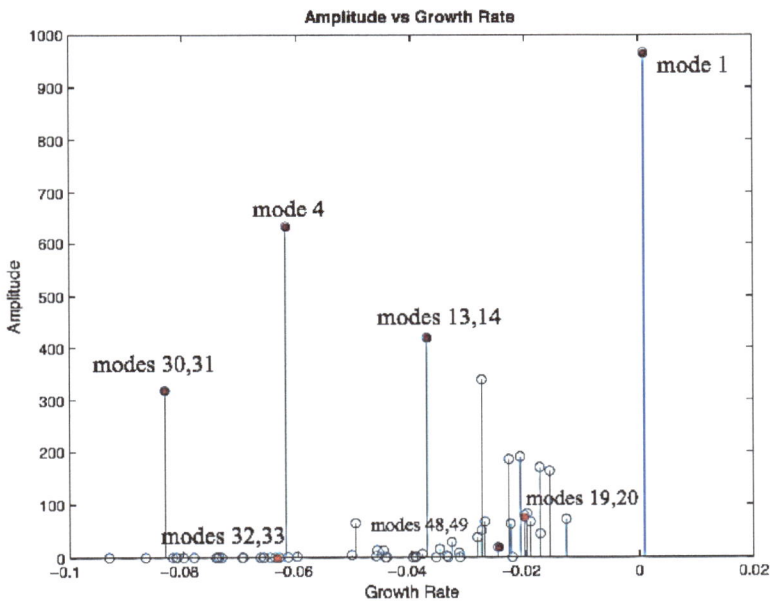

Fig. (**13**). Amplitude and growth rate in each mode.

Some DMD modes are illustrated in 3D by the Omega method with isosurfaces of $\Omega = 0.52$ ranging by the modal energy as shown in Fig. (**14**). The first and forth DMD modes are represented by the eigenvalues with only real parts. Thus the vortex structures of imaginary parts of modes 1 and 4 do not appear. In the other modes, the vortex structures in real parts and imaginary parts are similar. The same spatial mode shapes are demonstrated in complex conjugate modes such as in modes 13 and 14, modes 30 and 31, modes 19 and 20, modes 48 and 49. Moreover, the vortices of modes 27,32,33 cannot be captured by $\Omega = 0.52$ since their amplitudes are very low.

Mode	Real Part of DMD Mode	Imaginary Part of DMD Mode
1		None
4		None
13		
14		
30		

Fig. 14 cont.....

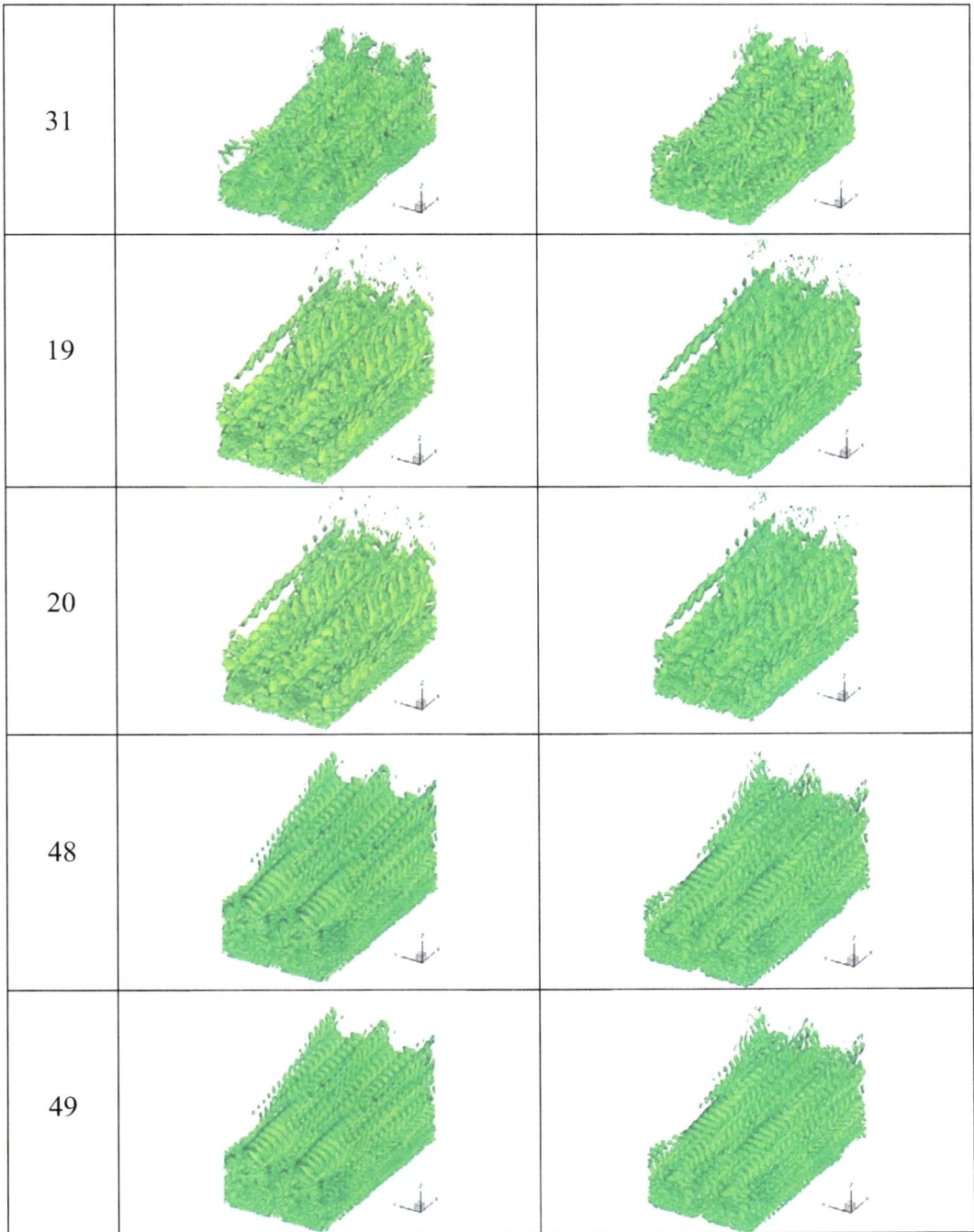

Fig. (14). Vortex structures of some DMD modes with isosurfaces of $\Omega = 0.52$.

The X-Z cross sections of vortex structures in real parts are presented in Fig. (**15**). It can be observed that more spanwise characteristics are clearly shown in the higher frequency modes. Also, lower amplitude modes get smaller vortex structures.

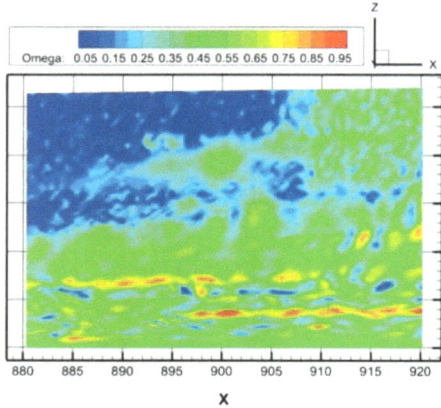

(a) DMD mode 1

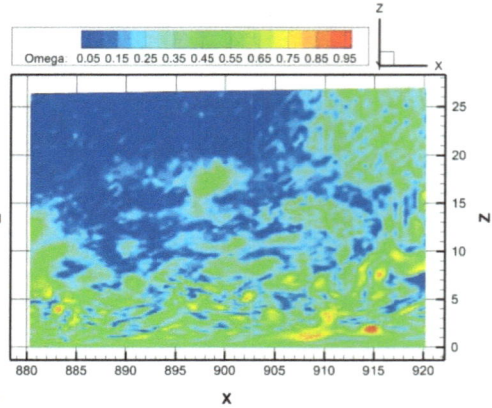

(b) DMD mode 4

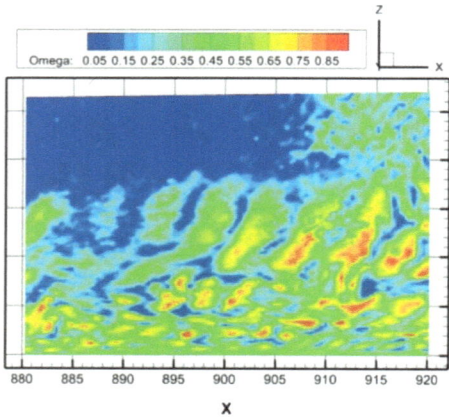

(c) DMD modes 13,14

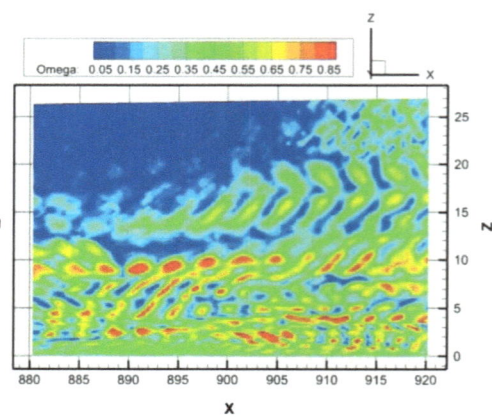

(d) DMD modes 19,20

Fig. 15 cont.....

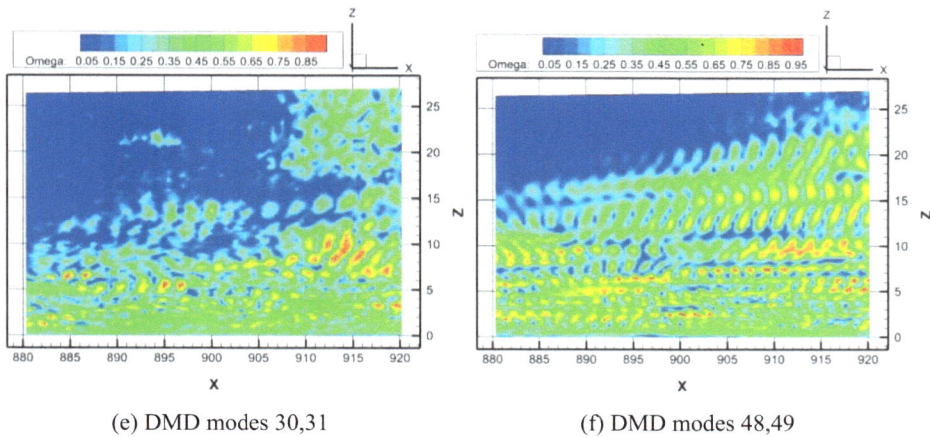

(e) DMD modes 30,31   (f) DMD modes 48,49

Fig. (15). Interior views of some modes by X-Z cross sections.

DISCUSSION AND CONCLUSION

POD and DMD are applied over 120 snapshots to study the vortex structure of flow on the flat plate. The ability of POD is to maximize the kinetic energy of the flow with a minimal number of modes and to extract coherent structures of flows. It is based on a spatial orthogonal framework. For the DMD, it is different from POD as it represents the dominant flow features with a temporal orthogonal framework. In POD, the modes are ranked by the energy content. POD mode 1 represents the most energetic mode and shaped in a streamwise structure. In DMD, the single dynamic feature is introduced to each mode. The dominant modal energy modes are presented in non-imaginary part of eigenvalue modes, which are modes 1 and 4. The complex conjugate modes show the same spatial and temporal features. They have the same spatial shapes, stability characteristics and frequencies (in a different sign of frequency). Higher frequency modes show more spanwise structures. DMD mode 1 has the most important dynamic features. It is shown in only real part and shaped in a streamwise characteristic similar to POD mode 1. Moreover, the mean flow is represented by the dominant mode based on highest energy content in POD, while DMD shows the mean flow by the dynamic features as the largest amplitude, lowest frequency and stability. The most dominant modes in both POD and DMD demonstrate the defined streamwise shape and constant time evolution. Therefore, it can be implied that the vortex structure in the late transition area of flow on the flat plate is dominated by the streamwise structure and stability with the lowest frequency oscillation.

CONSENT FOR PUBLICATION

Not applicable.

CONFLICT OF INTEREST

The author(s) confirm that this chapter contents have no conflict of interest.

ACKNOWLEDGEMENT

The authors are grateful to TACC (Texas Advanced Computation Center) for providing CPU hours to this research project. The computation is performed by using Code DNSUTA which was released by Dr. Chaoqun Liu at University of Texas at Arlington in 2009.

REFERENCES

[1]   J.L. Lumley, *The structure of inhomogeneous turbulent flows.* Atmospheric Turbulence and Radio Wave Propagation, 1967, pp. 166-178.

[2]   L. Sirovich, "Turbulence and the dynamics of coherent structures. Part I: Coherent structures", *Q. Appl. Math.,* vol. 45, no. 3, pp. 561-571, 1987.
[http://dx.doi.org/10.1090/qam/910462]

[3]   A. Duggleby, K.S. Ball, M.R. Paul, and P.F. Ficher, "Dynamical eigenfunction decomposition of turbulent pipe flow", *J. Turbul.,* vol. 8, pp. 1-28, 2007. [http://dx.doi.org/10.1080/14685240701376316]

[4]   A. Duggleby, K.S. Ball, and M.R. Paul, "The effect of spanwise wall oscilation on turbulent pipe flow structures resulting in drag reduction", *Phys. Fluids,* vol. 19, pp. 107-125, 2007. [http://dx.doi.org/10.1063/1.2825428]

[5]   J. Li, and W. Zhang, *The performance of proper orthogonal decomposition in discontinuous flows.* Theoretical and Applied Mechanics Letters, 2016, pp. 236-243.

[6]   B.A. Freno, and P. Cizmas, "A proper orthogonal decomposition method for nonlinear flows with deforming meshes", *Int. J. Heat Fluid Flow,* vol. 50, pp. 145-159, 2014.
 [http://dx.doi.org/10.1016/j.ijheatfluidflow.2014.07.001]

[7]   S. Laizet, S. Lardeau, and E. Lamballais, "Direct numerical simulation of a mixing layer downstream a thick splitter plate", *Phys. Fluids,* vol. 22, no. 1, pp. 1-15, 2010.
[http://dx.doi.org/10.1063/1.3275845]

[8]   L. Hellström, B. Ganapathisubramani, and A.J. Smits, "Coherent structures in transitional pipe flow", *Phys. Rev. Fluids,* vol. 1, 2016.024403 [http://dx.doi.org/10.1103/PhysRevFluids.1.024403]

[9]   L. Hellström, and A. J. Smits, "Structure identification in pipe flow using proper orthogonal decomposition", *Math.Phys. Eng. Sci,* vol. 375, no. 2089, 2017. [http://dx.doi.org/10.1098/rsta.2016.0086]

[10]   H. Gunes, "Proper orthogonal decomposition reconstruction of a transitional boundary layer with and without control", *Phys. Fluids,* vol. 16, p. 2763, 2004.
[http://dx.doi.org/10.1063/1.1758151]

[11]   S. Charkrit, X. Dong, and C. Liu, "POD analysis of losing symmetry in late flow transition", *AIAA Pap.,* pp. 2019-1870, 2019.
[http://dx.doi.org/10.2514/6.2019-1870]

[12]   X. Dong, S. Charkrit, X. Troung, and C. Liu, "POD study on vortex structures in MVG wake", *AIAA Pap.,* pp. 2019-1136, 2019.
[http://dx.doi.org/10.2514/6.2019-1136]

[13]   C. Jin, and H. Ma, "POD analysis of entropy generation in a laminar separation boundary layer", *Energies,* vol. 11, p. 3003, 2018. [http://dx.doi.org/10.3390/en11113003]

[14]   P.J. Schmid, and J. Sesterhenn, "Dynamic mode decomposition of numerical and experimental data", *In 61st Annual Meeting of the APS Division of Fluid Dynamics,* 2008 American Physical Society.

[15]   C.W. Rowley, I. Mezic, S. Bagheri, P. Schlatter, and D.S. Henningson, "Spectral analysis of nonlinear flows", *J. Fluid Mech.,* vol. 641, pp. 115-127, 2009.
[http://dx.doi.org/10.1017/S0022112009992059]

[16]   P. Premaratne, and H. Hu, "Analysis of turbine wake characteristics using dynamic mode decomposition", *35th AIAA Applied Aerodynamics Conference,* 2017 Denver, Colorado [http://dx.doi.org/10.2514/6.2017-4214]

[17]   B.D. Alina, and I.M. Navon, "The method of dynamic mode decomposition in shallow water and a swirling flow problem", *Int. J. Numer. Methods Fluids,* 2016.

[18]   A.T. Mohan, and D.V. Gaitondey, "Model reduction and analysis of deep dynamic stall on a plunging airfoil using dynamic mode decomposition", *AIAA Pap.,* pp. 2015-1058, 2015. [http://dx.doi.org/10.2514/6.2015-1058]

[19]   Q. Zhang, Y. Liu, and S. Wang, *The identification of coherent structures using proper orthogonal decomposition and dynamic mode decomposition.* J. Fluid Structure, 2014, pp. 53-72.

[20]   S. Tirunagari, V. Vuorinen, O. Kaario, and M. Larmi, "Analysis of proper orthogonal decomposition and dynamic mode decomposition on LES of subsonic jets", *CSI J. Computing,* vol. 1, pp. 19-24, 2012.

[21]   C. Liu, and P. Lu, "DNS Study on Physics of Late Boundary Layer Transition", *AIAA Pap.,* pp. 2012- 0083, 2012.
[http://dx.doi.org/10.2514/6.2012-83]

[22]   P. Lu, M. Thapa, and C. Liu, "Numerical investigation on chaos in late boundary layer transition", *Comput. Fluids,* vol. 91, pp. 68-76, 2014. [http://dx.doi.org/10.1016/j.compfluid.2013.11.027]

[23]   J.C. Hunt, A.A. Wray, and P. Moin, "Eddies, Streams, and Convergence Zones in Turbulent Flows", *Center for Turbulence Research: Proceedings of the Summer Program,* vol. N89-24555, 1988

[24]   J. Jeong, and F. Hussain, "On the identification of a vortex", *J. Fluid Mech.,* vol. 285, pp. 69-94, 1995.
[http://dx.doi.org/10.1017/S0022112095000462]

[25]   C. Liu, Y. Wang, Y. Yang, and Z. Duan, "New Omega Vortex Identification Method", *Sci. China Phys. Mech. Astron.,* vol. 59, no. 8, pp. 1-9, 2016.
[http://dx.doi.org/10.1007/s11433-016-0022-6]

[26]   X. Dong, S. Tian, and C. Liu, "Correlation analysis on volume vorticity and vortex in late boundary layer transition", *Phys. Fluids,* vol. 30, no. 1, 2018.014105.
[http://dx.doi.org/10.1063/1.5009115]

[27]   Y. Dong, Y. Yang, and C. Liu, "DNS Study on Three Vortex Identification Methods", *AIAA Pap.,* pp. 2017-0137, 2017.
[http://dx.doi.org/10.2514/6.2017-0137]

[28]   Y. Wang, Y. Yang, G. Yang, and C. Liu, "DNS study on vortex and vorticity in late boundary layer transition", *Commun. Comput. Phys.,* vol. 22, no. 2, pp. 441-459, 2017.
[http://dx.doi.org/10.4208/cicp.OA-2016-0183]

[29]   E. Abdel-Raouf, M.A. Sharif, and J. Baker, "Impulsively started, steady and pulsated annular inflows", *Fluid Dyn. Res.,* vol. 49, no. 2, 2017.025511 [http://dx.doi.org/10.1088/1873-7005/aa5add]

[30]   B. Epps, "Review of Vortex Identification Methods", *AIAA Pap.,* pp. 2017-0989, 2017.

[31]   X. Dong, Y. Wang, X. Chen, Y. Dong, Y. Zhang, and C. Liu, "Determination of epsilon for Omega vortex identification method", *J. Hydrodynam.,* vol. 30, pp. 541-548, 2018. [http://dx.doi.org/10.1007/s42241-018-0066-x]

[32]   M.A. Mendez, M. Raiola, A. Masullo, S. Discetti, A. Ianiro, R. Theunissen, and J.M. Buchlin, "POD- based background removal for particle image velocimetry", *Exp. Therm. Fluid Sci.,* 2017.
[http://dx.doi.org/10.1016/j.expthermflusci.2016.08.021]

[33]   P. J. Schmid, "Advanced Post-Processing of Experimental and Numerical Data", *VKI Lec-ture Series 2014-01.*

[34]   I. Mezic, ""Spectral properties of dynamical systems, model reduction and decompositions," Nonlinear Dyn", *Vpl.,* vol. 41, pp. 309-325, 2005.

CHAPTER 6

Comparison of Liutex and Eigenvalue-based Vortex Identification Criteria for Compressible Flows

Yisheng Gao[1] and Chaoqun Liu[2,*]

[1]*College of Aerospace Engineering, Nanjing University of Aeronautics and Astronautics, Nanjing 210016, China*

[2]*Department of Mathematics, University of Texas at Arlington, Arlington, Texas 76019, USA*

Abstract: Currently, the Q criterion, the Δ criterion and the λ_{ci} criterion are representative among the most widely used vortex identification criteria. These criteria can be categorized as eigenvalue-based criteria since they are exclusively determined by the eigenvalues or invariants of the velocity gradient tensor. However, these criteria are not always satisfactory and suffer from several defects, such as inadequacy of identifying the rotational axis and contamination by shearing. Recently, a novel concept of Liutex (previously named Rortex), including the scalar, vector and tensor form, was proposed to overcome the issues associated with the eigenvalue-based criteria. In the present paper, the comparison of Liutex and two eigenvalue-based criteria, namely the λ_{ci} criterion and the Q_D criterion, a modification of the Q criterion, is performed to assess these methods for compressible flows. According to the analysis of the deviatoric part of the velocity gradient tensor, all the scalar, vector and tensor forms of Liutex are valid for compressible flows without any modification, while two eigenvalue-based criteria, though applicable to compressible flows, will tend to be severely contaminated by shearing as for incompressible flows. Vortical structures induced by supersonic microramp vortex generator (MVG) at Mach 2.5 are examined to confirm the validity and superiority of Liutex for compressible flows.

Keywords: Compressible flows, Liutex/Rortex, Vortex identification, Vortex structures, Turbulence.

INTRODUCTION

Vortical structures, or formally referred to as coherent structures [1-3], are widely regarded as one of the most principal features of transitional and turbulent flows and serve a crucial role in turbulence generation and sustenance. During the last

*Corresponding author Chaoqun Liu: Department of Mathematics, University of Texas at Arlington, Arlington, Texas 76019, USA; Tel: +1-8172725151; Fax: +1-8172725802; E-mail: cliu@uta.edu

several decades, some typical elementary structures, such as hairpin vortices [4-7], vortex braids [8-9] and quasi-streamwise vortices [1, 10-11] *etc.*, have been identified and intensively studied. Unfortunately, despite the ubiquity and signify-cance of such spatially and temporally coherent vortical motions in transitional and turbulent flows, an unambiguous and rigorous definition of the vortex is yet to be achieved (without doubt, the concept of coherent structures is also somewhat ambiguous). The lack of a well-accepted definition has been perceived as one of the chief obstacles hindering the thorough understanding of vortical structures and the mechanism of turbulence generation and sustenance [12-13].

The classic vortex dynamics generally associate the vortex with the vorticity since the vorticity is mathematically mathewell-defined (the curl of the velocity vector). For example, Saffman [14] regards a vortex as a "finite volume of vorticity immersed in irrotational fluid." Nitsche [15] suggests that "a vortex is commonly associated with the rotational motion of fluid around a common centerline. It is defined by the vorticity in the fluid, which measures the rate of local fluid rotation." Wu *et al.* [16] declares that "a vortex is a connected region with high concentration of vorticity compared with its surrounding". However, the usage of vorticity-based methods is not always satisfactory, especially for turbulence flows. The prominent issue is that the vorticity cannot distinguish between a region with real vortical motion and a shear layer region. A typical example is the Blasius boundary layer where the magnitude of the vorticity is relatively large in the near-wall regions, but no real rotational motion will be observed. In addition, the vorticity can be somewhat misaligned with the direction of vortical structures [17]. It is not uncommon that the vorticity vector angle will be significantly larger than the local inclination of the vortex structure over almost the entire length of the quasi-streamwise vortex in the channel flow [18]. And the association between regions of strong vorticity and actual vortices can be rather weak in the turbulent boundary layer, especially in the near wall region [19]. In fact, it has been found by Wang *et al.* [20] that in the near-wall regions, the magnitude of the vorticity will be dramatically reduced along vorticity lines entering the vortex core.

To overcome the issues associated with vorticity-based methods for the identification and visualization of vortex structures, numerous vortex identification methods, including Eulerian non-local methods, Eulerian local region-type methods, Eulerian local line-type methods and Lagrangian methods have been proposed during the last three decades [21]. Most of the currently popular vortex identification criteria belong to Eulerian local region-type methods. Most of these criteria are exclusively determined by the eigenvalues or invariants of the velocity gradient tensor and thereby can be categorized as eigenvalue-based criteria. For example, the Q criterion [22] defines vortices as the regions where the vorticity magnitude prevails over the strain-rate magnitude. The Δ criterion [23-25] identifies the region where the velocity gradient tensor has complex eigenvalues by the discriminant of the characteristic equation. The λ_{ci} criterion [18] is an extension of the Δ criterion, which uses the (positive) imaginary part of the complex eigenvalue to determine the swirling strength. The λ_2 criterion [26] is a dynamical definition and based on the second-largest eigenvalue of $\boldsymbol{S}^2 + \boldsymbol{\Omega}^2$ ($\boldsymbol{S}$ and $\boldsymbol{\Omega}$ represent the symmetric and the antisymmetric parts of the velocity gradient tensor, respectively. Also, it should be noted that λ_2 can be exclusively determined by the eigenvalues only if the eigenvectors of the velocity gradient tensor are orthonormal). These methods can discriminate against shear layers, offering more detectable vortex structures than vorticity-based methods. Nevertheless, there exist several shortcomings of the eigenvalue-based criteria. One is the user-specified threshold. On one hand, the threshold is case-dependent and cannot be determined beforehand. On the other hand, no one can confirm if a threshold is appropriate or not, because different thresholds will present different vortical structures [27]. Therefore, the educed structures obtained from these criteria should be interpreted with care. As a remedy, relative values can be employed to avoid the usage of case-related thresholds, and one such example is the Omega method [28-30]. The Omega method (Ω) is originated from an idea that the vortex is a region where the vorticity overtakes the deformation and Ω is defined as a ratio of vorticity tensor norm squared over the sum of vorticity tensor norm squared and deformation tensor norm squared. Thus, the Omega method is robust to moderate threshold change and capable to capture both strong and weak vortices simultaneously. Another obvious drawback is the inadequacy of identifying the swirl axis or orientation. The existing eigenvalue-based criteria are scalar-valued criteria, which means that only iso-surfaces will be obtained, and no rotational axis can be identified by these criteria. In addition, eigenvalue-based criteria are prone to contamination by shearing [31-33]. The problem of contamination by shear motivates Kolář [34] to propose a triple decomposition from which the residual vorticity can be obtained after the extraction of an effective pure shearing motion and represents a direct and accurate measure of the pure rigid-body rotation of a fluid element. However, a basic reference frame (BRF) should be first determined and searching for BRF in 3D cases is very challenging, which severely limits the applicability of the triple decomposition. The extensive overview of the currently available vortex identification methods has been provided by several review papers [21, 35-36].

Recently, a novel eigenvector-based concept of Liutex (previously named Rortex) [33, 37-38] is introduced to address the abovementioned issues associated with the existing eigenvalue-based criteria. Compared to the existing eigenvalue-based methods, Liutex is not only Galilean invariant [39], but also a systematical definition of the local fluid rotation, including the scalar, vector and tensor forms [33, 40]. The scalar version (the magnitude) of Liutex represents the accurate local rotational strength without the contamination by shear. The direction of the Liutex vector, determined by the real eigenvector of the velocity gradient tensor, represents the local rotation axis, consistent with the analysis of the solution trajectories of ordinary differential equations [23]. And the tensor form of Liutex, rather than the vorticity tensor, represents the real rotational part of the velocity gradient tensor, which can be used for a universal decomposition of the velocity gradient tensor [40]. The comparison of Liutex and two eigenvalue-based methods for incompressible flows is provided [33] to indicate the superiority of Liutex over the eigenvalue-based methods.

Since many industrial flow problems, such as aerodynamics of aircraft, environmental flows and aerodynamics of rotors, propellers and blades, involve vortex structures in compressible flows, the applicability of vortex identification methods in compressible flows is of central importance. However, the very extensive literature on vortex identification or visualization contains only a minute portion of publications dealing with this topic. Kolář [41] investigates compressibility effect in several vortex identification methods, based on the analysis of the deviatoric part of the velocity gradient tensor. It is found that the Q criterion cannot distinguish between expansion and compression, thus as a whole is not extendable to compressible flows, while the Δ criterion and the λ_{ci} criterion retain the applicability to compressible flows. Later, a modification of the Q criterion named the Q_M criterion is proposed by Kolář *et al.* [42] to be applied in compressible flows, based on the concept of the corotation [43]. Yao *et al.* [44] propose an extension of the λ_2 criterion for compressible and variable density flows. In the present paper, the applicability of Liutex to compressible flows is examined. According to the analysis of the deviatoric part of the local velocity gradient tensor, all the scalar, vector and tensor forms of Liutex are valid for compressible flows without any modification. The analytical relations between Liutex and three eigenvalue-based vortex identification methods, namely the λ_{ci} criterion and the Q_D criterion [41], a modification of the Q criterion, are presented to indicate that eigenvalue-based methods will be prone to severe contamination by shear as for incompressible flows. And vortical structures induced by MVG at Mach 2.5 are studied to confirm the superiority of Liutex for compressible flows.

REVIEW OF THE DEFINITION OF LIUTEX

In this section, two equivalent expressions of Liutex will be briefly presented. The first version is based on the work of Liu *et al.* [37], Tian *et al.* [38] and Gao *et al.* [33]. The direction of the Liutex vector, representing the local rotational axis, is defined by the direction of $\vec{r}$ which fulfills

$$\mathrm{d}\vec{v} = \alpha d\vec{r} \tag{1}$$

when the velocity gradient tensor has one real eigenvalue and two complex eigenvalues. Here, $\vec{v}$ represents the velocity vector and α is a scalar. This definition is reasonable since it is expected that the velocity can only increase or decrease along the rotation axis and no cross-velocity gradient exists. Accordingly, if the z-axis is the rotational axis, the velocity can only increase or decrease along the z-axis, namely $dw \neq 0$, but $du = 0$ and $dv = 0$. When the velocity gradient tensor has three real eigenvalues, according to critical point theory [23], flow patterns indicate no swirling/rotational motion. In this case, Liutex should be defined as zero. It has been proven in Ref. [33] that the real eigenvector fulfills Eq. (1), thereby serving as the local rotational axis, expressed as:

$$\nabla\vec{v} \cdot \vec{r} = \lambda_r \vec{r} \tag{2}$$

$$\nabla\vec{v} = \begin{bmatrix} \dfrac{\partial u}{\partial x} & \dfrac{\partial u}{\partial y} & \dfrac{\partial u}{\partial z} \\ \dfrac{\partial v}{\partial x} & \dfrac{\partial v}{\partial y} & \dfrac{\partial v}{\partial z} \\ \dfrac{\partial w}{\partial x} & \dfrac{\partial w}{\partial y} & \dfrac{\partial w}{\partial z} \end{bmatrix} \tag{3}$$

where λ_r represents the real eigenvalue. After the direction of the Liutex vector is determined, the next step is to obtain the rotational strength (the magnitude of the Liutex vector). First, a coordinate rotation (Q rotation) is used to rotate the original z-axis to the direction of the local rotational axis $\vec{r}$ and the velocity gradient tensor $\nabla\vec{V}$ in the resulting XYZ frame will become

$$\nabla\vec{V} = Q\nabla\vec{v}Q^{\mathrm{T}} = \begin{bmatrix} \dfrac{\partial U}{\partial X} & \dfrac{\partial U}{\partial Y} & 0 \\ \dfrac{\partial V}{\partial X} & \dfrac{\partial V}{\partial Y} & 0 \\ \dfrac{\partial W}{\partial X} & \dfrac{\partial W}{\partial Y} & \dfrac{\partial W}{\partial Z} \end{bmatrix} \tag{4}$$

where $\boldsymbol{Q}$ is a (proper) rotation matrix and U, V, W represent the velocity components in the XYZ frame, respectively. The components of $\boldsymbol{Q}$ can be found in Ref. [13]. And then, a second rotation (P rotation) is applied to rotate the reference frame around the Z-axis and the corresponding velocity gradient tensor $\nabla \vec{V}_\theta$ can be written as

$$
\nabla \vec{V}_\theta =
\begin{bmatrix}
\frac{\partial U}{\partial X}|_\theta & \frac{\partial U}{\partial Y}|_\theta & 0 \\
\frac{\partial V}{\partial X}|_\theta & \frac{\partial V}{\partial Y}|_\theta & 0 \\
\frac{\partial W}{\partial X}|_\theta & \frac{\partial W}{\partial Y}|_\theta & \frac{\partial W}{\partial Z}|_\theta
\end{bmatrix}
= P \nabla \vec{V} P^{-1}
\tag{5}
$$

where $\boldsymbol{P}$ is a (proper) rotation matrix around the Z-axis and can be written as

$$
\boldsymbol{P} =
\begin{bmatrix}
cos\theta & sin\theta & 0 \\
-sin\theta & cos\theta & 0 \\
0 & 0 & 1
\end{bmatrix}
\tag{6}
$$

The components of 2×2 upper left submatrix of $\nabla \vec{V}_\theta$ are

$$
\frac{\partial U}{\partial Y}|_\theta = \alpha si\, n(2\theta + \varphi) - \beta
\tag{7a}
$$

$$
\frac{\partial V}{\partial X}|_\theta = \alpha si\, n(2\theta + \varphi) + \beta
\tag{7b}
$$

$$
\frac{\partial U}{\partial X}|_\theta = -\alpha \cos(2\theta + \varphi) + \frac{1}{2}\left(\frac{\partial U}{\partial X} + \frac{\partial V}{\partial Y}\right)
\tag{7c}
$$

$$
\frac{\partial V}{\partial Y}|_\theta = \alpha \cos(2\theta + \varphi) + \frac{1}{2}\left(\frac{\partial U}{\partial X} + \frac{\partial V}{\partial Y}\right)
\tag{7d}
$$

where

$$
\alpha = \frac{1}{2}\sqrt{\left(\frac{\partial V}{\partial Y} - \frac{\partial U}{\partial X}\right)^2 + \left(\frac{\partial V}{\partial X} + \frac{\partial U}{\partial Y}\right)^2}
\tag{8}
$$

$$
\beta = \frac{1}{2}\left(\frac{\partial V}{\partial X} - \frac{\partial U}{\partial Y}\right)
\tag{9}
$$

$$\varphi = \begin{cases} acos\left(\dfrac{\frac{1}{2}\left(\frac{\partial V}{\partial Y}-\frac{\partial U}{\partial X}\right)}{\alpha}\right), & \dfrac{\partial V}{\partial X}+\dfrac{\partial U}{\partial Y} \geq 0 \\[3mm] asin\left(\dfrac{\frac{1}{2}\left(\frac{\partial V}{\partial X}+\frac{\partial U}{\partial Y}\right)}{\alpha}\right), & \dfrac{\partial V}{\partial X}+\dfrac{\partial U}{\partial Y} < 0, \dfrac{\partial V}{\partial Y}-\dfrac{\partial U}{\partial X} \geq 0 \\[3mm] asin\left(\dfrac{-\frac{1}{2}\left(\frac{\partial V}{\partial X}+\frac{\partial U}{\partial Y}\right)}{\alpha}\right)+\pi, & \dfrac{\partial V}{\partial X}+\dfrac{\partial U}{\partial Y} < 0, \dfrac{\partial V}{\partial Y}-\dfrac{\partial U}{\partial X} < 0 \end{cases} \tag{10}$$

(Note: If $\alpha = 0$, we have $\dfrac{\partial V}{\partial Y}-\dfrac{\partial U}{\partial X}=0, \dfrac{\partial V}{\partial X}+\dfrac{\partial U}{\partial Y}=0, \dfrac{\partial V}{\partial X}=\beta, \dfrac{\partial U}{\partial Y}=-\beta$ for any θ. In this case, φ is not needed.)

The rotational strength is defined as twice the minimal absolute value of the off-diagonal component of the 2×2 upper left submatrix when θ varies from 0 to 2π and is given by

$$R = 2(\beta - \alpha) \tag{11}$$

The study [45] provides a justification of the above definition of the rotational strength. In fact, the minimal absolute value is the only option which will not cause a contradiction. Therefore, the Liutex vector $\vec{R}$ can be expressed as follows

$$\vec{R} = R\vec{r} \tag{12}$$

The corresponding tensor form of Liutex can be written as

$$\overset{\Rightarrow}{R} = \begin{bmatrix} 0 & -\phi & 0 \\ \phi & 0 & 0 \\ 0 & 0 & 0 \end{bmatrix} \tag{13a}$$

$$\phi = \beta - \alpha = R/2 \tag{13b}$$

Recently, a second expression of the Liutex vector, which is a simple and explicit version, is proposed by Wang *et al.* [46]. The magnitude of Liutex can be directly calculated by the following explicit expression

$$R = \langle \vec{\omega}, \vec{r} \rangle - \sqrt{\langle \vec{\omega}, \vec{r} \rangle^2 - 4\lambda_{ci}^2} \qquad (14)$$

where $\vec{\omega}$ represents the vorticity vector, $\langle \cdot, \cdot \rangle$ the inner product and λ_{ci} the imaginary part of the complex eigenvalue. It should be noted that since the direction of the eigenvector is unique up to the $\pm$ sign, an additional condition

$$\vec{\omega} \cdot \vec{r} > 0 \qquad (15)$$

should be satisfied to assure the uniqueness of the direction. By using the explicit formula, the successive rotations given by Eqs. (4) and (5) can be totally avoided, which implies a substantial efficiency improvement [46].

EIGENVALUE-BASED VORTEX IDENTIFICATION CRITERIA AND COMPRESSIBLE EXTENSIONS

As pointed out in the study [33], most of the currently popular Eulerian vortex identification methods are exclusively determined by the eigenvalues or invariants of the velocity gradient tensor. If λ_1, λ_2 and λ_3 are used to denote three eigenvalues of the velocity gradient tensor, the characteristic equation can be written as

$$\lambda^3 + \tilde{P}\lambda^2 + \tilde{Q}\lambda + \tilde{R} = 0 \qquad (16)$$

where

$$\tilde{P} = -(\lambda_1 + \lambda_2 + \lambda_3) = -tr(\nabla \vec{v}) \qquad (17)$$

$$\tilde{Q} = \lambda_1\lambda_2 + \lambda_2\lambda_3 + \lambda_3\lambda_1 = -\frac{1}{2}(\text{tr}(\nabla \vec{v}^2) - \text{tr}(\nabla \vec{v})^2) \qquad (18)$$

$$\tilde{R} = -\lambda_1\lambda_2\lambda_3 = -\det(\nabla \vec{v}) \qquad (19)$$

$\tilde{P}$, $\tilde{Q}$ and $\tilde{R}$ are three invariants. And the velocity gradient tensor is usually decomposed to a symmetric part (strain rate tensor) and an antisymmetric part (vorticity tensor) as

$$\nabla\vec{v} = \begin{bmatrix} \dfrac{\partial u}{\partial x} & \dfrac{\partial u}{\partial y} & \dfrac{\partial u}{\partial z} \\[2mm] \dfrac{\partial v}{\partial x} & \dfrac{\partial v}{\partial y} & \dfrac{\partial v}{\partial z} \\[2mm] \dfrac{\partial w}{\partial x} & \dfrac{\partial w}{\partial y} & \dfrac{\partial w}{\partial z} \end{bmatrix} = S + W$$

$$S = \frac{1}{2}(\nabla\vec{v} + \nabla\vec{v}^{T}) = \begin{bmatrix} \dfrac{\partial u}{\partial x} & \dfrac{1}{2}\left(\dfrac{\partial u}{\partial y} + \dfrac{\partial v}{\partial x}\right) & \dfrac{1}{2}\left(\dfrac{\partial u}{\partial z} + \dfrac{\partial w}{\partial x}\right) \\[2mm] \dfrac{1}{2}\left(\dfrac{\partial v}{\partial x} + \dfrac{\partial u}{\partial y}\right) & \dfrac{\partial v}{\partial y} & \dfrac{1}{2}\left(\dfrac{\partial v}{\partial z} + \dfrac{\partial w}{\partial x}\right) \\[2mm] \dfrac{1}{2}\left(\dfrac{\partial w}{\partial x} + \dfrac{\partial u}{\partial z}\right) & \dfrac{1}{2}\left(\dfrac{\partial w}{\partial y} + \dfrac{\partial v}{\partial z}\right) & \dfrac{\partial w}{\partial z} \end{bmatrix} \quad (20)$$

$$W = \frac{1}{2}(\nabla\vec{v} - \nabla\vec{v}^{T}) = \begin{bmatrix} 0 & \dfrac{1}{2}\left(\dfrac{\partial u}{\partial y} - \dfrac{\partial v}{\partial x}\right) & \dfrac{1}{2}\left(\dfrac{\partial u}{\partial z} - \dfrac{\partial w}{\partial x}\right) \\[2mm] \dfrac{1}{2}\left(\dfrac{\partial v}{\partial x} - \dfrac{\partial u}{\partial y}\right) & 0 & \dfrac{1}{2}\left(\dfrac{\partial v}{\partial z} - \dfrac{\partial w}{\partial x}\right) \\[2mm] \dfrac{1}{2}\left(\dfrac{\partial w}{\partial x} - \dfrac{\partial u}{\partial z}\right) & \dfrac{1}{2}\left(\dfrac{\partial w}{\partial y} - \dfrac{\partial v}{\partial z}\right) & 0 \end{bmatrix} \quad (21)$$

For incompressible flows, the continuous equation implies the vanishing divergence, namely

$$\tilde{P} = -(\lambda_1 + \lambda_2 + \lambda_3) = -tr(\nabla\vec{v}) = -\left(\frac{\partial u}{\partial x} + \frac{\partial v}{\partial y} + \frac{\partial w}{\partial z}\right) = 0 \quad (22)$$

But for compressible flows, the divergence of the velocity is no longer equal to zero. Therefore, Kolář [41] and Epps [21] suggest that if the divergence does not play a role, the criterion should only depend on the deviatoric part of the velocity gradient tensor which is expressed as

$$\nabla\vec{v}_D \equiv \nabla\vec{v} - kI = \begin{bmatrix} \dfrac{\partial u}{\partial x} - k & \dfrac{\partial u}{\partial y} & \dfrac{\partial u}{\partial z} \\[2mm] \dfrac{\partial v}{\partial x} & \dfrac{\partial v}{\partial y} - k & \dfrac{\partial v}{\partial z} \\[2mm] \dfrac{\partial w}{\partial x} & \dfrac{\partial w}{\partial y} & \dfrac{\partial w}{\partial z} - k \end{bmatrix} \quad (23)$$

$$k = \frac{1}{3}tr(\nabla\vec{v}) = -\frac{1}{3}\tilde{P} = \frac{1}{3}(\lambda_1 + \lambda_2 + \lambda_3) \quad (24)$$

Here, $k > 0$ implies isentropic expansion while $k < 0$ implies compression. And the deviatoric part of the strain rate tensor can be written as

$$S_D \equiv S - kI = \begin{bmatrix} \frac{\partial u}{\partial x} - k & \frac{1}{2}\left(\frac{\partial u}{\partial y} + \frac{\partial v}{\partial x}\right) & \frac{1}{2}\left(\frac{\partial u}{\partial z} + \frac{\partial w}{\partial x}\right) \\ \frac{1}{2}\left(\frac{\partial v}{\partial x} + \frac{\partial u}{\partial y}\right) & \frac{\partial v}{\partial y} - k & \frac{1}{2}\left(\frac{\partial v}{\partial z} + \frac{\partial w}{\partial y}\right) \\ \frac{1}{2}\left(\frac{\partial w}{\partial x} + \frac{\partial u}{\partial z}\right) & \frac{1}{2}\left(\frac{\partial w}{\partial y} + \frac{\partial v}{\partial z}\right) & \frac{\partial w}{\partial z} - k \end{bmatrix} \tag{25}$$

According to Eqs. (23), (24) and (25), we have

$$tr(\nabla \vec{v}_D) = tr(S_D) = \left(\frac{\partial u}{\partial x} - k\right) + \left(\frac{\partial v}{\partial y} - k\right) + \left(\frac{\partial w}{\partial z} - k\right) = 0 \tag{26}$$

In the following, two representatives of eigenvalue-based criteria and their compressible extensions are discussed.

(1) Q criterion

The Q criterion proposed by Hunt *et al.* [22] detects vortices as fluid regions with the vorticity magnitude in excess of the strain-rate magnitude, which can be written as

$$Q = \frac{1}{2}(\|W\|^2 - \|S\|^2) > 0 \tag{27}$$

where $\|\cdot\|^2$ represents the Frobenius norm. A second condition which requires the pressure in the vortical regions to be lower than the ambient pressure is often omitted in practice. According to Eqs. (17) and (18), Q can be expressed in terms of invariants as

$$Q = \tilde{Q} - \frac{1}{2}\tilde{P}^2 \tag{28}$$

For incompressible flows, Eq. (22) implies $Q = \tilde{Q}$. But for compressible flows, Q will be expressed as [21]

$$Q = \frac{1}{2}(\|W\|^2 - \|S_D\|^2) - \frac{3}{2}k^2 \tag{29}$$

Obviously, from Eq. (29), the Q criterion cannot distinguish between expansion ($k > 0$) and compression ($k < 0$), which means that the original Q criterion suffers from the ambiguity and is not extendable to compressible flows. Thus, Kolář [41] proposes a modification of the Q criterion in terms of a deviatoric part of S which reads

$$Q_D = \frac{1}{2}(\|W\|^2 - \|S_D\|^2) \tag{30}$$

According to Eqs. (24), (29) and (30), Q_D can be also expressed in terms of invariants as

$$Q_D = \tilde{Q} - \frac{1}{3}\tilde{P}^2 \tag{31}$$

(2) λ_{ci} criterion

Based on critical point theory [23], in a reference frame moving with a fluid particle, the instantaneous trajectory of the neighboring fluid particle is governed by a linear system of first-order ordinary differential equations which can be written as

$$\frac{dy}{dt} = \nabla\vec{v} \cdot y \tag{32}$$

where y represents the position relative of the neighboring fluid particle relative to the position of the reference fluid particle. The characteristic equation for the velocity gradient tensor $\nabla\vec{v}$ is given by Eq. (16) and the corresponding discriminant is

$$\Delta = \left(\frac{\tilde{Q}}{3} - \frac{\tilde{P}^2}{9}\right)^3 + \left(\frac{\tilde{R}}{2} + \frac{\tilde{P}^3}{27} - \frac{\tilde{P}\tilde{Q}}{6}\right)^2 \tag{33}$$

If $\Delta \leq 0$, three eigenvalues of $\nabla\vec{v}$ are real; if $\Delta > 0$, there exist one real eigenvalue and two conjugate complex eigenvalues. The fact that the trajectory will exhibit a swirling flow pattern if $\nabla\vec{v}$ has two complex eigenvalues prompts Chong *et al.* [23] to propose the Δ criterion which defines a vortex core to be the region where $\nabla\vec{v}$ has complex eigenvalues.

The λ_{ci} criterion [18] can be regarded as an enhanced version of the Δ criterion and identical to the Δ criterion when zero threshold is applied. The solution of Eq. (32) can be facilitated by eigendecomposition

$$\nabla\vec{v} = [\vec{r} \quad \vec{v}_{cr} \quad \vec{v}_{ci}]\begin{bmatrix} \lambda_r & 0 & 0 \\ 0 & \lambda_{cr} & \lambda_{ci} \\ 0 & -\lambda_{ci} & \lambda_{cr} \end{bmatrix}[\vec{r} \quad \vec{v}_{cr} \quad \vec{v}_{ci}]^{-1} \tag{34}$$

Here, $(\lambda_r, \vec{r})$ is the real eigenpair and $(\lambda_{cr} \pm i\lambda_{ci}, \vec{v}_{cr} \pm i\vec{v}_{ci})$ the complex conjugate eigenpair. In the local curvilinear coordinate system (c_1, c_2, c_3) spanned by the eigenvector $(\vec{r}, \vec{v}_{cr}, \vec{v}_{ci})$, the instantaneous streamlines are the same as pathlines and can be written as

$$c_1(t) = c_1(0)e^{\lambda_r t} \tag{35a}$$

$$c_2(t) = [c_2(0)\cos(\lambda_{ci}t) + c_3(0)\sin(\lambda_{ci}t)]e^{\lambda_{cr}t} \tag{35b}$$

$$c_3(t) = [c_3(0)\cos(\lambda_{ci}t) - c_2(0)\sin(\lambda_{ci}t)]e^{\lambda_{cr}t} \tag{35c}$$

where t represents the time-like parameter and the constants $c_1(0)$, $c_2(0)$ and $c_3(0)$ are determined by the initial conditions. From Eq. (35b) and (35c), the period of orbit of a fluid particle is $2\pi/\lambda_{ci}$, so the imaginary part of the complex value λ_{ci} is called swirling strength.

Kolář [41] has pointed out that there is no compressibility effect on the Δ criterion and the λ_{ci} criterion stems from the Δ criterion and therefore retains the applicability to compressible flows. Here, a simple proof of the applicability of the λ_{ci} criterion to compressible flows is presented.

According to the definition of eigenvalue and Eq. (34), we have

$$\nabla\vec{v} \cdot (\vec{v}_{cr} + i\vec{v}_{ci}) = (\lambda_{cr} + i\lambda_{ci})(\vec{v}_{cr} + i\vec{v}_{ci}) \tag{36}$$

Therefore,

$$\nabla\vec{v}_D \cdot (\vec{v}_{cr} + i\vec{v}_{ci}) = (\nabla\vec{v} - k\mathbf{I}) \cdot (\vec{v}_{cr} + i\vec{v}_{ci}) = \nabla\vec{v} \cdot (\vec{v}_{cr} + i\vec{v}_{ci}) - k\mathbf{I} \cdot (\vec{v}_{cr} + i\vec{v}_{ci}) = (\lambda_{cr} + i\lambda_{ci})(\vec{v}_{cr} + i\vec{v}_{ci}) - k(\vec{v}_{cr} + i\vec{v}_{ci}) = (\lambda_{cr} - k + i\lambda_{ci})(\vec{v}_{cr} + i\vec{v}_{ci}) \tag{37}$$

Eq. (37) indicates that the uniform dilatation given by k does not alter the imaginary part of the complex eigenvalue λ_{ci} and λ_{ci} remains the same for the deviatoric part of the velocity gradient tensor. Thus, the applicability of the λ_{ci} criterion to compressible flows has been confirmed.

THE APPLICABILITY OF LIUTEX TO COMPRESSIBLE FLOWS

Since Liutex is based on the real eigenvector, the real eigenvector of the deviatoric part of the velocity gradient tensor is considered first. Similar to Eq. (36), we have

$$\nabla \vec{v} \cdot \vec{r} = \lambda_r \vec{r} \tag{38}$$

Therefore,

$$\nabla \vec{v}_D \cdot \vec{r} = (\nabla \vec{v} - kI) \cdot \vec{r} = \nabla \vec{v} \cdot \vec{r} - kI \cdot \vec{r} = \lambda_r \vec{r} - k\vec{r} = (\lambda_r - k)\vec{r} \tag{39}$$

Eq. (39) implies that the real eigenvalue is altered but the real eigenvector of the original velocity gradient tensor retains the real eigenvector of the deviatoric part of the velocity gradient tensor. Thus, compressibility effect will not change the direction of Liutex.

To examine the compressibility effect on the magnitude of Liutex, Eq. (14) can be used. Since the vorticity vector $\vec{\omega}$ only depends on the off-diagonal terms of the velocity gradient, $\vec{\omega}$ remains the same if the deviatoric part of the velocity gradient tensor is applied. And as proven in the previous section, λ_{ci} also remains the same. Consequently, the magnitude of Liutex given by Eq. (14) remains the same, independent of the compressibility effect. In addition, through Eq. (13), it can be concluded that the tensor form of Liutex will be not altered as well. So, all the scalar, vector and tensor forms of Liutex are valid for compressible flows without any modification.

ANALYTICAL RELATIONS BETWEEN LIUTEX, Q_D AND λ_{ci}

According to studies. [33, 40], when $2\theta + \varphi = \pi/2$, the velocity gradient tensor given by Eq. (5) becomes

$$\nabla \vec{V} = \begin{bmatrix} \lambda_{cr} & -\phi & 0 \\ \phi + s & \lambda_{cr} & 0 \\ \xi & \eta & \lambda_r \end{bmatrix} \tag{40}$$

where λ_{cr} is the real part of the complex eigenvalues and s, ξ, η represent the shearing components along the transformed axes. Since the 2×2 upper left submatrix in Eq. (40) has the same complex eigenvalues as the original velocity gradient tensor, the relation between the magnitude of Liutex and λ_{ci} can be immediately obtained as

$$\lambda_{ci} = \sqrt{\frac{R^2}{4} + \frac{Rs}{2}} \tag{41}$$

For compressible flows, the deviatoric part of Eq. (40) is

$$\nabla \vec{V}_D = \begin{bmatrix} \lambda_{cr} - k & -\phi & 0 \\ \phi + s & \lambda_{cr} - k & 0 \\ \xi & \eta & \lambda_r - k \end{bmatrix} \tag{42}$$

In this case, the real part of the complex eigenvalue becomes $\lambda_{cr} - k$, but the the imaginary part remains the same. So, the relation given by Eq. (41) is still valid for compressible flows.

The analytical relation between Liutex and Q_D can be obtained as

$$Q_D = \tilde{Q} - \frac{1}{3}\tilde{P}^2 = (\lambda_{cr} + i\lambda_{ci})(\lambda_{cr} - i\lambda_{ci}) + (\lambda_{cr} - i\lambda_{ci})\lambda_r + \lambda_r(\lambda_{cr} + i\lambda_{ci}) -$$
$$\frac{1}{3}(\lambda_{cr} + i\lambda_{ci} + \lambda_{cr} - i\lambda_{ci} + \lambda_r)^2 = \lambda_{ci}^2 + \frac{2}{3}\lambda_{cr}\lambda_r - \frac{1}{3}(\lambda_{cr}^2 + \lambda_r^2) = \frac{R^2}{4} + \frac{Rs}{2} + \frac{2}{3}\lambda_{cr}\lambda_r -$$
$$\frac{1}{3}(\lambda_{cr}^2 + \lambda_r^2) \tag{43}$$

As in the case of incompressible flows [33], from Eq. (43), it can be found that the shearing effect s always exists in the imaginary part of the complex eigenvalues. Therefore, if eigenvalue-based criteria are dependent on the complex eigenvalues, they will be inevitably contaminated by shearing. Eqs. (41) and (43) indicate the shearing effect s on λ_{ci} and Q_D, respectively.

TEST CASE

Here the implicit large-eddy simulation (ILES) data of MVG at Mach 2.5 and $Re_\theta = 5760$ (Reynolds number based on momentum thickness) [47, 48] are examined to compare Liutex with the λ_{ci} and Q_D criteria. The ILES data are generated by a LES code called LESUTA [49]. A fifth-order bandwidth-optimized WENO scheme is applied for the convective terms and the traditional fourth-order central scheme is used twice to compute the second-order derivatives for viscous terms. Without explicitly using the subgrid scale (SGS) model, the intrinsic dissipation of the numerical method is utilized to dissipate the turbulent energy accumulated at the unresolved scales with high wave numbers. For time integration, the explicit third-order TVD-type Runge-Kutta scheme is employed. The grid number for the whole field is $n_{spanwise} \times n_{normal} \times n_{streamwise} = 137 \times 192 \times 1600$. For the detailed case setup [47, 48].

Fig. (**1**) illustrates iso-surfaces obtained by Q_D, λ_{ci} and Liutex with small thresholds. It can be found that vortical structures identified by three methods are very similar (the structures identified by λ_{ci} and Liutex are not so smooth as Q_D since Q_D is a smooth function while λ_{ci} and Liutex are truncated according to their definitions). This implies that all methods present a similar (approximated) vortex boundary for compressible flows. However, λ_{ci} and Q_D would be severely contaminated by shearing. To investigate the effect of the contamination by shearing, three points A, B and C on the iso-surface of Liutex ($R = 1.0$), as shown in Fig. (**2**), are examined. Point A is located immediately after MVG. Point B is located in the middle part and Point C is located after the weakened shock wave. Table **1** lists the eigenvalues, Q_D, Liutex and shearing components of Points A, B, C obtained by Liutex-based tensor decomposition given by Eq. (40), which demonstrate the local rotation, expansion or compression and shearing effects. It is obvious that the shearing components s in the plane perpendicular to the local rotational axis are relatively large and can be much larger than the magnitude of Liutex, which implies that the shearing effect is so strong that a large portion of Q_D and λ_{ci} are severely contaminated by shearing. It should be also noted that the real eigenvalue and the real part of the complex eigenvalue, which indicate the expansion or compression effects, are no longer small and may be important for compressible flows. The Liutex-based tensor decomposition given by Eq. (40) can provide the complete rotation, expansion or compression and shearing effects of any point, manifesting the superiority over the scalar-based criteria which can only present iso-surfaces to detect vortical structures.

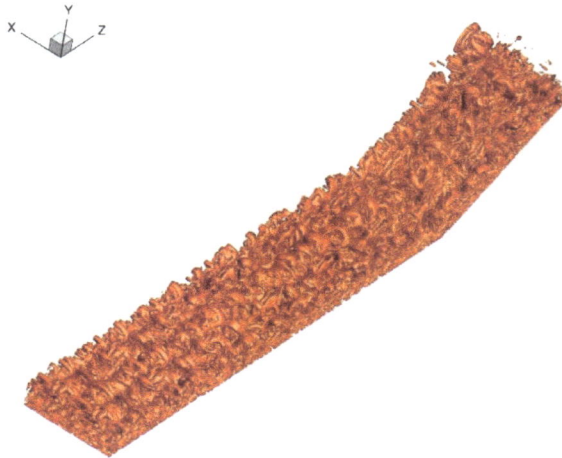

Global view

Fig. 1 cont.....

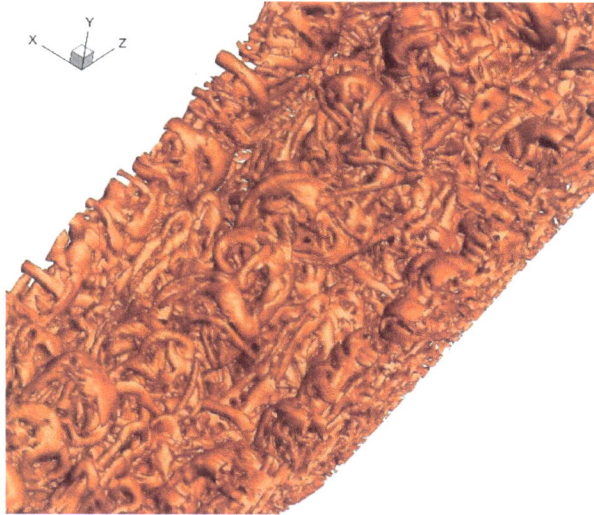

Local view

(a) $Q_D = 0.01$

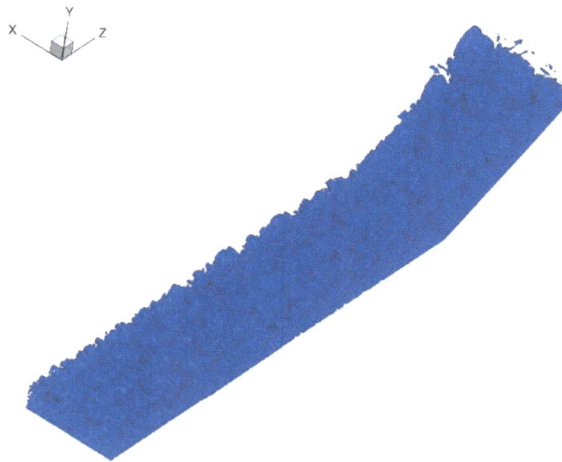

Global view

Fig. 1 cont.....

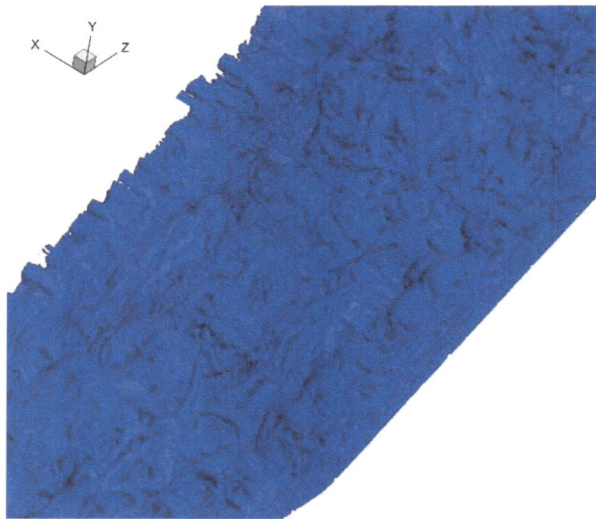

Local view

(b)　　　$\lambda_{ci} = 0.1$

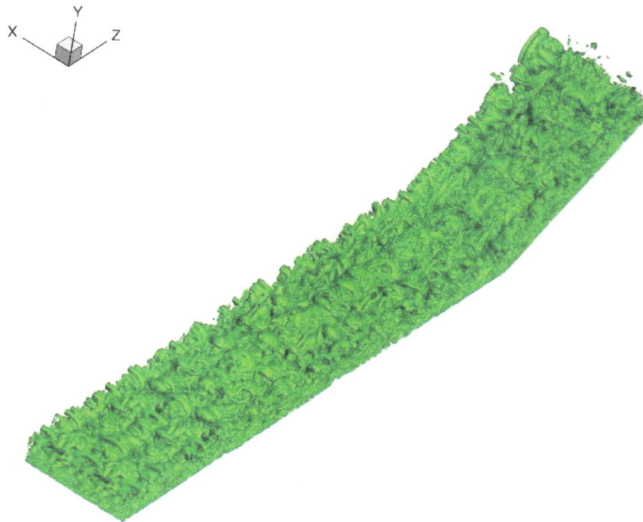

Global view

Fig. 1 cont.....

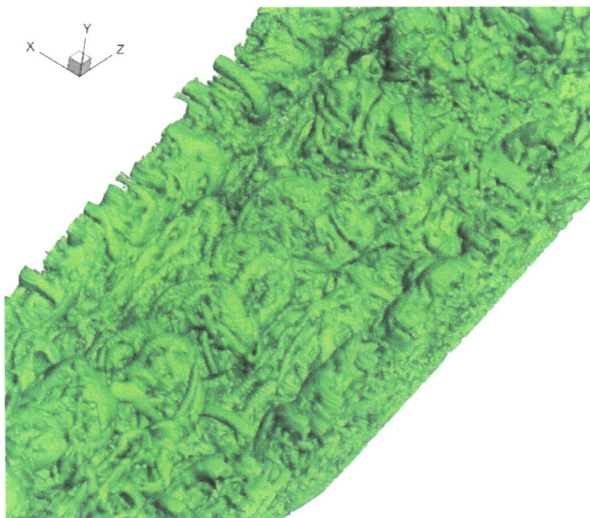

Local view

(c)   Liutex=0.15

Fig. (1). Iso-surfaces of Q_D, λ_{ci} and Liutex for the MVG case with small thresholds.

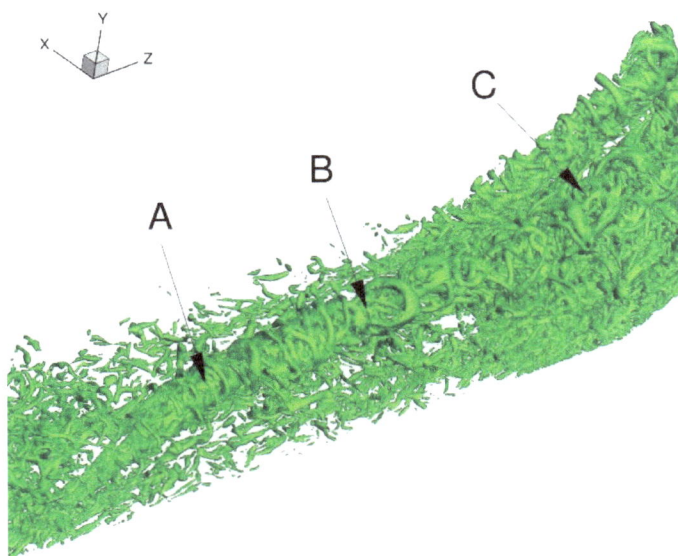

Fig. (2). Iso-surfaces of Liutex with $R = 1.0$. Three points are examined to study the contamination by shearing.

Table 1. Eigenvalues (λ_r, λ_{cr}, λ_{ci}), Q_D, Liutex and shearing components of Point A, B, C.

Variable	Point A	Point B	Point C
λ_r	0.671	0.258	-0.076
λ_{cr}	-0.191	-0.123	-0.523
λ_{ci}	1.716	1.046	1.391
Q_D	2.807	1.778	1.853
R	1.0	1.0	1.0
s	5.610	1.776	3.454
ξ	1.247	0.031	1.082
η	-0.372	-0.158	-0.924

CONCLUDING REMARKS

In the present study, the applicability of Liutex to compressible flows is examined. All the scalar, vector and tensor forms of Liutex remains the same if the original velocity gradient tensor is substituted by the deviatoric part, which implies there is no compressibility effect on Liutex and Liutex is valid for compressible flows without any modification. The analytical relations between Liutex and two eigenvalue-based criteria, namely the λ_{ci} criterion and the Q_D criterion, are presented to indicate eigenvalue-based criteria will be severely contaminated by shearing in compressible flows if the imaginary part of the complex eigenvalue is involved. The data from ILES of MVG confirms the validity and superiority of Liutex for compressible flows.

CONSENT FOR PUBLICATION

Not applicable.

CONFLICT OF INTEREST

The authors confirm that this chapter contents have no conflict of interest.

ACKNOWLEDGEMENTS

This work was supported by the Department of Mathematics at University of Texas at Arlington and AFOSR grant MURI FA9559-16-1-0364. The work was also supported by the Priority Academic Program Development of Jiangsu Higher Education Institutions. The authors are grateful to Texas Advanced Computing Center (TACC) for providing computation hours. This work was accomplished by using Code LESUTA which was developed by Drs. Qin Li and Chaoqun Liu at the University of Texas at Arlington.

REFERENCES

[1]   S.K. Robinson, "Coherent motion in the turbulent boundary layer", *Annu. Rev. Fluid Mech.,* vol. 23, pp. 601-639, 1991.
[http://dx.doi.org/10.1146/annurev.fl.23.010191.003125]

[2]   A.K.M.F. Hussain, "Coherent structures and turbulence", *J. Fluid Mech.,* vol. 173, pp. 303-356, 1986.
[http://dx.doi.org/10.1017/S0022112086001192]

[3]   L. Sirovich, "Turbulence and the dynamics of coherent structures. Part I: Coherent structures", *Q. Appl. Math.,* vol. 45, no. 3, pp. 561-571, 1987.
[http://dx.doi.org/10.1090/qam/910462]

[4]   T. Theodorsen, "Mechanism of turbulence", *Proceedings of the Midwestern Conference on Fluid Mechanics,* 1952

[5]   R.J. Adrian, "Hairpin vortex organization in wall turbulence", *Phys. Fluids,* vol. 19, no. 041301, 2007.

[6]   C. Liu, Y. Yan, and P. Lu, "Physics of turbulence generation and sustenance in a boundary layer", *Comput. Fluids,* vol. 102, pp. 353-384, 2014. [http://dx.doi.org/10.1016/j.compfluid.2014.06.032]

[7]   X. Wu, and P. Moin, "Direct numerical simulation of turbulence in a nominally zero-pressure-gradient flat-plate boundary layer", *J. Fluid Mech.,* vol. 630, pp. 5-41, 2009.
[http://dx.doi.org/10.1017/S0022112009006624]

[8]   M.M. Rogers, and R.D. Moser, "Direct simulation of a self-similar turbulent mixing layer", *Phys. Fluids,* vol. 6, no. 2, pp. 903-923, 1993.
[http://dx.doi.org/10.1063/1.868325]

[9]   J.E. Martin, and E. Meiburg, "Numerical investigation of three-dimensionally evolving jets subject to axisymmetric and azimuthal perturbations", *J. Fluid Mech.,* vol. 230, pp. 271-318, 1991.
[http://dx.doi.org/10.1017/S0022112091000794]

[10]   J.W. Brooke, and T.J. Hanratty, "Origin of turbulence-producing eddies in a channel flow", *Phys. Fluids A Fluid Dyn.,* vol. 5, no. 4, pp. 1011-1022, 1993. [http://dx.doi.org/10.1063/1.858666]

[11]   J. Jeong, F. Hussain, W. Schoppa, and J. Kim, "Coherent structures near the wall in a turbulent channel flow", *J. Fluid Mech.,* vol. 332, pp. 185-214, 1997. [http://dx.doi.org/10.1017/S0022112096003965]

[12]   C. Liu, and X. Cai, "New theory on turbulence generation and structure—DNS and experiment", *Sci. China Phys. Mech. Astron.,* vol. 60, 2017.084731 [http://dx.doi.org/10.1007/s11433-017-9047-2]

[13]   C. Liu, Y. Gao, X. Dong, Y. Wang, J. Liu, Y. Zhang, X. Cai, and N. Gui, "Third generation of vortex identification methods: Omega and Liutex/Rortex based systems", *J. Hydrodynam.,* vol. 31, no. 2, pp. 205-223, 2019.
[http://dx.doi.org/10.1007/s42241-019-0022-4]

[14] P. Saffman, *Vortices dynamics.* Cambridge university press: Cambridge, 1992.

[15] M. Nitsche, Vortex Dynamics.*Encyclopedia of Mathematics and Physics.* Academic Press: New York, 2006.
[http://dx.doi.org/10.1016/B0-12-512666-2/00254-6]

[16] J-Z. Wu, H-Y. Ma, and M-D. Zhou, *Vorticity and vortices dynamics.* Springer-Verlag: Berlin, Heidelberg, 2006.
[http://dx.doi.org/10.1007/978-3-540-29028-5]

[17] Q. Gao, C. Ortiz-Dueñas, and E.K. Longmire, "Analysis of vortex populations in turbulent wall-bounded flows", *J. Fluid Mech.,* vol. 678, pp. 87-123, 2011.
[http://dx.doi.org/10.1017/jfm.2011.101]

[18] J. Zhou, R. Adrian, S. Balachandar, and T. Kendall, "Mechanisms for generating coherent packets of hairpin vortices in channel flow", *J. Fluid Mech.,* vol. 387, pp. 353-396, 1999.
[http://dx.doi.org/10.1017/S002211209900467X]

[19] S.K. Robinson, A review of vortex structures and associated coherent motions in turbulent boundary layers.*Structure of Turbulence and Drag Reduction.* Springer-Verlag: Berlin, Heidelberg, 1990.
[http://dx.doi.org/10.1007/978-3-642-50971-1_2]

[20] Y. Wang, Y. Yang, G. Yang, and C. Liu, "DNS study on vortex and vorticity in late boundary layer transition", *Commun. Comput. Phys.,* vol. 22, pp. 441-459, 2017.
[http://dx.doi.org/10.4208/cicp.OA-2016-0183]

[21] B. Epps, "Review of Vortex Identification Methods", *55th AIAA Aerospace Sciences Meeting, AIAA SciTech Forum,* 2017pp. 2017-0989

[22] J. Hunt, A. Wray, and P. Moin, "Eddies, streams, and convergence zones in turbulent flows",

[23] M. Chong, A. Perry, and B. Cantwell, "A general classification of three-dimensional flow fields", *Phys. Fluids A Fluid Dyn.,* vol. 2, no. 5, pp. 765-777, 1990. [http://dx.doi.org/10.1063/1.857730]

[24] U. Dallmann, "Topological structures of three-dimensional flow separation", *16th AIAA Fluid and Plasma Dynamics conference,* 1983 Danvers, MA

[25] H. Vollmers, H.P. Kreplin, and H.U. Meier, "Separation and vortical-type flow around a prolate spheroid-evaluation of relevant parameters", *Proceedings of the AGARD Symposium on Aerodynamics of Vortical Type Flows in Three Dimensions,* 1983

[26] J. Jeong, and F. Hussain, "On the identification of a vortices", *J. Fluid Mech.,* vol. 285, pp. 69-94, 1995.
[http://dx.doi.org/10.1017/S0022112095000462]

[27] C. Liu, "Numerical and theoretical study on 'vortex breakdown'", *Int. J. Comput. Math.,* vol. 88, no. 17, pp. 3702-3708, 2011.
[http://dx.doi.org/10.1080/00207160.2011.617438]

[28] C. Liu, Y. Wang, Y. Yang, and Z. Duan, "New omega vortex identification method", *Sci. China Phys. Mech. Astron.,* vol. 59, no. 8, 2016.684711 [http://dx.doi.org/10.1007/s11433-016-0022-6]

[29] X. Dong, Y. Wang, X. Chen, Y. Dong, Y. Zhang, and C. Liu, "Determination of epsilon for Omega vortex identification method", *J. Hydrodynam.,* vol. 30, no. 4, pp. 541-548, 2018.
[http://dx.doi.org/10.1007/s42241-018-0066-x]

[30] J. Liu, Y. Wang, Y. Gao, and C. Liu, "Galilean invariance of Omega vortex identification method", *J. Hydrodynam.,* vol. 31, no. 2, pp. 249-255, 2019.
[http://dx.doi.org/10.1007/s42241-019-0024-2]

[31] Y. Maciel, M. Robitaille, and S. Rahgozar, "A method for characterizing cross-sections of vortices in turbulent flows", *Int. J. Heat Fluid FL.,* vol. 37, pp. 177-188, 2012.
[http://dx.doi.org/10.1016/j.ijheatfluidflow.2012.06.005]

[32] H. Chen, R.J. Adrian, Q. Zhong, and X. Wang, "Analytic solutions for three dimensional swirling strength in compressible and incompressible flows", *Phys. Fluids,* vol. 26, no. 8, 2014.081701
[http://dx.doi.org/10.1063/1.4893343]

[33] Y. Gao, and C. Liu, "Rortex and comparison with eigenvalue-based vortex identification criteria", *Phys. Fluids,* vol. 30, no. 8, 2018.085107 [http://dx.doi.org/10.1063/1.5040112]

[34] V. Kolář, "Vortex identification: New requirements and limitations", *Int. J. Heat Fluid FL.,* vol. 28, no. 4, pp. 638-652, 2007.
[http://dx.doi.org/10.1016/j.ijheatfluidflow.2007.03.004]

[35] Y. Zhang, K. Liu, H. Xian, and X. Du, "A review of methods for vortex identification in hydroturbines", *Renew. Sustain. Energy Rev.,* vol. 81, no. 1, pp. 1269-1285, 2018. [http://dx.doi.org/10.1016/j.rser.2017.05.058]

[36] Y. Zhang, X. Qiu, P. Chen, K. Liu, X. Dong, and C. Liu, "A selected review of vortex identification methods with applications", *J. Hydrodynam.,* vol. 30, no. 5, pp. 767-779, 2018. [http://dx.doi.org/10.1007/s42241-018-0112-8]

[37] C. Liu, Y. Gao, S. Tian, and X. Dong, "Rortex—A new vortex vector definition and vorticity tensor and vector decompositions", *Phys. Fluids,* vol. 30, no. 3, 2018.035103 [http://dx.doi.org/10.1063/1.5023001]

[38] S. Tian, Y. Gao, X. Dong, and C. Liu, "Definitions of vortex vector and vortex", *J. Fluid Mech.,* vol. 849, pp. 312-339, 2018. [http://dx.doi.org/10.1017/jfm.2018.406]

[39] Y. Wang, Y. Gao, and C. Liu, "Letter: Galilean invariance of Rortex", *Phys. Fluids,* vol. 30, no. 11, 2018.111701. [http://dx.doi.org/10.1063/1.5058939]

[40] Y. Gao, and C. Liu, "Rortex based velocity gradient tensor decomposition", *Phys. Fluids,* vol. 31, no. 1, 2019.011704. [http://dx.doi.org/10.1063/1.5084739]

[41] V. Kolář, "Compressibility effect in vortex identication", *AIAA J.,* vol. 47, no. 2, pp. 473-475, 2009. [http://dx.doi.org/10.2514/1.40131]

[42] V. Kolář, and J. Šístek, "Corotational and compressibility aspects leading to a modication of the vortex-identication Q-Criterion", *AIAA J.,* vol. 53, no. 8, pp. 2406-2410, 2015. [http://dx.doi.org/10.2514/1.J053697]

[43] V. Kolář, J. Šístek, F. Cirak, and P. Moses, "Average corotation of line segments near a point and vortex identification", *AIAA J.,* vol. 51, no. 11, pp. 2678-2694, 2013. [http://dx.doi.org/10.2514/1.J052330]

[44] J. Yao, and F. Hussain, "Toward vortex identification based on local pressure-minimum criterion in compressible and variable density flows", *J. Fluid Mech.,* vol. 850, pp. 5-17, 2018. [http://dx.doi.org/10.1017/jfm.2018.465]

[45] J. Liu, Y. Deng, Y. Gao, S. Charkrit, and C. Liu, "Mathematical foundation of turbulence generation—From symmetric to asymmetric Liutex", *J. Hydrodynam.,* vol. 31, no. 3, pp. 632-636, 2019. [http://dx.doi.org/10.1007/s42241-019-0049-6]

[46] Y. Wang, Y. Gao, J. Liu, and C. Liu, "Explicit formula for the Liutex vector and physical meaning of vorticity based on the Liutex-Shear decomposition", *J. Hydrodynam.,* vol. 31, no. 3, pp. 464-474, 2019. [http://dx.doi.org/10.1007/s42241-019-0032-2]

[47] Y. Yang, Y. Yan, and C. Liu, "ILES for mechanism of ramp-type MVG reducing shock induced flow separation", *Sci. China Phys. Mech. Astron.,* vol. 59, no. 12, 2016.124711 [http://dx.doi.org/10.1007/s11433-016-0348-2]

[48] Y. Yan, C. Chen, X. Wang, and C. Liu, "LES and analyses on the vortex structure behind supersonic MVG with turbulent inflow", *Appl. Math. Model.,* vol. 38, no. 1, pp. 196-211, 2014. [http://dx.doi.org/10.1016/j.apm.2013.05.048]

[49] Q. Li, and C. Liu, *Implicit LES for Supersonic Microramp Vortex Generator: New Discoveries and New Mechanisms*, 2011. [http://dx.doi.org/10.21236/ADA564811]

CHAPTER 7

Observation of Coherent Structures of Low Reynolds Number Turbulent Boundary Layer by DNS and Experiment

Panpan Yan[1,2], Chaoqun Liu[2,*], Yanang Guo[3] and Xiaoshu Cai[3]

[1]Shenyang Aircraft Design and Research Institute, Aviation Industry of China, Shenyang 110035, China

[2]Department of Mathematics, University of Texas at Arlington, Arlington, Texas 76019, USA

[3]Institute of Particle and Two-phase Flow Measurement, University of Shanghai for Science and Technology, Shanghai 200093, China

Abstract: An elaborate direct numerical simulation (DNS) for late boundary layer transition has been conducted. The DNS results are qualitatively compared with a new Lagrangian property experimental technique named the moving single-frame and long-exposure (MSFLE) imaging method to obtain a deeper understanding on the coherent structures of a transitional and turbulent boundary layer at low Reynolds number. Multilevel vortex structures are clearly observed by both experiment and DNS. This study found that there are multilevel co-rotating vortices, showing how energy is transported from the main flow to the bottom of the boundary layer and how the streaks or white-black strips are formed. The results also show that the lower level vortices cannot simply be produced by the upper-level vortex inducement. There are multilevel hairpin vortex ejections and sweeps inside the boundary layer of the transitional and low Reynolds number turbulent flows. The ejections and sweeps are much stronger around the hairpin legs and necks than those in the ring area. This clearly shows that the ring-like vortices are the production of the strong vortex neck rotation. In conjunction with ejections and sweeps, a lot of strong shear layers are produced. Because fluid cannot tolerate the strong shear, the shear layer must turn to rotation and form many vortices. This would help reveal the mechanism of multilevel vortices and turbulence generation. Although the upper-level vortices could be larger than the lower ones, the lower level vortices sometimes have the same size as the neighboring upper-level vortices. The vortex cascade and large vortex breakdown are not observed by either DNS or experiment.

Keywords: DNS, Ejection and sweep, Hairpin vortices, Multilevel vortices, experiment, Turbulent boundary layer.

---

*Corresponding author Chaoqun Liu: Department of Mathematics, University of Texas at Arlington, Arlington, Texas 76019, USA; Tel: +1-8172725151; Fax: +1-8172725802; E-mail: cliu@uta.edu

INTRODUCTION

Wall-bounded turbulent flow is a fundamental scientific topic and has received a lot of attention for over a century due to its importance to both scientific research and many industrial applications such as transition control and drag reduction in aerospace engineering [1, 2]. Many efforts have been made for turbulent boundary layer flow [3-5]. However, the mechanism of wall-bounded turbulent flow still remains a puzzle. Nearly four decades ago, Falco [6] gave a well-known visualization of a low Reynolds number turbulent boundary layer that illustrates several known types of coherent structures. Theodorsen [7] had identified the horseshoe vortex in wall-bounded turbulent flow by experimental observation. Robinson [8] believed that the coherent structure is responsible for the production and dissipation of turbulence in a boundary layer, and the study of turbulence structures is of fundamental importance to understand and control the turbulent boundary layer. Kline *et al.* [9] found that the long streamwise streaks of hydrogen bubbles in the near-wall region by experiment and that the spanwise spacing of these streaks were about 100 wall units. They believed that the instability of these streaks plays a vital role in turbulence generation. Working with Kline, Robinson [10] gave a summary of the structures. He observed that quasi-streamwise vortices are located close to the wall, arches or horseshoe vortices in the wake region, and there is a mixture of quasi-streamwise vortices and arches in the logarithmic layer. Liu *et al.* [11-13], Rist *et al.* [14], and Wu and Moin [15] obtained transitional and low Reynolds number turbulent flow with a forest of hairpin vortices through DNS while Eitel-Amor found that the hairpin vortex structures are not a feature in fully developed turbulence [38]. Scaling theories coupled with the notion of coherently organized motions were first proposed by Townsend [16, 17] and advanced by Perry and Chong [18]. They gave a model of individual hairpin vortices scattered randomly in the flow and found that the vortices are statistically independent of each other. They imagined the wall layer as a forest of single layer hairpin vortices which can be modeled with simplified shapes in a hierarchy of scales above the wall. Yan *et al.* [19] found that the hairpin vortex is a combination of the Λ-vortex roots and vortex rings. Λ-vortex roots and vortex rings are formed separately and independently, and the mechanism of the Λ-vortex self-deformation to hairpin vortex does not exist. Liu *et al.* [20, 21] believed that turbulence is an inherent property of fluid flow, although the external disturbance is needed. The nature of turbulence generation is that fluids, away from the wall, cannot tolerate the shear, and shear must transfer to rotation. Therefore, we believe shear layer instability is the mother of turbulence. However, physics of turbulence is very complex and a number of articles have given a variety of theories [22-27]. Adrian [28] believed hairpins could be auto-generated to form packets that populate a significant fraction of the boundary layer, and he addressed the important role of hairpin vortex ejections and sweeps. Marusic's review paper [3] mentioned that there is still a dichotomy on whether the hairpin vortex exists or not. Schoppa and Hussain [29] thought complete hairpin vortices do not exist in wall bounded turbulence. Therefore, more work must be conducted to get a deep understanding about the vortex structure of the low Reynolds number turbulent boundary layer.

A new Lagrangian property experiment technique, which is named as the moving single-frame and long-exposure (MSFLE) imaging method, is proposed for measuring the coherent structure in a turbulent boundary layer. Meanwhile, a high order direct numerical simulation with nearly 60 million grid points with 400,000 time steps is carried out in order to get a deep understanding of the coherent structure for the low Reynolds number turbulent boundary layer flow.

This chapter is organized in the following way. In Section 2, the DNS case setup and validation and the experiment setup are described. Section 3 provides the comparison of the DNS results with the experiment and describes our new DNS observations. Finally, some conclusions are presented in Section 4.

CASE SET UP

DNS Case Setup and Code Validation

Fig. (**1**) shows the computational domain where x, y, z represent the streamwise, spanwise and wall-normal directions respectively. There are 1920×128×241 grid points in the computation domain. The points are distributed uniformly in the streamwise and spanwise directions and stretched in the wall normal direction to ensure the grid has enough resolution to capture all small length scales. The first grid interval is set to 0.43 in wall units (z^+=0.43) in the normal direction. As shown in Fig. (**1b**), the whole domain is decomposed in the x-direction to implement parallel computation by using the Message Passing Interface (MPI) technique. Table I gives the details of the flow conditions including Mach number, Reynolds number, *etc.* Here δ_{in} is the inflow displacement thickness, and other parameters are non-dimensionalized by δ_{in} as reference length. L_x and L_y are the length of computational domain in the x and y directions, and $L_{z_{in}}$ is the height of the inlet. x_{in} is the distance between the leading edge and inlet. T_w and T_∞ represent the wall temperature and freestream temperature. The Reynolds number of the inlet is defined as $Re = \rho_\infty U_\infty \delta_{in}/\mu_\infty$.

The following equations give the inflow boundary conditions:

$$Q = Q_{Blasius} + A_{2d}q'_{2d}e^{i(\alpha x - \omega t)} + A_{3d}q'_{3d}e^{i(\alpha x \pm \beta y - \omega t)} \tag{1}$$

where Q contains three velocity components, pressure, and temperature, and $Q_{Blasius}$ represents the Blasius solution. The second and third terms denote the 2D and 3D T-S waves, and more details can be found [30]. The non-slipping and adiabatic condition is applied to the wall and non-reflecting condition is applied for the outflow boundaries. Periodic conditions are used at the spanwise boundaries.

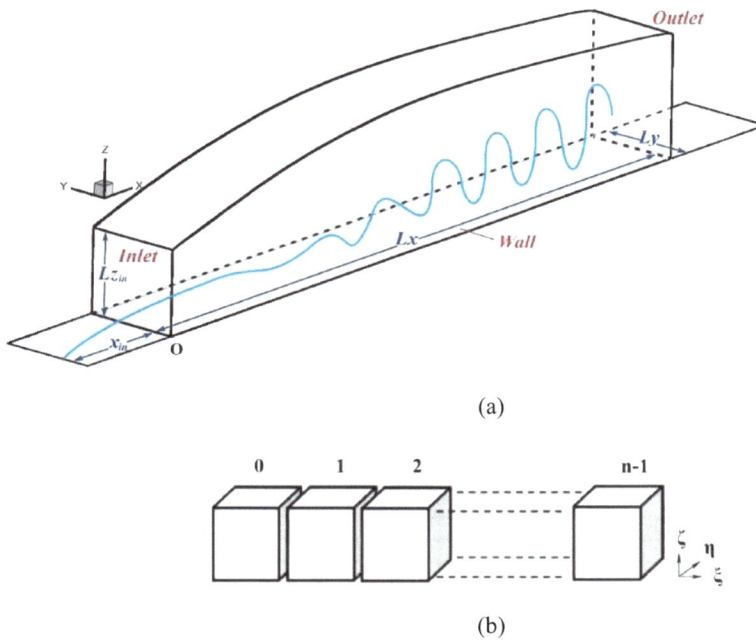

(a)

(b)

Fig. (1). Computational domain sketch.

Table 1. Flow parameters.

M_∞	Re	x_{in}	L_x	L_y	$L_{z_{in}}$	T_w	T_∞
0.5	1000	$300.79\delta_{in}$	$780.03\delta_{in}$	$22\delta_{in}$	$40\delta_{in}$	273.15K	273.15K

Numerical Methods

The flow is governed by the 3D compressible N-S equations [31] that can be written in the curvilinear coordinates as follows:

$$\frac{\partial \hat{\mathbf{Q}}}{\partial \tau} + \frac{\partial \hat{\mathbf{F}}}{\partial \xi} + \frac{\partial \hat{\mathbf{G}}}{\partial \eta} + \frac{\partial \hat{\mathbf{H}}}{\partial \zeta} = \frac{1}{Re}\left(\frac{\partial \hat{\mathbf{F}}_v}{\partial \xi} + \frac{\partial \hat{\mathbf{G}}_v}{\partial \eta} + \frac{\partial \hat{\mathbf{H}}_v}{\partial \zeta} \right) \tag{2}$$

where

$$\hat{\mathbf{Q}} = J^{-1}\mathbf{Q}$$

$$\hat{\mathbf{F}} = J^{-1}\left(\xi_x \mathbf{F} + \xi_y \mathbf{G} + \xi_z \mathbf{H} \right)$$

$$\hat{\mathbf{G}} = J^{-1}\left(\eta_x \mathbf{F} + \eta_y \mathbf{G} + \eta_z \mathbf{H} \right)$$

$$\hat{\mathbf{H}} = J^{-1}\left(\zeta_x \mathbf{F} + \zeta_y \mathbf{G} + \zeta_z \mathbf{H} \right)$$

$$\hat{\mathbf{F}}_v = J^{-1}\left(\xi_x \mathbf{F}_v + \xi_y \mathbf{G}_v + \xi_z \mathbf{H}_v \right)$$

$$\hat{\mathbf{G}}_v = J^{-1}\left(\eta_x \mathbf{F}_v + \eta_y \mathbf{G}_v + \eta_z \mathbf{H}_v \right)$$

$$\hat{\mathbf{H}}_v = J^{-1}\left(\zeta_x \mathbf{F}_v + \zeta_y \mathbf{G}_v + \zeta_z \mathbf{H}_v \right)$$

$$J^{-1} = \begin{vmatrix} 1 & 0 & 0 & 0 \\ 0 & x_\xi & x_\eta & x_\zeta \\ 0 & y_\xi & y_\eta & y_\zeta \\ 0 & z_\xi & z_\eta & z_\zeta \end{vmatrix} \tag{3}$$

$$\begin{pmatrix} \xi_x & \xi_y & \xi_z \\ \eta_x & \eta_y & \eta_z \\ \zeta_x & \zeta_y & \zeta_z \end{pmatrix} = J \begin{pmatrix} y_\eta z_\zeta - y_\zeta z_\eta & z_\eta x_\zeta - z_\zeta x_\eta & x_\eta y_\zeta - x_\zeta y_\eta \\ y_\zeta z_\xi - y_\xi z_\zeta & z_\zeta x_\xi - z_\xi x_\zeta & x_\zeta y_\xi - x_\xi y_\zeta \\ y_\xi z_\eta - y_\eta z_\xi & z_\xi x_\eta - z_\eta x_\xi & x_\xi y_\eta - x_\eta y_\xi \end{pmatrix}$$

$$\mathbf{Q} = \begin{bmatrix} \rho & \rho u & \rho v & \rho w & \rho E \end{bmatrix}^{T}$$

$$\mathbf{F} = \begin{bmatrix} \rho u & \left(\rho u^2 + p \right) & \rho u v & \rho u w & (e+p)u \end{bmatrix}^{T} \qquad (4)$$

$$\mathbf{G} = \begin{bmatrix} \rho v & \rho u v & \left(\rho v^2 + p \right) & \rho v w & (e+p)v \end{bmatrix}^{T}$$

$$\mathbf{H} = \begin{bmatrix} \rho w & \rho u w & \rho v w & \left(\rho w^2 + p \right) & (e+p)w \end{bmatrix}^{T}$$

$$\mathbf{F}_v = \begin{bmatrix} 0 & \sigma_{xx} & \sigma_{xy} & \sigma_{xz} & u\sigma_{xx} + v\sigma_{xy} + w\sigma_{xz} + \dfrac{1}{(\gamma-1)PrM_\infty^2}\kappa(T)\dfrac{\partial T}{\partial x} \end{bmatrix}^{T}$$

$$\mathbf{G}_v = \begin{bmatrix} 0 & \sigma_{xy} & \sigma_{yy} & \sigma_{yz} & u\sigma_{xy} + v\sigma_{yy} + w\sigma_{yz} + \dfrac{1}{(\gamma-1)PrM_\infty^2}\kappa(T)\dfrac{\partial T}{\partial y} \end{bmatrix}^{T}$$

$$\mathbf{H}_v = \begin{bmatrix} 0 & \sigma_{xz} & \sigma_{yz} & \sigma_{zz} & u\sigma_{xz} + v\sigma_{yz} + w\sigma_{zz} + \dfrac{1}{(\gamma-1)PrM_\infty^2}\kappa(T)\dfrac{\partial T}{\partial z} \end{bmatrix}^{T}$$

The reference values for length, density, velocity, temperature, and pressure are δ_{in}, ρ_∞, U_∞, T_∞ and $\rho_\infty U_\infty^2$, respectively.

We employ a six-order compact scheme [32] for the spatial discretization in the streamwise and normal directions, while in the spanwise direction, pseudo-spectral method is used. A high-order spatial scheme filtering is used to eliminate the spurious numerical oscillations caused by the central difference scheme,

$$\frac{1}{3}f'_{j-1} + f'_j + \frac{1}{3}f'_{j+1} = \frac{1}{h}\left(-\frac{1}{36}f_{j-2} - \frac{7}{9}f_{j-1} + \frac{7}{9}f_{j+1} + \frac{1}{36}f_{j+2} \right), j = 3,\ldots N-2 \qquad (5)$$

A third order Total Variation Diminishing Runge-Kutta scheme [33] is employed.

$$\begin{aligned} Q^{(0)} &= Q^{(n)} \\ Q^{(1)} &= Q^{(0)} + \Delta t R^{(0)} \\ Q^{(2)} &= \frac{3}{4}Q^{(0)} + \frac{1}{4}Q^{(1)} + \frac{1}{4}\Delta t R^{(1)} \\ Q^{(n+1)} &= \frac{1}{3}Q^{(0)} + \frac{2}{3}Q^{(2)} + \frac{2}{3}\Delta t R^{(2)} \end{aligned} \qquad (6)$$

Code Validation

Since the code "DNSUTA" developed at the University of Texas at Arlington has been carefully validated by NASA Langley [34] and UTA researchers [35, 36], only a short description of the validation will be addressed here, and readers are encouraged to refer to our previous papers for details.

The simulation is performed for about 600,000 timesteps. Fig. (**2**) gives the time- and spanwise-averaged skin-friction coefficient along the streamwise direction obtained by our DNS, which is represented by the solid line. In Fig. (**2**), the dashed line presents the result of the laminar flow, and the dash-dotted line is the result given by Ducros *et al.* [37]. As seen, before $x \approx 450\delta_{in}$ the result of our DNS is consistent with the laminar flow result. The skin-friction then grows sharply, and after this the result is coincident with the turbulent flow result. This means that $x \approx 500\delta_{in}$ is the "onset point" of the flow transition. The time and spanwise-averaged streamwise velocity profile are shown in 3. As seen, the velocity profile at $x \approx 300.79\delta_{in}$ is a typical laminar boundary layer velocity profile and the velocity profile at $x \approx 632.33\delta_{in}$ becomes a turbulent flow velocity profile (Log law). These two comparisons show that our DNS result is reliable and the simulation from laminar flow to turbulent flow has been achieved. For more discussion of the convergence criteria for DNS [38].

Utilizing the new Omega method (Ω) proposed by Liu *et al.* [39], which is found to be quite suitable for the identification of vortex structures [40]. The currently popular vortex identification methods are based on the analysis of the velocity gradient tensor∇V, which can be divided into symmetric and antisymmetric parts, as:

$$\nabla V = \frac{1}{2}(\nabla V + \nabla V^T) + \frac{1}{2}(\nabla V - \nabla V^T) = \mathbf{A} + \mathbf{B} \qquad (7)$$

where A is a symmetric matrix, B is an antisymmetric matrix.

Q – criterion

The definition of the Q – criterion is described in [41]

$$Q = \frac{1}{2}(b - a) \qquad (1)$$

with

$$a = \text{trace}\left(A^T A\right) = \sum_{i=1}^{3} \sum_{j=1}^{3} \left(A_{ij}\right)^2 \qquad (9)$$

$$b = \text{trace}\left(B^T B\right) = \sum_{i=1}^{3} \sum_{j=1}^{3} \left(B_{ij}\right)^2 \qquad (10)$$

And the zone with $Q \geq 0$ indicates the existence of the vortex.

Ω method

Liu *et al.* [39] presented the Ω method as follows:

$$\Omega = \frac{b}{a+b+\varepsilon} \qquad (2)$$

where definitions of *a, b* are the same as those in Eq. (8). The Omega method shows the relative vortex strength and is able to capture both weak and strong vortices simultaneously without much threshold adjustment. For details in comparison with Q and other vortex identification methods, one can refer to their recent publication in [42]. In Fig. (4), $\Omega = 0.52$ and $\varepsilon = 0.001 * (b - a)$ were adopted. Note that our experiences of vortex identification and many other citations have shown that one does not need to adjust the threshold and $\Omega = 0.52$ can be satisfied for most of our cases which we studied.

The evolution of coherent structures in the transition process are shown in Fig. (4). The formation of ring-like vortices chains is consistent with the experiments conducted by Lee and Li [43] as shown in Fig. (5).

All these verifications and validations above show that our code is correct and our DNS results are reliable. In Fig. (3), a Log Law of the velocity profile of the boundary layer has been achieved, which is $u^+ = \frac{1}{k}\ln y^+ + C^+$, where we pick k=0.42 and $C^+ = 0.5$.

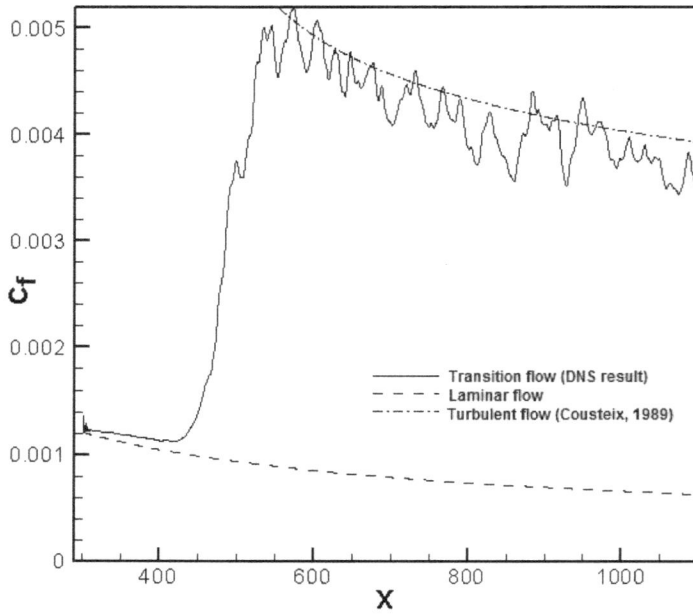

Fig. (2). Time-and spanwise-averaged skin-friction coefficient along the streamwise direction **[36]**.

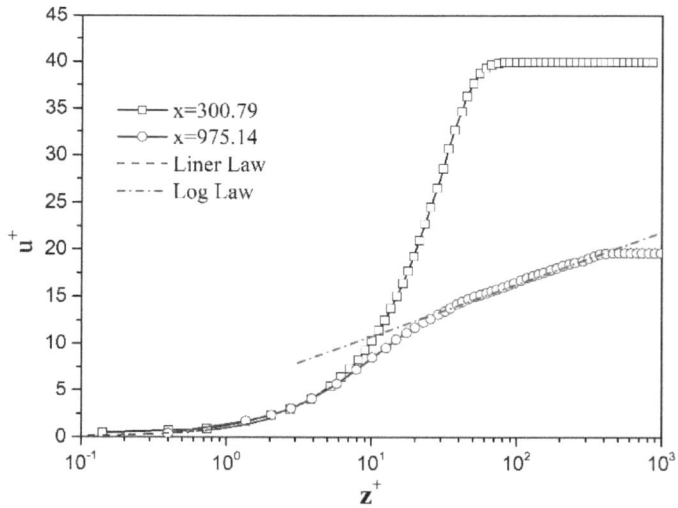

Fig. (3). Log-linear plots of the time- and spanwise-averaged velocity in wall unit.

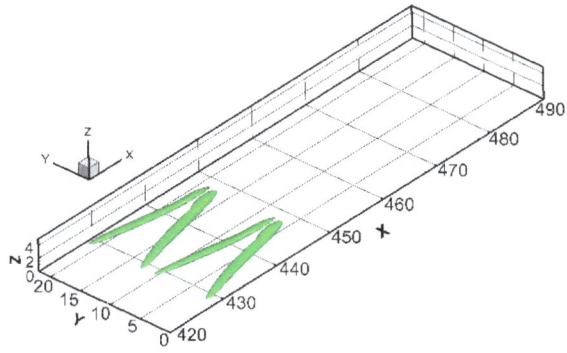

(a)

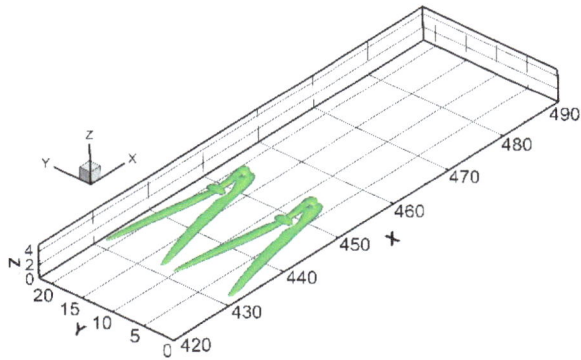

(b)

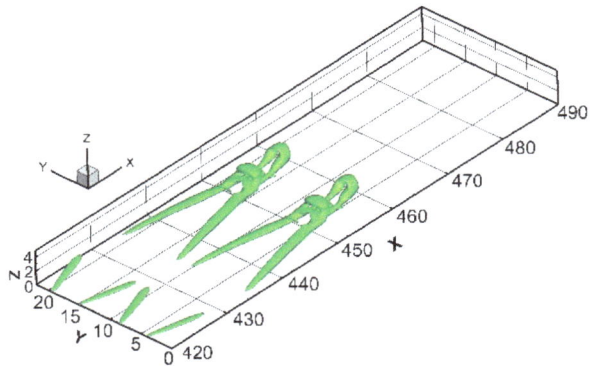

(c)

Fig. 4 cont.....

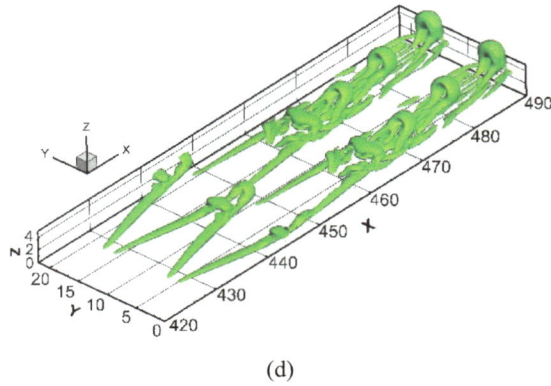

(d)

Fig. (4). Evolution of the vortex structures during the late transition by DNS [36].

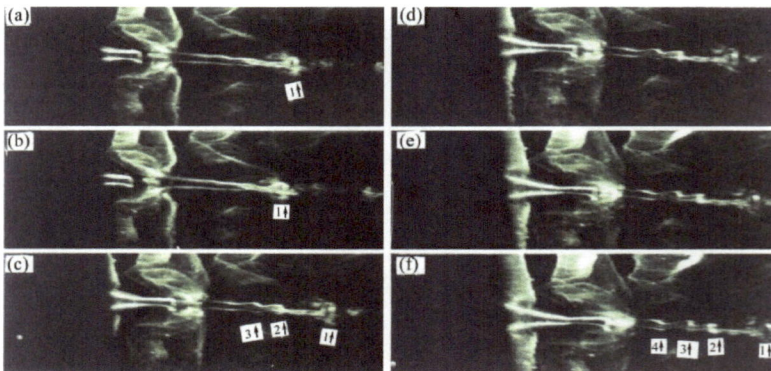

Fig. (5). Evolution of the ring-like vortex experiment [36].

Experiment Setup

Although the DNS results for the late boundary layer transition is well validated, we still want to compare the vortex structure qualitatively with experiment for early turbulent flow with low Reynolds number. According to the velocity profile, flow in the last section of the DNS domain after $x=900$ has become turbulent (see Fig. **3**) and is comparable with the experiment work for turbulent flow with the low Reynolds number. Please refer to chapter 11 for detailed description about experiment.

Introduction to MSFLE Method

One way to study the flow field is to intersperse tracer particles in the flow, and then we can obtain the flow field characteristics by observing the motion of tracer particles. The moving single-frame and long-exposure (MSFLE) imaging method is based on the above concept and is developed from the single-frame and long-exposure (SFLE) method [44]. SFLE can record the trajectory of tracer particles clearly by setting proper exposure time. Fig. (**6**) shows a typical motion trajectory image captured by SFLE method. One can obtain the particle velocity by the formula:

$$V = \frac{S-D}{M\Delta t} \qquad (12)$$

Where D represents the diameter of the particle, S is the total length of trajectory, M is the magnification factor of lens, and Δt is exposure time.

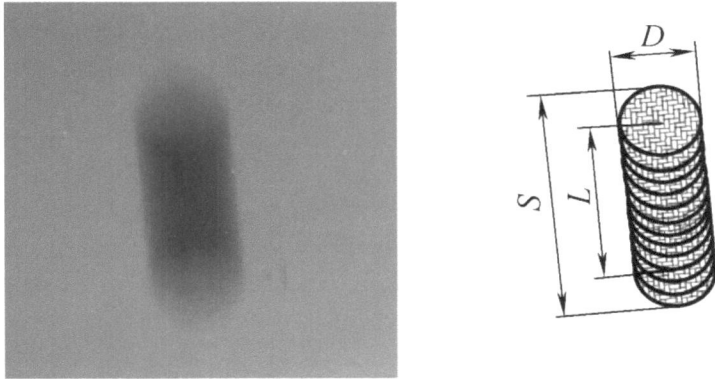

Fig. (6). Typical motion trajectory image and calculation model.

It should be noted that the view field and resolution of the camera are contradictory when the SFSE method is used. If we want to obtain the elaborated structure of the vortex by using the high-resolution camera, the view field is narrowed as a consequence, and this will make it hard to capture the movement of coherent structures. To solve this problem, we propose the MSFLE method. In this method, we allow the camera to move at the same speed as the speed of the focused part of the coherent structure, so that the trajectory of the particle can be recorded properly by the long exposure method.

The coherent structure of the turbulent boundary layer travels downstream with the flow and has streamwise migration speed. When observing the boundary layer by the Euler method, we can only get the instantaneous information of the flow field for a fixed location. Thus, it is difficult to obtain the coherent structure due to the impact of migration speed. By using the MSFLE method, which is a Lagrangian-type measurement technique, the coherent structure can be obtained once the speed of the camera is set to be the same as the speed of the focused part of the coherent structure. In this way, we can observe the evolution process of coherent structures with time and space, and there is no need for the vortex identification and Galilean decomposition.

Experimental Facility

The experiment is conducted in a low-speed circulating water tunnel. The size of the water tunnel is 2500mm×104mm×90mm. The definition of the coordinate system and measurement system is shown in Fig. (7). We define x, y, z as the streamwise, spanwise and wall-normal directions respectively, and the origin point is located at the front of the water tunnel. A plexiglass plate is placed in the middle of the water tunnel, and the size of the plate is 1500mm×78mm×4.7mm, which is located from x=700mm to x=2200mm. The leading edge of the plate is shaped as an ellipse with a ratio of 4:1 for the major axis to the minor axis as shown in Fig. (7). Tracer particles are hollow glass microspheres with an average diameter of 20μm, and the density of tracer particles is 0.6g/ml. The light source is 450nm wavelength continuous laser diode, which produces a 1 mm thick laser sheet. Images are captured by a CMOS camera which has a resolution of 1280 × 1024 pixels. The entire measurement system is installed on a horizontal guide rail, which can move along the streamwise direction.

The laser sheet is placed in two different positions as shown in Fig. (7). In Fig. (7a), the angle between the laser sheet and flow direction is 53 degrees; the camera is perpendicular to the laser sheet. In Fig. (7b), the angle is increased to 135 degrees, but the camera is kept perpendicular to the laser sheet. Fig. (7c) shows the picture of the experimental equipment with laser sheet placed at 135 degrees to the streamwise direction.

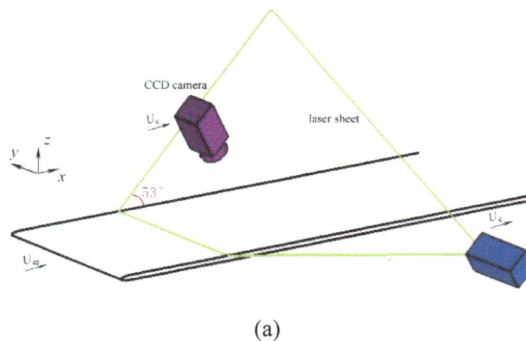

(a)

Fig. 7 cont.....

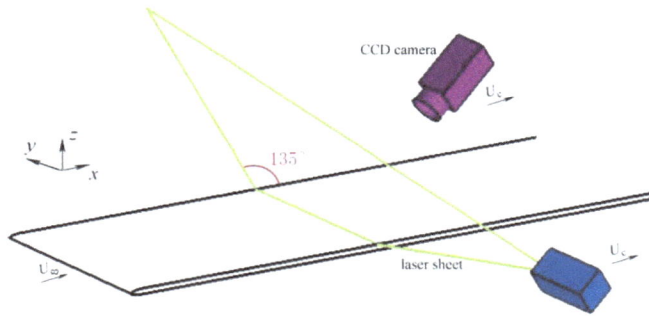

(b)

(c)

Fig. (7). Definition of the coordinate system and measurement system. (**a**) Laser sheet placed with 53 degrees inclination to the flow direction. (**b**) Laser sheet placed with 135 degrees inclination to the flow direction. (**c**) Picture of experimental equipment with Laser sheet placed at 135 degrees.

The test conditions can be described as follows. Inflow speed is $U_\infty = 85mm \cdot s^{-1}$. Because the velocity inside the boundary layer is slower than the main flow, the speed of the camera is set as $U_c = 0.94U_\infty$. Measuring range is from $x = 1440\ mm$ to $x = 2190\ mm$, and the Reynolds number of these two positions are $Re_\theta = \theta U_\infty / v = 429$ and 750, respectively, where θ is the momentum thickness of the boundary layer and v is the coefficient of the kinematic viscosity.

During these experiments, the camera captured the coherent structure of flow continuously with the MSFLE method. The single frame exposure time is 200ms, and the time interval between two frames is less than 70μs. The magnification factor of the camera is 0.14, the sensor resolution is 1280×1024 pixels, and the size of the pixel is 4.8×4.8 μm². Therefore, the active area is 6.144×4.915mm², and the field of view is 43.89×35.11mm².

Experimental Validation

We measured the boundary layer velocity profile by SFLE method at $x = 1440\ mm$ and $x = 2190\ mm$ separately. The laser sheet is placed on the central plane parallel to the flow direction, and the camera is orientated to the side view perpendicular to the laser sheet. The velocity profile is shown in Fig. (**8**), u^+ and z^+ are calculated by $z^+ = zu_\tau/v$, $u^+ = u/u_\tau$, and the solid line is calculated by the Spalding [45] velocity profile equation

$$z^+ = u^+ + e^{-KB}\left[e^{Ku^+} - 1 - Ku^+ - \frac{\left(Ku^+\right)^2}{2!} - \frac{\left(Ku^+\right)^3}{3!}\right] \tag{13}$$

with K=0.4, B=5.5. As seen, the experimental result coincides with the Spalding theory, which means our experiment method is quite accurate. The results of two different locations are similar to each other, which demonstrates that the turbulent boundary layer is fully developed in the test area. Our DNS results for the mean velocity are also extracted and calculated in the same frame, showing a good agreement with experimental results and the Spalding formula. The profile of the mean velocity also shows that a fully developed turbulent flow with low Reynolds number is obtained by both experiment and DNS.

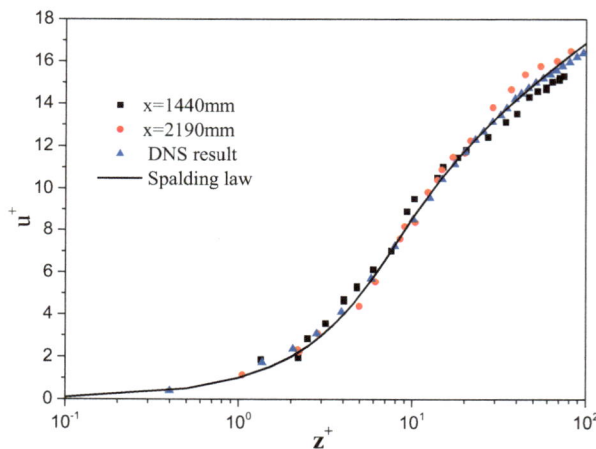

Fig. (8). Dimensionless velocity profile of experimental result, our DNS result, and Spalding law at x=1440mm and x=2190mm.

RESULT AND DISCUSSION

Comparison Between DNS and Experiment Results

Note that the experimental setup is not exactly the same as DNS's setup especially in the inflow conditions, but both (experiment and DNS at downstream section after x=900) can be classified as a "low Reynolds number turbulent flow". The side wall influence in the experiment can be ignored in the middle of the test section where the flow can still be considered as a turbulent boundary layer while DNS assumes a periodic boundary condition in the spanwise direction. Therefore, the middle section of experiment and the downstream section of DNS are still qualitatively comparable in vortex structure.

Since turbulent flow is fluctuated, it is difficult to ensure the flow initial condition and inflow boundary condition are the same for both experiment and DNS. The recently proposed diagnostic-plot method [46] may be a promising tool to design the setup of turbulence boundary layer experiment for the comparison of the numerical and experimental results but will not be used here. Quantitatively instantaneous comparison is not realistic. However, from Fig. (**8**), we can find that the velocity profile of DNS coincides with the experimental result, which means that the flow states of both the experiment and the DNS last section are the same, which can be classified as a low Reynolds number but still developed turbulent flow, and the turbulence coherent structure in both cases should be very similar. Therefore, the qualitative comparison in vortex structure between the DNS and experiment is still reliable and convincing. A more reliable and efficient method to design the set-up for Fig. (**9a**) shows the coherent structures in a low Reynolds number but fully developed turbulent boundary layer obtained by the MSFLE experimental method; the laser sheet is placed with 53 degrees inclination to the flow direction which is parallel to x-axis. Fig. (**9b**) gives the particle trace lines calculated by our DNS data. Nearly 30,000 seed points were placed, and the points were picked from one plane, which has the same direction as the laser sheet. In Fig. (**9a**), the bright dots indicate the tracer particles moving mainly in the streamwise direction, and the long bright lines denote the tracer particles moving in the spanwise or wall-normal direction since a dot means the particle moving direction is orthogonal to the laser sheet. The structure we obtained in Fig. (**9a**) is a horseshoe structure. To demonstrate this, we divided the horseshoe vortex into five different sections: sections A and B representing left and right legs, sections C and D for left and right necks and section E for the center of the horseshoe vortex as shown in Fig. (**9c**). Fig. (**9d**) shows the relative position between the laser sheet and the horseshoe vortex. The laser sheet intersects through the upper part of the horseshoe vortex. If a horseshoe vortex shown in Fig. (**9c**) is cut by a plane shown in Fig. (**9d**), then the flow directions of these five different sections are determined. It is easy to see that the local flow direction in section A is toward to the left, to the right in section B, to the upper left in section C, to the upper right in section D, and opposite to the streamwise direction in section E.

From Fig. (**9a**), we can find that the bright lines in section A are nearly horizontal, which shows that the tracer particles in section A move to the left. In the same way, we can find that particles in section B move to the right. The bright lines in section C are moving toward to the upper left, which means particles in section C move to the upper left. Similarly, we can see that particles in section D move to the upper right. In section E, there is a lot of bright dots instead of long bright lines, so that the particles in section E move in the opposite of the streamwise direction. The trajectories of tracer particles of five different sections obtained by experiment have the similar trend as analyzed above. Apparently, the structure we obtained in Fig. (**9a**) is a horseshoe structure proved by experiment and DNS.

In addition, we can clearly find two vortex cores at the bottom of section A and B in Fig. (**9a**). This is because the laser sheet cut through the left and right legs of the horseshoe. The horseshoe-like structure is also obtained by our DNS result as shown in Fig. (**9b**). The result is similar to the one obtained by the MSFLE experiment.

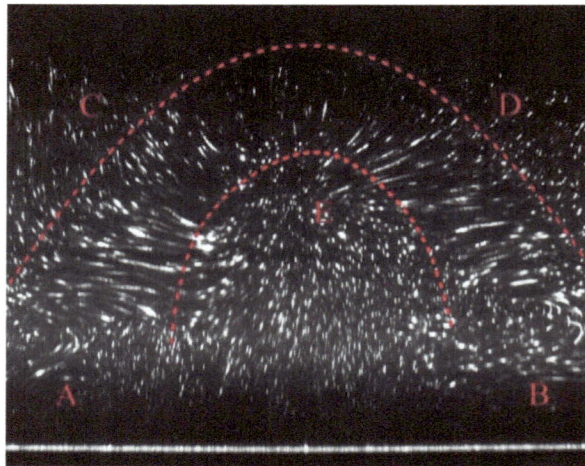

(a)

Fig. 9 cont.....

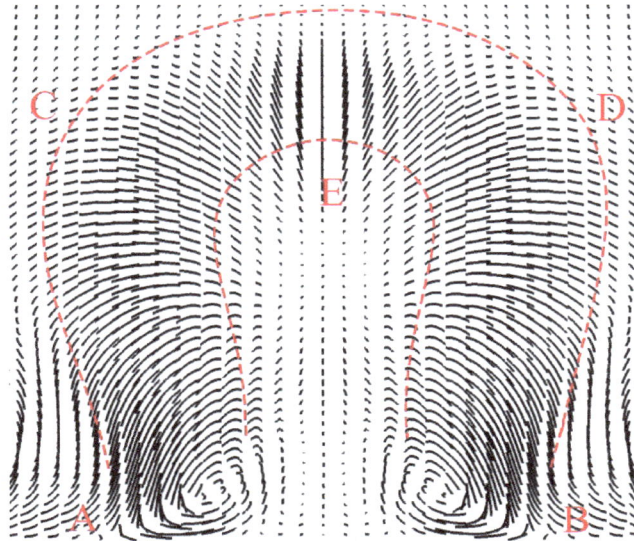

(b)

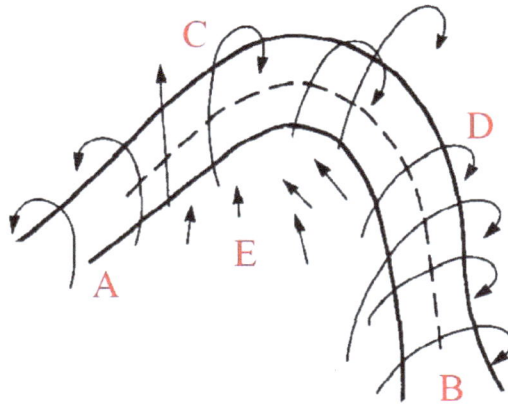

(c)

Fig. 9 cont.....

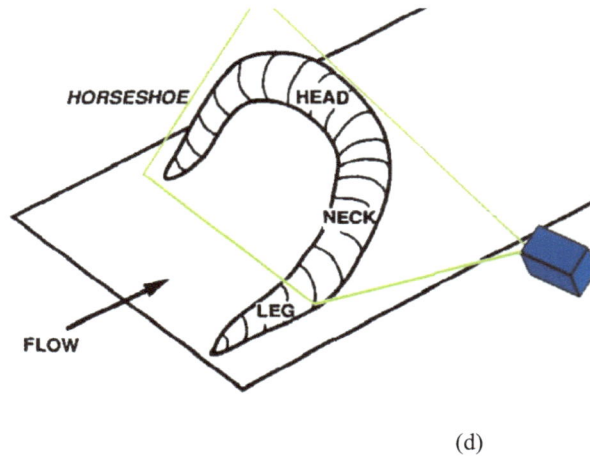

(d)

Fig. (9). (a) Horseshoe vortex structure obtained by MSFLE method. **(b)** Horseshoe vortex structure obtained by DNS data. **(c)** Sketch of five different sections of the horseshoe vortex and velocity directions (from Ref. 7). **(d)** Sketch of the relative position of the laser sheet and horseshoe vortex **[8]**.

Fig. **(10a)** gives the MSFLE experimental result with the laser sheet placed at 135 degrees inclination to the flow direction as illustrated in Fig. **(10b)** gives the particle trace calculated by our DNS data with seed points picked from a plane that has the same direction as the laser sheet. Fig. **(10b)** is colored by Omega (See Ref. 34), and the red area represents a large value of Omega, which most likely captures the vortex cores. The vortices denoted by the particle trace lines are in accordance with the vortex cores represented by the Omega contour. It shows that both Omega method and particle trace lines are appropriate to identify vortex structures. As shown in Fig. **(10a)**, seven streamwise vortices are captured by the MSFLE experiment, and these vortices are believed to be the legs of the hairpin vortices. We can find that the vortices are located in different layers. A total of three layers of vortices can be observed: vortices 1, 4, and 7 lie in the first layer near to the wall; vortices 2 and 5 are observed in the middle layer; and vortices 3 and 6 are found on the top layer. This phenomenon is also observed by our DNS result as shown in Fig. **(10b)**. Similar multilevel vortex structures are obtained by calculating particle trace lines. Totally 8 vortices are captured. Vortices 1 and 8 lie in the first layer near the wall; vortices 2, 4, 5, and 7 appear in the middle layer; and vortices 3 and 6 are in the top layer.

In conclusion, MSFLE method was able to capture the coherent vortex structures of turbulent flow, including the horseshoe vortex and hairpin vortex legs successfully. There are also multilevel hairpin vortices inside a low Reynolds number turbulent boundary layer in our DNS results, which qualitatively agree with the experimental results.

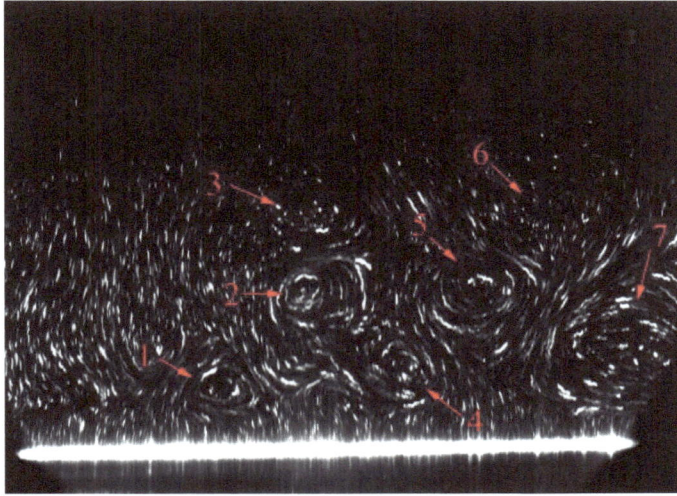

(a)

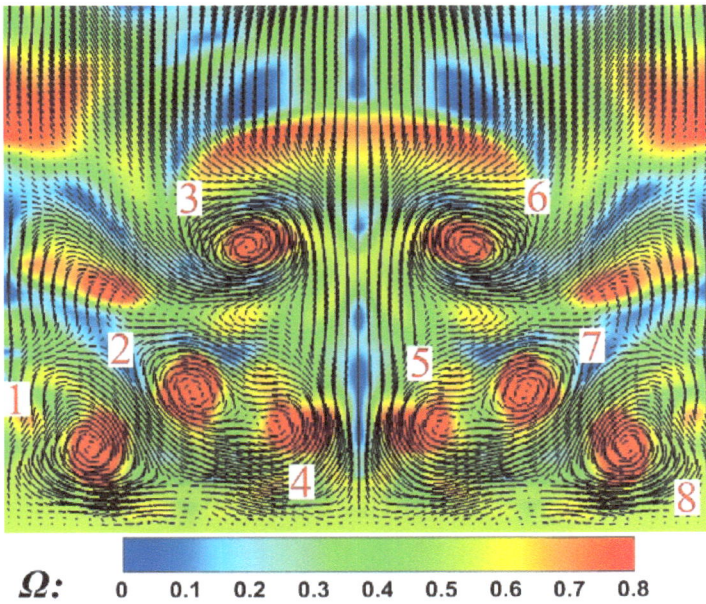

(b)

Fig. (10). (a) Multilevel hairpin vortex legs obtained by the MSFLE method with laser sheet placed at 135 degrees to intersect the streamwise direction. (b) Multilevel hairpin vortex legs obtained by our DNS data colored by the Ω contour.

Discussion on Multilevel Vortex Structures

Multilevel vortex structures are observed in the low Reynolds number turbulent boundary layer flow. Fig. (**11**) shows a DNS result of multilevel vortex structures identified by an Omega iso-surface ($\Omega = 0.52$). It is found that the downstream area has more levels of vortices than the upstream region.

Fig. (**12a**) shows the particle trace lines calculated by the DNS data with the seed points picked from a vertical plane located at x=913. As seen, a total of 4 levels of vortices can be observed. In the middle of the plane, there are four pairs of hairpin vortex legs denoted by A_1 to D_1 and A_2 to D_2. In the following, we only focus on the left legs of these vortices denoted by A_1-D_1. Fig. (**12b**) is the trace line colored by Omega. The particle trace lines coincide with the vortex cores indicated by the Omega method (the red areas). It is easy to find that the rotating directions of the two vortex legs A_1 and A_2 are opposite. The other three pairs of vortices show the same trend. However, it is difficult to distinguish the vortex rotating direction by the trace line. So, we use vorticity to distinguish the clockwise vortices and counterclockwise vortices. Fig. (**12c**) is the trace line colored by the x-direction component of the vorticity. The red color denotes that the vortex rotates counterclockwise, and the blue color means that the vortex rotates clockwise. It is evident that the same pair of vortex legs, A_1 and A_2 for example, have different colors, which means they are counter-rotating vortices. Apparently, the results coincide with our analysis above. So, it is appropriate to identify the hairpin or Lambda vortex rotational direction by the x-direction component of the vorticity.

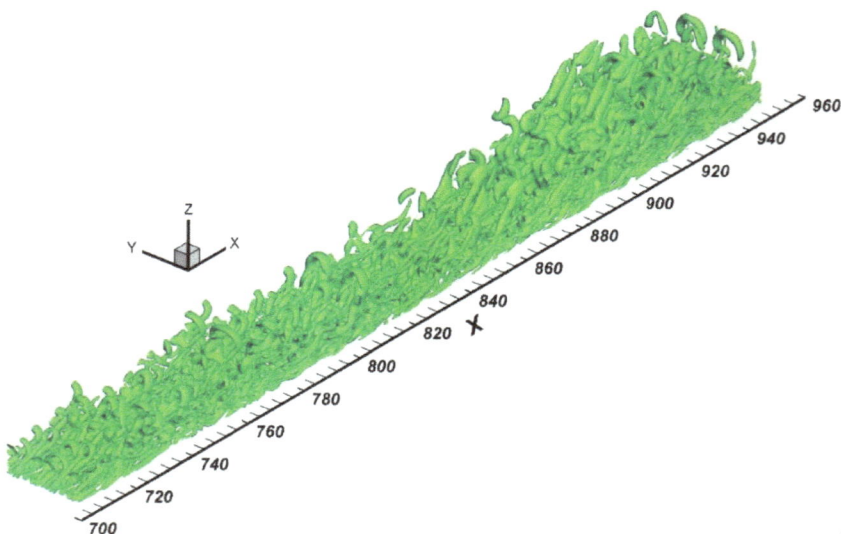

(a)

Fig. 11 cont.....

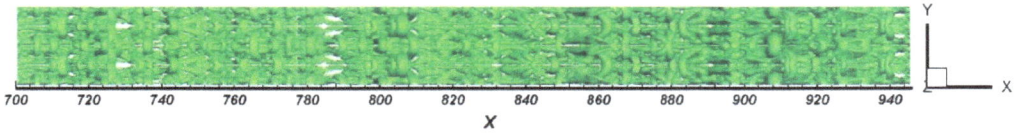

(b)

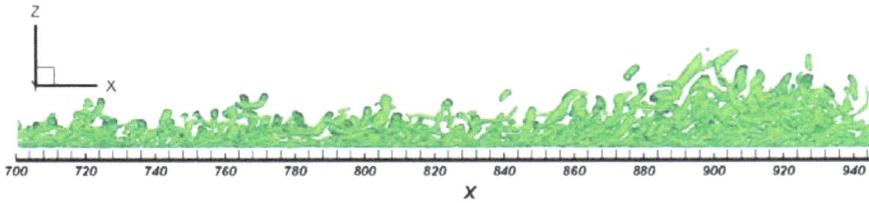

(c)

Fig. (11). View of multilevel vortices in the low Reynolds number boundary layer transitional and early turbulent flow, (**a**) Angle view (**b**) Top view (**c**) Side view (provided by our DNS).

Fig. (**12c**) shows that the rotating directions of the first level vortex A_1 and second level vortex B_1 are opposite in the rotation direction. However, the upper three-levels of the vortices B_1, C_1, and D_1 have the same rotating direction[s], and they are all counterclockwise vortices. It means that the lower vortices are not simply induced by the upper-level vortices as proposed by Adrian [**22**]. If the lower-level vortices are simply induced by upper-level vortices, the rotating direction should be opposite between two levels, which is clearly different from what we have observed. Thus, further research must be conducted for the mechanism of the multilevel vortex generation. Note that Omega can represent the strength of the vortex rotation, but cannot distinguish the rotating directions of the vortices.

From both Figs. (**11** and **12**) we observe that in the low Reynolds number turbulent boundary layer flow, the size of the upper-level vortex is not twice big of the lower level vortex, and sometimes they have almost the same size. The eddy cascade phenomenon indicated by Richardson [47] is not observed in the low Reynolds number turbulent boundary layer flow. Therefore, it is hard to conclude that the energy is transported from the large vortex to small vortices through vortex breakdown process that we did not find by either experiment or DNS.

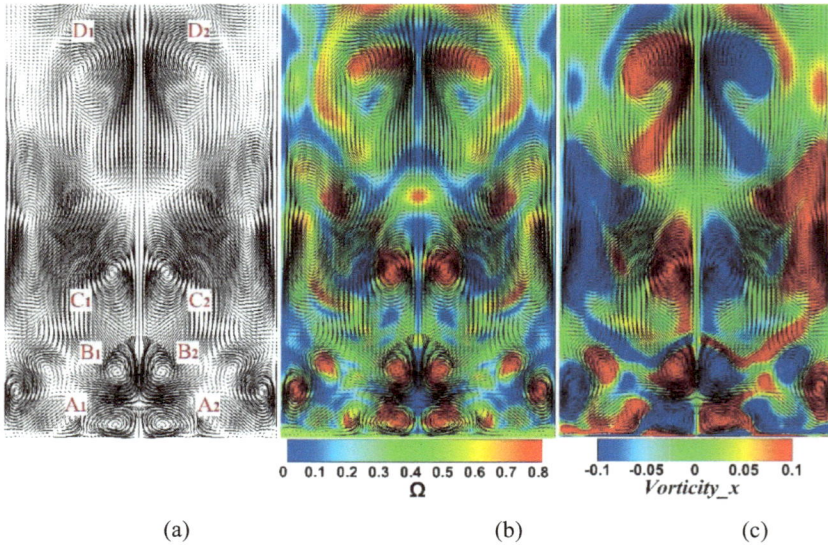

Fig. (12). (a) Multilevel vortex structures identified by trace lines for x=913 in a wall-normal and spanwise plane. **(b)** Multilevel vortex structures colored by Omega. **(c)** Multilevel vortex structures colored by the x-direction component of vorticity.

Multilevel Vortices Ejections and Sweeps

The observation here shows that the energy of vortices inside the boundary layer comes from the main flow. The question is how the energy is transported from the main flow to the boundary layer and how the new vortices are generated. Kolmogorov [48] proposed that the large eddies pass energy to small eddies through "vortex breakdown". However, it has never been observed in the transitional and turbulent boundary layer yet. In our experiment and DNS results, there is no vortex breakdown phenomenon observed in the low Reynolds number turbulent boundary layer flow. Adrian [22] demonstrated that there are ejections and sweeps motions inside the boundary layer, which are induced by the hairpin vortices. Ejection represents the negative streamwise fluctuations, which lifts flow away from the wall. Thus, the low-speed flow near the wall is ejected into the high-speed upper zone. Sweep represents the positive streamwise fluctuations, which brings the high-speed fluid toward the wall. Thus, the high-speed flow sweeps into the low-speed area near the wall. This phenomenon has been observed by our DNS data as shown in Fig. (**13**).

Fig. (**13a**) displays the sketch of a single hairpin vortex structure (identified by Omega iso-surface with $\Omega=0.52$) colored by wall-normal velocity, and the vector field denotes the sweeps and ejections around the vortex legs. Fig. (**13b**) shows the sweeps and ejections around the vortex neck. The red areas mean that the fluid is going up to the main flow corresponding to the ejection event. The red areas mainly locate at the inboard region of the vortex legs. The velocity vector clearly shows the hairpin vortex ejection event exists near the inboard region of the vortex legs and necks. The blue areas indicate that the fluid is going down to the wall corresponding to the sweep events. The sweep events mainly happen in the outboard region of the hairpin vortex legs.

We picked 5 monitoring points from the hairpin vortex as shown in Fig. (**13c**), and the wall-normal velocities of these monitoring points are listed in Table **2**: P_1 is placed in the outboard region of the vortex legs, showing the sweep strength of vortex leg; P_2 and P_3 are placed in the inboard region of the hairpin vortex, representing the ejection strength of vortex leg and neck; P_4 and P_5 represent the ejection and sweep strength of the hairpin vortex rings. As seen, the wall-normal velocities of vortex legs and necks are $-0.21U_\infty$, $0.19U_\infty$ and $0.22U_\infty$, respectively. These magnitudes are about twice as large as the wall normal velocities in the area of the vortex ring, which are $0.1U_\infty$ and $-0.06U_\infty$. The ejection or sweep strength is much stronger in the hairpin legs and necks than in the ring area.

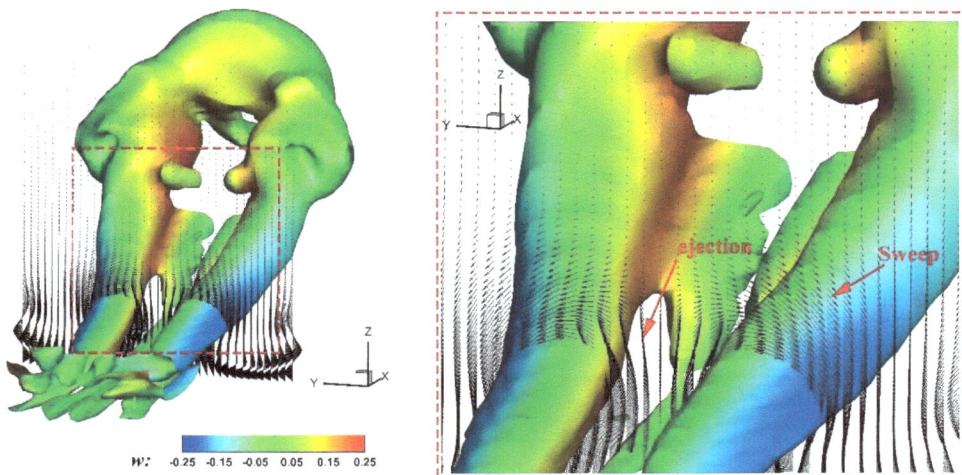

(a)

Fig. 13 cont.....

(b)

(c)

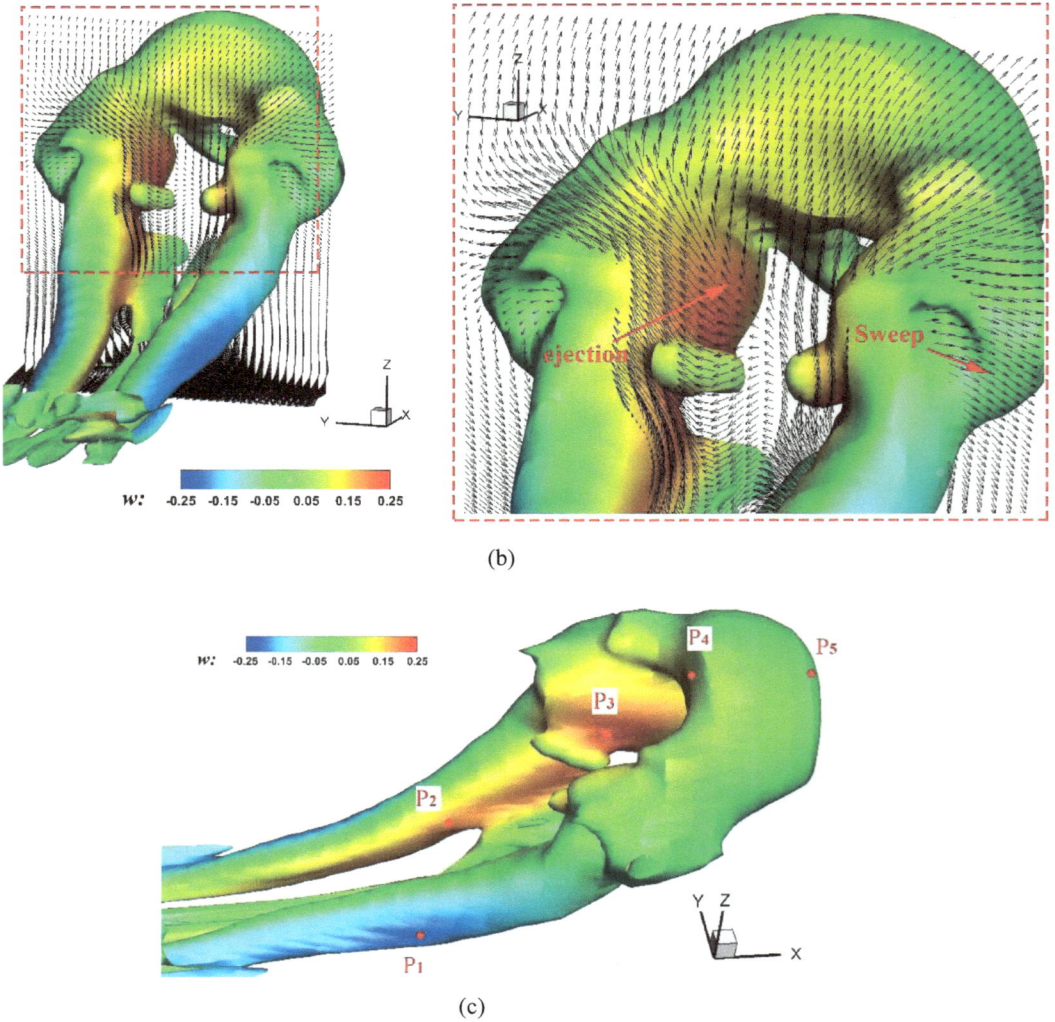

Fig. (13). Hairpin vortex is identified by iso-surface with $\Omega = 0.52$ and colored by wall-normal velocity. (**a**) Ejection and sweeps near the hairpin vortex legs. (**b**) Ejection and sweeps near the hairpin vortex neck. (**c**) Sketch of five monitor points.

Table 2. Wall-normal velocity of five monitoring points.

	P_1	P_2	P_3	P_4	P_5
V_z/U_∞	-0.21	0.19	0.22	0.1	-0.06

A total of 7 monitoring points is placed as shown in Fig. (**14a**): monitoring points P_1-P_3 are placed on the inboard region of the three vortex legs and monitor points P_4-P_7 are placed in the rings of vortex A and B respectively. The wall-normal velocities of these monitoring points are listed in Table III. We can see that the wall-normal velocities of P_2 and P_3 are positive and wall-normal velocity of P_1 is negative, which means the rotational directions of the left legs of vortex A and B are counterclockwise, and the rotating direction of the left leg of vortex C is clockwise as denoted by the black arrow in Fig. (**14b**). The rotational directions of the first level vortex C and the second level vortex A are opposite, which is coincident with the result shown in Fig. (**12**). From Table **3**, we can also find that the wall-normal velocity strengths of vortex legs are $-0.33U_\infty$, $0.22U_\infty$ and $0.12U_\infty$, respectively. These are about three to five times stronger than the rings, which are $0.04U_\infty$, $-0.06U_\infty$, $0.1U_\infty$ and $-0.08U_\infty$. The observation clearly shows the ejection or sweep strength is much stronger in the hairpin leg and neck areas than in the ring area.

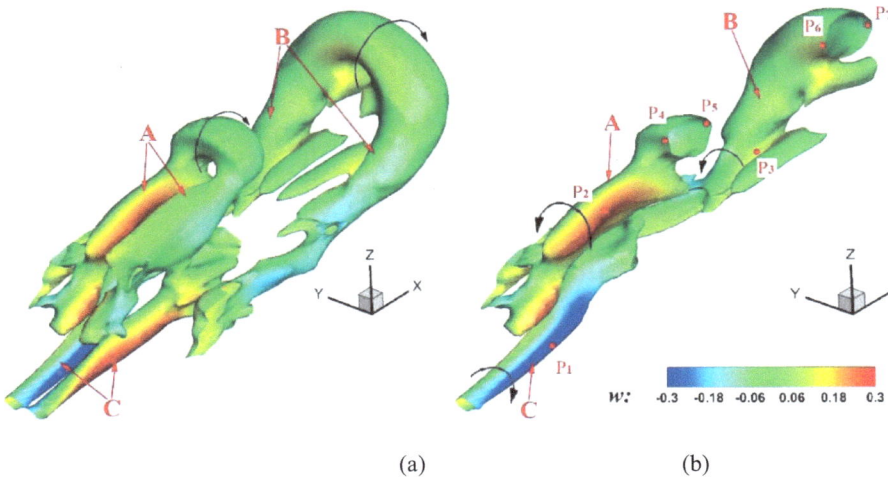

(a)   (b)

Fig. (14). (**a**) Two-level Hairpin vortex structures colored by wall-normal velocity (**b**) half of the vortex structure cut by the central plane.

Table 3. Wall-normal velocity of seven monitor points.

	P_1	P_2	P_3	P_4	P_5	P_6	P_7
V_z/U_∞	-0.33	0.22	0.12	0.04	-0.06	0.1	-0.08

Vortex C does not have a vortex ring because vortex C is a clockwise rotational vortex, it cannot form the ring head must be a counterclockwise vortex due to the profile of the Blasius velocity. The rotating directions of the rings of vortices A and B are denoted by the black arrow in Fig. (**14a**). The upper ring head is higher and its velocity must be larger than the lower part, and the vortex ring can only rotate in the direction shown in Fig. (**14a**). If the legs rotate in the opposite direction just like vortex C, the vortex ring cannot rotate in an opposite direction as a part of the hairpin vortex, which would be impossible for vortex C to have a vortex ring. This would lead to a velocity of the upper ring area smaller than the one in the lower ring area. Therefore, clockwise rotation vortex legs cannot form a vortex ring. It also proves that the hairpin vortex is not a single vortex, but composed of two different parts: the vortex legs and the vortex ring. Our previous paper has discussed this conclusion in detail [19].

Fig. (**15a**) shows the same vortex structures as shown in Fig. (**12**), which are colored by the wall-normal velocity. The red area means the fluid is going up to the main flow, and the blue area indicates the fluid is going down to the wall. Fig. (**15b**) is colored by the streamwise velocity and Fig. (**15c**) shows the positions of 8 monitoring points. The wall-normal velocity and streamwise velocity of the eight monitoring points are listed in Table IV.

As seen in Fig. (**15a**), the vortex center is located between the red and blue area, and it is easy to find that the vortices in the upper three level B_1, C_1 and D_1 are counterclockwise vortices, but the first level vortex A_1 is a clockwise vortex. The monitoring points P_1, P_2, P_3, and P_4 are located in the inboard region of vortex legs. From Table **4**, we can see that the wall-normal velocities of P_2, P_3, and P_4 are upward. Thus, the low-speed flow in the inboard region of second level vortex legs (B_1 and B_2) is ejected upward. Therefore, because the rotation direction of the third level vortices (C_1 and C_2) is the same as the second level, and the low-speed flow lifted by the second level vortex will continue to be lifted up by the ejection of the third level vortices, and then be lifted up a third time by the fourth level vortices (D_1 and D_2). Therefore, the low-speed flow between the vortex legs near the wall is lifted up to the main flow by the multilevel vortices which are co-rotating with the clockwise rotation. In Fig. (**15b**) we can observe the low-speed zone inboard of the two vortex legs. The streamwise velocities of points P_2, P_3 and P_4 are $0.43U_\infty$, $0.64 U_\infty$ and $0.77 U_\infty$, respectively, which are obviously smaller than the neighboring regions. This is a key point to understand why the flow streaks or white and black stripes can be formed, which is caused by the multilevel ejections. The black/white strips strictly require multilevel vortices have the same rotation direction (clockwise for example), challenging the simple inducement theory [22] because the simple inducement theory must lead to vortices with opposite rotation directions.

Now let us consider outboard regions of the hairpin vortex legs. Points P_5, P_6, P_7, and P_8 are located in the outboard region of vortex legs. From Table **4**, we can see that the fluids at points P_2, P_3 and P_4 are going down to the wall and represent the sweep event. The high-speed main flow is brought toward the wall by the top level vortex D_1; then the third level vortex C_1 continues to drive the high-speed flow toward the wall, and so does the second level vortex B_1. In this way, the high-speed flow near the outboard hairpin vortex legs is brought down three times to the near-wall area by the multilevel vortices. In Fig. (**15b**) we can observe the high-speed zone outboard of the second and third level vortices. The streamwise velocities of point P_5, P_6 and P_7 are $0.995 U_\infty$, $0.85 U_\infty$ and $0.8 U_\infty$, respectively, which are obviously greater than their neighboring regions. In this way, the energy is transported by the multiple sweeps from the main flow into the turbulent boundary layer. Again, these new observations about the sweeps also challenge the simple vortex inducement theory.

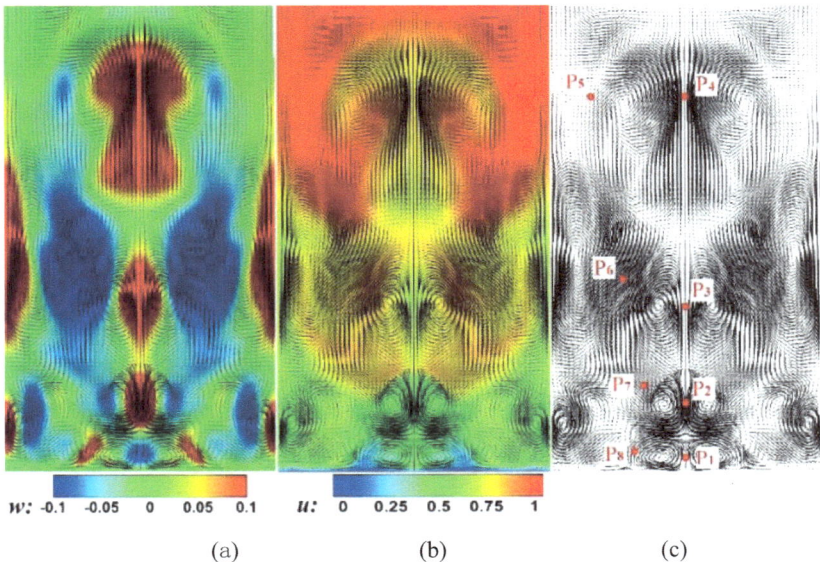

Fig. (15). Multilevel vortex structures colored by (**a**) wall-normal velocity. (**b**) Streamwise velocity. (**c**) Sketch of eight monitoring points.

The multilevel vortices ejections and sweeps can also explain the streamwise streaks of the hydrogen bubbles reported by Kline *et al.* [**9**] as shown in Fig. (**16**).

These streaks are shown as regions of the low streamwise momentum. Fig. (**17**) gives the DNS results of the streaks in a streamwise-spanwise plane at different heights. The height is the same as the position of monitoring points P_1, P_3 and P_4 shown in Fig. (**15c**). We can observe that the low-speed streaks are located in the inboard area of the hairpin legs and they are formed by the ejection events mentioned above. The streamwise streaks are the result of the multilevel vortex ejection events rather than the driving force of the formation of vortices.

Table 4. Wall-normal velocity and streamwise velocity of monitoring points.

	P_1	P_2	P_3	P_4	P_5	P_6	P_7	P_8
V_z/U_∞	-0.11	0.27	0.18	0.23	-0.02	-0.15	-0.06	0.17
V_x/U_c	0.39	0.43	0.64	0.77	0.995	0.85	0.8	0.22

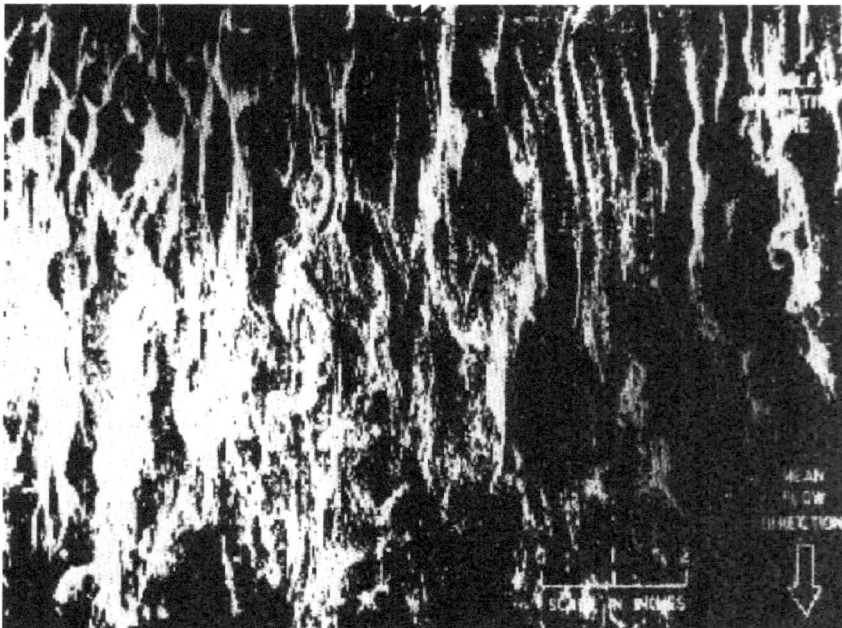

Fig. (16). H_2 bubble visualization of low-speed streaks in a streamwise-spanwise plane [28].

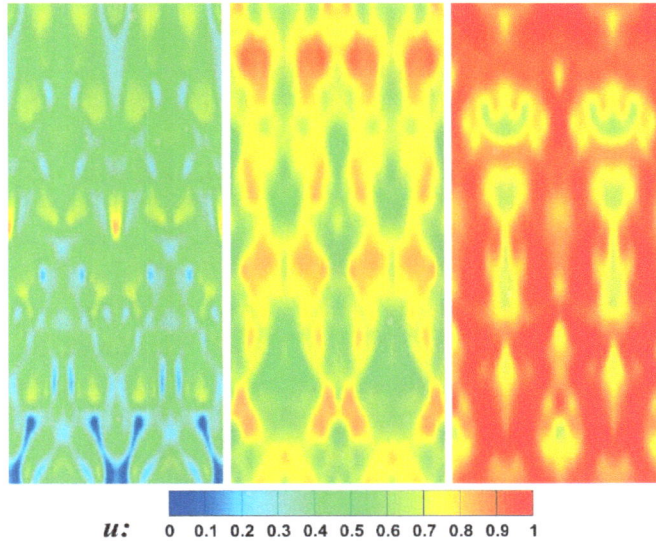

Fig. (17). Streaks in a streamwise-spanwise plane by DNS data for different height.

The multilevel ejections and sweeps mentioned above not only transport the energy from the main flow to the boundary layer but also form a lot of strong shear layers. The fluid cannot tolerate high shear and the shear layer will turn into the vortices. So, more small vortices will be developed due to multilevel hairpin vortices ejections and sweeps. It is confirmed that small-scale vortices are not created by large vortex breakdown and the energy is not transported by the eddy cascade. Both eddy cascade and vortex breakdown are not observed by our experiment or DNS results in the turbulent boundary-layer flow with a low Reynolds number. Therefore, none of them can be confirmed.

One thing must be explained here that although the last section of the DNS domain can be considered as the fully developed turbulent flow based on the velocity profile (see Fig. **3**), the DNS results provided here are mostly in the late flow transition stage and, therefore, are still symmetric, which are more organized and easy to do analyses.

CONCLUSIONS

It is very encouraging that the experimental observations match the DNS visualization closely in the coherent vortex structure of a fully developed turbulent boundary at a low Reynolds number. The DNS visualization matching the experimental observation demonstrates that our results are reliable in a transient sense.

Based on the DNS simulation and MSFLE experiment, the following conclusions can be made:

1. The MSFLE method can capture the vortex structures in a turbulent boundary layer at a low Reynolds number, and the results agree closely with our DNS as the camera moves forward with the vortex at the same speed.

2. Multilevel hairpin vortex structures are observed by both the experiment and DNS results. The upper three-layer vortices rotate in the same direction. These new findings on co-rotating vortices provide evidence that the lower level vortices cannot be simply induced by the higher level vortices.

3. In the low Reynolds number turbulent boundary layer flow, the size of the upper layer vortices are not twice as big as the lower layer vortices; sometimes the upper and lower level vortices are nearly equal in size. The Kolmogorov's vortex breakdown process and Richardson's eddy cascade phenomena are not observed.

4. There are multilevel hairpin vortex ejections and sweeps inside the turbulent boundary layer. The strength of ejections or sweeps is much stronger near the hairpin legs and necks than in the ring area. The low-speed flow near the wall is lifted up by multi-ejection events, and the high-speed flow is brought toward the wall by multilevel sweeps. The energy is transported from the main flow to the boundary layer by the multilevel sweep process.

5. Multilevel hairpin vortex ejections and sweeps are responsible for the formation of streamwise streaks; these low-speed streaks are the result of the multilevel vortex ejection process rather than the driving force of the formation of hairpin vortices.

6. Multilevel vortex ejections and sweeps will produce many strong shear layers. However, fluid cannot tolerate high shear and the shear layers will turn into vortices, which could reveal the mechanism of multilevel vortices generation.

NOMENCLATURE

V	Particle velocity
S	total length of particle trajectory
D	diameter of the particle
M	magnification factor

Δt	exposure time
M_∞	Mach number
R_e	Reynolds number
δ_{in}	inflow displacement thickness
ρ_∞	free stream velocity
T_w	wall temperature
T_∞	free stream temperature
$L_{z_{in}}$	height at inflow boundary
L_x	length of computational domain along the x direction
L_y	length of computational domain along the y direction
x_{in}	distance between leading edge of flat plate and up-stream boundary of computational domain
A_{2d}	amplitude of 2D inlet disturbance
A_{3d}	amplitude of 3D inlet disturbance
ω	frequency of inlet disturbance
α	streamwise wave number of inlet disturbance
β	spanwise wave number of inlet disturbance
μ_∞	viscosity
x, y, z	streamwise, spanwise, wall-normal directions
u, v, w	streamwise, spanwise, wall-normal velocity

CONSENT FOR PUBLICATION

Not applicable.

CONFLICT OF INTEREST

The authors confirm that this chapter contents have no conflict of interest.

ACKNOWLEDGEMENTS

Declared none.

REFERENCES

[1]　A. Kurz, S. Grundmann, and C. Tropea, "Boundary layer transition control using DBD plasma actuators", *AerospaceLab,* vol. 6, p. 1, 2013.

[2]   W.P. Li, and H. Liu, "On the mechanism of turbulent darg reduction with riblets", In: *16th European turbulence Conference*, Stockholm, Sweden, 2017.

[3]   I. Marusic, B.J. McKeon, P.A. Monkewitz, H.M. Nagib, A.J. Smits, and K.R. Sreenivasan, "Wall- bounded turbulent flows at high Reynolds numbers: recent advances and key issues", *Phys. Fluids,* vol. 22, 2010.065103. [http://dx.doi.org/10.1063/1.3453711]

[4]   Y.D. Chashechkin, *"Visualization and identification of vortex structures in stratified wakes,"* in *Fluid Mechanics and its Applications, Eddy Structure Identification in Free Turbulent Shear Flows.,* J.P. Bonnet, M.N. Glauser, Eds., vol. Vol. 21. Springer: Dordrecht, 1993, pp. 393-403. [http://dx.doi.org/10.1007/978-94-011-2098-2_32]

[5]   W. Cheng, D.I. Pullin, and R. Samtaney, "Large-eddy simulation of separation and reattachment of a flat plate turbulent boundary layer", *J. Fluid Mech.,* vol. 785, p. 78, 2015. [http://dx.doi.org/10.1017/jfm.2015.604]

[6]   R.E. Falco, "Coherent motions in the outer region of turbulent boundary layers", *Phys. Fluids,* vol. 20, p. 124, 1977. [http://dx.doi.org/10.1063/1.861721]

[7]   T. Theodorsen, "Mechanism of turbulence", *Proceedings of the Midwestern Conference on Fluid Mechanics,* 1952

[8]   S.K. Robinson, "Coherent motions in the turbulent boundary layer", *Annu. Rev. Fluid Mech.,* vol. 23, p. 601, 1991. [http://dx.doi.org/10.1146/annurev.fl.23.010191.003125]

[9]   S.J. Kline, W.C. Reynolds, R.A. Schraub, and P.W. Runstadler, "The structure of turbulent boundary layers", *J. Fluid Mech.,* vol. 30, p. 741, 1967. [http://dx.doi.org/10.1017/S0022112067001740]

[10]   S.J. Kline, and S.K. Robinson, *"Quasi-coherent structures in the turbulent boundary layer, Part 1: Status report on a community-wide summary of the data,"* in *Near Wall Turbulence.,* S.J. Kline, N.H. Afgan, Eds., Hemisphere: Washington, DC, 1989, p. 218.

[11]   C. Liu, and Z. Liu, "Multigrid mapping and box relaxation for simulation of the whole process of flow transition in 3-D boundary layers", *J. Comput. Phys.,* vol. 119, p. 325, 1995. [http://dx.doi.org/10.1006/jcph.1995.1138]

[12]   Z. Liu, G. Xiong, and C. Liu, "Direct numerical simulation for the whole process of transition on 3-D airfoils", *Proceedings of Fluid Dynamics Conference,* 1995 New Orleans, LA

[13]   C. Liu, and Z. Liu, "Direct numerical simulation for flow transition around airfoils", *Proceedings of first AFOSR international conference on DNS/LES,* 1997 Ruston, Louisiana

[14]   S. Bake, D. Meyer, and U. Rist, "Turbulence mechanism in Klebanoff transition: a quantitative comparison of experiment and direct numerical simulation", *J. Fluid Mech.,* vol. 459, p. 217, 2002. [http://dx.doi.org/10.1017/S0022112002007954]

[15]   X. Wu, and P. Moin, "Direct numerical simulation of turbulence in a nominally zeropressure gradient flat-plate boundary layer", *J. Fluid Mech.,* vol. 630, p. 5, 2009. [http://dx.doi.org/10.1017/S0022112009006624]

[16]   *A. A. Townsend, The Structure of Turbulent Shear Flow.* Cambridge University Press: Cambridge, 1956.

[17]   A.A. Townsend, *The Structure of Turbulent Shear Flow.* 2nd ed. Cambridge University Press: Cambridge, 1976.

[18]   A.E. Perry, and M.S. Chong, "On the mechanism of wall turbulence", *J. Fluid Mech.,* vol. 119, p. 173, 1982. [http://dx.doi.org/10.1017/S0022112082001311]

[19]   Y. Yan, and C. Chen, "DNS study on Λ-vortex and vortex ring formation in flow transition at Mach number 0.5", *J. Turbul.,* vol. 15, p. 1, 2014. [http://dx.doi.org/10.1080/14685248.2013.871023]

[20]   C. Liu, Y. Yan, and P. Lu, "Physics of turbulence generation and sustenance in a boundary layer", *Comput. Fluids,* vol. 102, p. 353, 2014. [http://dx.doi.org/10.1016/j.compfluid.2014.06.032]

[21]   C. Liu, and X. Cai, "New theory on turbulence generation and structure – DNS and experiment", *Sci. China Phys. Mech. Astron.,* vol. 60, p. 31, 2017. [http://dx.doi.org/10.1007/s11433-017-9047-2]

[22]   G. Eitel-Amor, R. Örlü, P. Schlatter, and O. Flores, "Hairpin vortices in turbulent boundary layers", *Phys. Fluids,* vol. 27, 2015.025108 [http://dx.doi.org/10.1063/1.4907783]

[23]   J.C. del Álamo, J. Jiménez, P. Zandonade, and R.D. Moser, "Self-similar vortex clusters in the turbulent logarithmic region", *J. Fluid Mech.,* vol. 561, pp. 329-358, 2006. [http://dx.doi.org/10.1017/S0022112006000814]

[24]   A. Lozano-Durán, O. Flores, and J. Jiménez, "The three-dimensional structure of momentum transfer in turbulent channels", *J. Fluid Mech.,* vol. 694, no. 10, pp. 100-130, 2012. [http://dx.doi.org/10.1017/jfm.2011.524]

[25]   A. Lozano-Duran, and J. Jimenez, "Time-resolved evolution of coherent structures in turbulent channels: Characterization of eddies and cascades", *J. Fluid Mech.,* vol. 759, pp. 432-471, 2014. [http://dx.doi.org/10.1017/jfm.2014.575]

[26]   M. Atzori1, R. Vinuesa1, A. Lozano-Duran, and P. Schlatter, "Characterization of turbulent coherent structures in square duct flow", *IOP Conf. Series: Journal of Physics: Conf. Series,* p. 012008, 2018.

[27]   R. Vinuesa, R. Örlü, and P. Schlatter, "Characterisation of backflow events over a wing section", *J. Turbul.*, vol. 18, no. 2, pp. 170-185, 2017.
[http://dx.doi.org/10.1080/14685248.2016.1259626]

[28]   R.J. Adrian, "Hairpin vortex organization in wall turbulence", *Phys. Fluids,* vol. 19, 2007.041301
[http://dx.doi.org/10.1063/1.2717527]

[29]   W. Schoppa, and F. Hussain, "Coherent structure generation in near-wall turbulence", *J. Fluid Mech.*, vol. 453, p. 57, 2002.
[http://dx.doi.org/10.1017/S002211200100667X]

[30]   M.R. Malik, "Numerical methods for hypersonic boundary layer stability", *J. Comput. Phys.*, vol. 86, p. 376, 1990.
[http://dx.doi.org/10.1016/0021-9991(90)90106-B]

[31]   R. Temam, *Navier-Stokes Equations.* vol. Vol. 2. North-Holland: Amsterdam, 1984.

[32]   S.K. Lele, "Compact finite difference schemes with spectral-like resolution", *J. Comput. Phys.*, vol. 103, p. 16, 1992.
[http://dx.doi.org/10.1016/0021-9991(92)90324-R]

[33]   C.W. Shu, and S. Osher, "Efficient implementation of essentially nonoscillatory shock-capturing schemes", *J. Comput. Phys.*, vol. 77, p. 439, 1988.
[http://dx.doi.org/10.1016/0021-9991(88)90177-5]

[34]   J. Li, C. Liu, M. Choudhari, and C.L. Chang, "Cross-validation of DNS and PSE results for instability- wave propagation in compressible boundary layers past curvilinear surfaces", *16th AIAA Computational Fluid Dynamics Conference,* 2003 Orlando, FL

[35]   C. Liu, and L. Chen, "Parallel DNS for vortex structure of late stages of flow transition", *Comput. Fluids,* vol. 45, p. 129, 2011.
[http://dx.doi.org/10.1016/j.compfluid.2010.11.006]

[36]   X. Dong, S. Tian, and C. Liu, "Correlation analysis on volume vorticity and vortex in late boundary layer transition", *Phys. Fluids,* vol. 30, p. 14105, 2018.
[http://dx.doi.org/10.1063/1.5009115]

[37]   F. Ducros, P. Comte, and M. Lesieur, "Large-eddy simulation of transition to turbulence in a boundary layer developing spatially over a flat plate", *J. Fluid Mech.,* vol. 326, p. 1, 1996.
[http://dx.doi.org/10.1017/S0022112096008221]

[38]   R. Vinuesa, C. Prus, P. Schlatter, and H.M. Nagib, "Convergence of numerical simulations of turbulent wall-bounded flows and mean cross-flow structure of rectangular ducts, Meccanica, December 2016, Volume 51, Issue 12, pp 3025–3042, 2016",

[39]   C. Liu, Y. Wang, Y. Yang, and Z. Duan, "New omega vortex identification method", *Sci. China Phys. Mech. Astron.,* vol. 59, p. 1, 2016.
[http://dx.doi.org/10.1007/s11433-016-0022-6]

[40]   Y. Zhang, K. Liu, H. Xian, and X. Du, "A review of methods for vortex identification in hydroturbines", *Renew. Sustain. Energy Rev.,* vol. 81, p. 1269, 2017. [http://dx.doi.org/10.1016/j.rser.2017.05.058]

[41]   J. Hunt, A. Wary, and P. Moin, "Eddies, streams, convergence zones in turbulent flows [C", *Proceedings of the Summer Program 1988 in its Studying Turbulence Using Numerical Simulation Databases,* 1988pp. 193-208 Stanford, California, USA

[42]   C. Liu, Y. Gao, X. Dong, Y. Wang, J. Liu, Y. Zhang, X. Cai, and N. Gui, "Third generation of vortex identification methods: Omega and Liutex/Rortex based systems", *J. Hydrodynam.,* vol. 31, no. 2, pp. 205-223, 2019.
[http://dx.doi.org/10.1007/s42241-019-0022-4]

[43]   C. Lee, and R. Li, "Dominant structure for turbulent production in a transitional boundary layer", *J. Turbul.,* vol. 8, p. 1, 2007. [http://dx.doi.org/10.1080/14685240600925163]

[44]   J. Chen, and X. Cai, "Research on the single frame imaging method for measuring multi-parameter fields in flow field", *J. Exp. Fluid Mech.,* vol. 29, p. 67, 2015.

[45]   D.B. Spalding, "A single formula for the "law of the wall,"", *J. Appl. Mech.,* vol. 28, p. 455, 1961.
[http://dx.doi.org/10.1115/1.3641728]

[46]   S. Vila, R. Vinuesa, S. Discetti, A. Ianiro, R. Schlatter, and P. Örlü, "2017, On the identification of well-behaved turbulent boundary layers", *J. Fluid Mech.,* vol. 822, pp. 109-138, 2017. [http://dx.doi.org/10.1017/jfm.2017.258]

[47]   L.F. Richardson, "The supply of energy from and to atmospheric eddies", *Proc. R. Soc. Lond., A Contain. Pap. Math. Phys. Character,* vol. 97, p. 354, 1920. [http://dx.doi.org/10.1098/rspa.1920.0039]

[48]   A.N. Kolmogorov, "The local structure of turbulence in incompressible viscous fluid for very large Reynolds numbers", *Dokl. Akad. Nauk,* vol. 30, p. 299, 1941.

CHAPTER 8

Direct Numerical Simulation of Incompressible Flow in a Channel with Rib Structures

Ting Yu, Duo Wang, Heng Li, Hongyi Xu*

Aeronautics and Astronautics Department, Fudan University, Shanghai, PR China

Abstract: This chapter applied the state-of-the-art flow simulation method, *i.e.* the Direct Numerical Simulation (DNS), and strongly coupled the DNS with the heat-transfer governing equation to solve the thermal turbulence in both 2-dimensional (2D) and 3-dimensional (3D) channels with the rib tabulator structures. An innovative approach was applied to the simulations. The surface roughness effects of the cooling vane were directly tackled by including the roughness geometry in the DNS and applying the immersed-boundary method to handle the geometry complexities due to the roughness. Two inlet conditions, namely the uniform flow and full-developed turbulence, were applied at the inflow surface of the channel. Half height of the channel was used as the scale length. The Prandtl (Pr) number was set at Pr = 0.7. Five Reynolds (Re) number of 1000, 2500, 5000, 7500 and 10000 were calculated in the 2D cases and the Reynolds numbers of 2500 and 5000 were applied in 3D cases where a periodical condition was applied in the span-wise direction. Additionally, Reynolds number of 10000 was set in the case with roughened surface. The stream-wise velocity, turbulence intensity, and the Nusselt (Nu) number were analyzed. Results in the 2D and 3D cases presented a significant difference on flow structure. At the same time, with increasing Reynolds number, the length of recirculation zone and the enhancement of heat transfer showed a decreasing trend.

Keywords: Direct Numerical Simulation, Heat Transfer, Rib Tabulator, Roughness, Rortex.

INTRODUCTION

Surface structures, such as pin-fins, ribs and dimples, are effective cooling methods in modern aero-engine design. These structures are commonly used to enhance heat exchange by increasing the heat-exchange surfaces and the turbulence level in flow.

*Corresponding author Hongyi Xu: Aeronautics and Astronautics Department, Fudan University, Shanghai, PR China; Tel: 86-021-55665062; Email: hongyi_xu@fudan.edu.cn

Chaoqun Liu and Yisheng Gao (Eds.)

The rib tabulator is one of the important structures applied in aero-engine components, for example, the combustion ducts and the internal cooling channel of turbine blade. Therefore, it is important to study the flow around a rib and to analyze its effects on the flow field and closely-related heat transfer process. As well known, the flow separation occurs in front of a rib and then reattachment can be found at the bottom wall after rib. Many researches have been conducted to investigate the interactions of these structures with the strongly-coupled flow and heat transfer effects.

The experiments from Eaton *et al.* [1], Chun *et al.* [2], and Liu *et al.* [3] showed the strong unsteadiness of these flow structures. Bergeles and Athanassiadis [4] found that the structures included the combination of complex phenomena, such as flow separation and reattachment. Liu *et al.* [5] provided their experiment results about the flow over a rib, which tried to elucidate the unsteady behaviors of the separation and reattachment over a 2D square rib.

As the computational capabilities grew, more numerical results of flow over the rib were given. Leonardi *et al.* [6] presented the periodic flow in channel with square ribs, the relationship between the flow structure and the ratio of rib height to gap of two ribs was studied. LES method was applied by Cui *et al.* [7], which focused on two types of flows between the ribs. Matsubara *et al.* [8] and Miura *et al.* [9] simulated the flow structure and heat transfer of 3D rib with different Reynolds numbers and aspect ratios.

The current chapter numerically analyzed the flow and heat transfer processes in both 2D and 3D single-rib geometries. Ten cases were investigated including seven 2D cases and three 3D cases. The Reynold numbers (Re) of 2500, 5000 and 7500 were applied in the 2D cases, and the Re of 2500 and 5000 were applied in the 3D cases. Fully-developed turbulent inlet conditions were used in the case with Re=2500. Additionally, most previous work focused on the smooth boundary wall. Since these studies were not able to reflect the effects of the roughness, a 2D case with roughened boundary surfaces was simulated in this study. This case had six dis-symmetric ribs. The Reynolds number of 10000 was applied. Fully-developed turbulent inlet condition was used. X-ray scanning technologies were applied to obtain the geometric contours of the roughness on the cooling-vane surface. An immersed boundary (IB) method coupled with the adaptive mesh refinement technology was implemented to tackle the extremely complex geometries of the surface roughness [10]. In all cases, Prandtl number (Pr) was fixed at 0.7 in these studies to represent air flow. The state-of-the-art simulation method in fluid mechanics, namely Direct Numerical Simulation (DNS), was applied to obtain the flow solution, and multi-grid method was applied to increase the solution efficiency. In order to capture the turbulence and the associated vortical structure, a new vortex identification method was applied, namely, newly defined Rortex proposed by Liu *et al.* [10].

The data are useful and helpful in pushing the nowadays frontiers in each of these relevant science fields, such as the fluid mechanics, heat transfer as well as computations. Also, the database will help transforming the design and manufacture of the turbine blade from the current one based on engineering experiments and empirical data to the future one intelligently-guided by reliable simulation databank.

MATHEMATICAL-PHYSICAL MODELS AND METHODS

Governing Equations

The mathematical models included the conservations of mass, momentum and energy, which are written in the non-dimensional form under Cartesian coordinators:

$$\frac{\partial \bar{u}_j}{\partial x_j} = 0 \tag{1}$$

$$\frac{\partial \bar{u}_j}{\partial t} + \frac{\partial \bar{u}_j \bar{u}_i}{\partial x_j} = -\frac{\partial p}{\partial x_j} + \frac{1}{Re} \frac{\partial}{\partial x_j} \left(\frac{\partial \bar{u}_i}{\partial x_j} + \frac{\partial \bar{u}_j}{\partial x_i} \right) \tag{2}$$

$$\frac{\partial \theta}{\partial t} + \frac{\partial \bar{u}_j \theta}{\partial x_j} = \frac{1}{RePr} \frac{\partial^2 \theta}{\partial x_j \partial x_j} + \frac{Ec}{Re} \frac{1}{2} \left(\frac{\partial \bar{u}_i}{\partial x_j} + \frac{\partial \bar{u}_j}{\partial x_i} \right) \left(\frac{\partial \bar{u}_i}{\partial x_j} + \frac{\partial \bar{u}_j}{\partial x_i} \right) \tag{3}$$

where the subscripts $i, j, k = 1,2,3$ represented the three spatial directions x, y, z, with x being the stream-wise and y, z being the cross-streamwise directions; the repeated subscripts followed the Einstein summation; $\bar{u}_j$ and θ are the non-dimension forms of velocities and the temperature, respectively, and p is the static pressure. The criterion numbers of Re, Pr and Ec are defined as $Re = \rho U h/\mu$, $Pr = \mu C_p/k$ and $Ec = U^2/C_p(Th - Tc)$ with ρ, μ, C_p being the fluid density, molecular viscosity and specific heat capacity, U being the fluid incoming velocity, Th *and* Tc being the hot(wall) and cool(incoming) temperatures, respectively, and h being the height of rib as seen in Fig. (**2**).

Computation Scheme

The second-order ifnite volume schemes were applied in the spatial discretization. □ In temporal direction, the convection and diff usion terms were handled by the second-order time marching schemes of Adams-Bashforth and Adams-Moulton, respectively. The velocity and pressure ifelds were decoupled by the fractional step □ method [12]. To achieve a high-performance solution efifciency, the robust □ Flexible-Cycle Additive-Correction (FCAC)-multigrid (MG) in [13] was applied to tackle the unsteady multi-scale turbulence motions.

Vortex Identification Method

The new vortex identiifcation method wa. applied to analyze the lfow structures. It □ was called Rortex method [10]. Rortex method was a new vortex identification presented by Liu *et al.* [10]. A newly-defined quantity of Rortex, was used to represent the vortex. The definition of Rortex was given below.

The gradient of velocity was given as following.

$$\begin{vmatrix} \dfrac{\partial u}{\partial x} & \dfrac{\partial u}{\partial y} & \dfrac{\partial u}{\partial z} \\[2mm] \dfrac{\partial v}{\partial x} & \dfrac{\partial v}{\partial y} & \dfrac{\partial v}{\partial z} \\[2mm] \dfrac{\partial w}{\partial x} & \dfrac{\partial w}{\partial y} & \dfrac{\partial w}{\partial z} \end{vmatrix} \qquad\qquad (4)$$

Firstly, origin reference frame xyz is transformed to a new reference frame XYZ. In the new reference frame, the fluid rotation axis is parallel to the axis Z. And according to the real Schur decomposition, the gradient of the velocity presented by the new reference frame and the transformation can be written as below.

$$
\begin{vmatrix}
\dfrac{\partial U}{\partial X} & \dfrac{\partial U}{\partial Y} & 0 \\[2mm]
\dfrac{\partial V}{\partial X} & \dfrac{\partial V}{\partial Y} & 0 \\[2mm]
\dfrac{\partial W}{\partial X} & \dfrac{\partial W}{\partial Y} & \dfrac{\partial W}{\partial Z}
\end{vmatrix}
\tag{5}
$$

$$
\nabla \vec{V} = Q \nabla \vec{v} Q^{-1}
\tag{6}
$$

The matrix Q can then be found by real Schur decomposition. Thus, two variables can be calculated:

$$
\alpha = \frac{1}{2}\sqrt{\left(\frac{\partial V}{\partial Y} - \frac{\partial U}{\partial X}\right)^2 + \left(\frac{\partial V}{\partial X} + \frac{\partial U}{\partial Y}\right)^2}
\tag{7}
$$

$$
\beta = \frac{1}{2}\left(\frac{\partial V}{\partial X} - \frac{\partial U}{\partial Y}\right)
\tag{8}
$$

The definition of Rortex is given as following

$$
R = \begin{cases}
\beta - \alpha, & \text{if } \alpha^2 - \beta^2 \text{ less than } 0 \text{ and } \beta \text{ greater than } 0 \\
\beta + \alpha, & \text{if } \alpha^2 - \beta^2 \text{ less than } 0 \text{ and } \beta \text{ greater than } 0 \\
0, & \text{if } \alpha^2 - \beta^2 \text{ less than } 0
\end{cases}
\tag{9}
$$

The vorticity is then decomposed as

$$
\Omega = \vec{R} + \vec{S}
\tag{10}
$$

with, $\vec{S}$ being the term of shear.

Surface Roughness Geometry

The X-ray scanning technologies were applied to obtain the geometric contours of the roughness on the cooling-vane surface and the computational grids near the roughened surfaces were generated using the IB method. Fig. (**1**) demonstrated the roughness distributions on surface and the computational grids generated near the surfaces with the rib-tabulator structures.

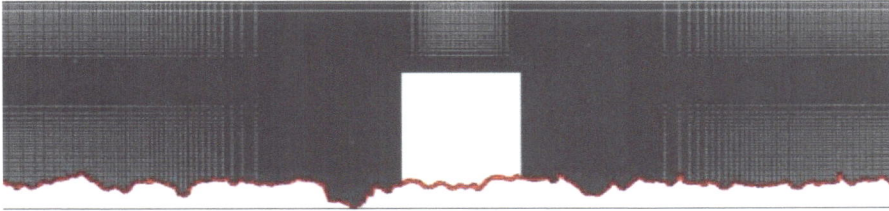

Fig. (1). Computational Grid Generated using IB Method.

RESULTS OF CASES WITH SINGLE RIB

Computational Cases

The structure and parameters of cases with single rib were given first. Fig. (**2**) presents the profile of structure for the cases with single rib. In Fig. (**2**), w was the width of the rib, h presented the height of the rib, H was the height of the channel, $L1$ and $L2$ were the length of the channel before and behind the rib respectively. In cases, $H = 10h$, $w = 1h$, $L1 = 12h$, $L2 = 24h$, $Pr = 0.7$ and the grid points in x direction and y direction were 320 and 256. The scale length was half height of the channel. 3D cases extended the grid in z direction. $W = 2h$ stands for the length of the channel in z direction $W = 2h$, periodic condition was given as the boundary condition in this direction.

Fig. (2). Grid Structure.

In seven 2D cases, five had uniform flow inlet condition and their Reynold numbers were 1000, 2500, 5000, 7500 and 10000 separately. The fully developed turbulence inlet condition was applied for the cases with Reynold number 2500 and 5000. Two of 3D cases had uniform inlet condition, the Reynolds numbers of these two cases were 2500 and 5000. The fully developed turbulence inlet condition and Reynolds number 2500 were applied in another 3D case.

Table 1. Computed validation cases.

Inlet Condition	Re	Dimension
Uniform	1000	2D
Uniform	2500	2D
Turbulent	2500	2D
Uniform	5000	2D
Turbulent	5000	2D
Uniform	7500	2D
Uniform	10000	2D
Uniform	2500	3D
Turbulent	2500	3D
Uniform	5000	3D

Results of Validation

In order to validate the flow model and the code, the results of validation cases are presented below and are compared with the reference data.

Fig. (**3**) shows the profiles of streamwise velocity in two 3D cases and these two cases are compared with the experiment results from Liu *et al.* [5]. The x-axis is the no-dimensional stream-wise velocity and y-axis is the position in y direction. The location $x = 0h$ is fixed at the leading edge of the rib. The presented results of velocity profiles generally match with the reference results. In particular, the green symbols are more accurate than the red symbols when comparing to the results from Liu *et al.* [5], which is obviously due to the effects of the incoming flow boundary conditions. The fully-developed turbulence incoming conditions are definitively more realistic than the uniform inlet and therefore, are closer to the results obtained in Liu's research. The approaching flow separates from the wall in front of the rib and forms a separated zone. Then, the flow detaches immediately at the leading edge of the rib and a thin layer of reversed flow forms above the rib [5]. The turbulent flow develops downstream of the rib and a slender recirculation zone generates behind the rib.

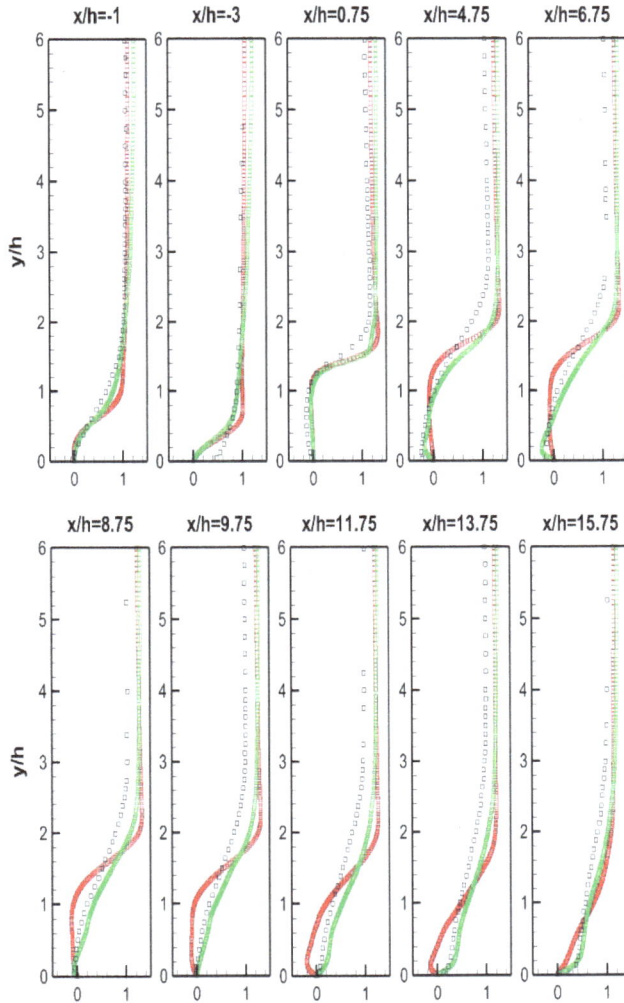

Fig. (3). Profiles of Stream-Wise Velocity (□:Liu *et al*. [5] ▪:3D Case, Re=2500, Uniform Flow Inlet , ▪:3D Case, Re=2500, Turbulence Inlet).

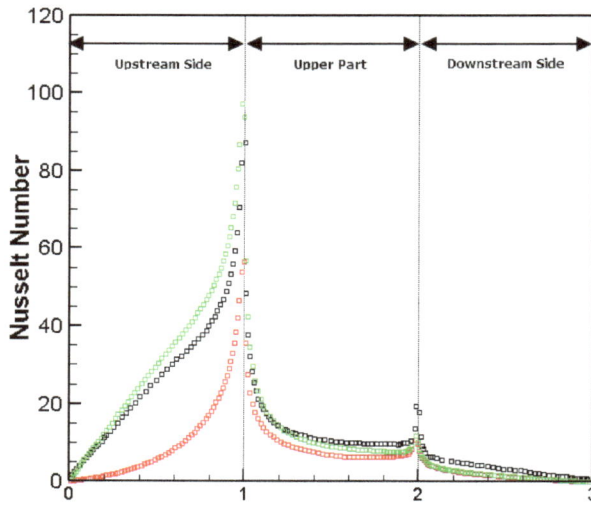

Fig. (4). Nusselt Number (□:Liu *et al.* [5] □:3D Case, Re=2500, Uniform Flow Inlet , □:3D Case, Re=2500, Turbulence Inlet).

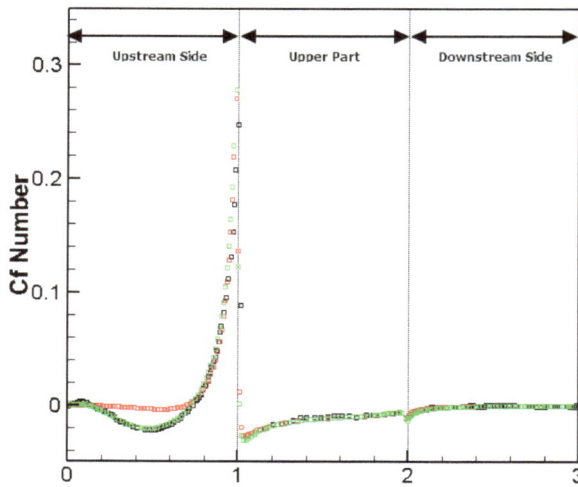

Fig. (5). Cf Number (□:Liu *et al.* [5] □:3D Case, Re=2500, Uniform Flow Inlet , □:3D Case, Re=2500, Turbulence Inlet).

Figs. (**4** and **5**) present Nu number and Cf number separately. The reference data of Nu number come from Matsubara *et al.* [8] and the reference data of Cf number is from Miura *et al.* [9]. Both Nu and Cf number reach the maximum point at the leading edge of the rib and then change smoothly on the top face of the rib. Both the absolute values of Nu and Cf number get a local maximum point at the top right corner of the rib. These two physical variables match well with the reference data.

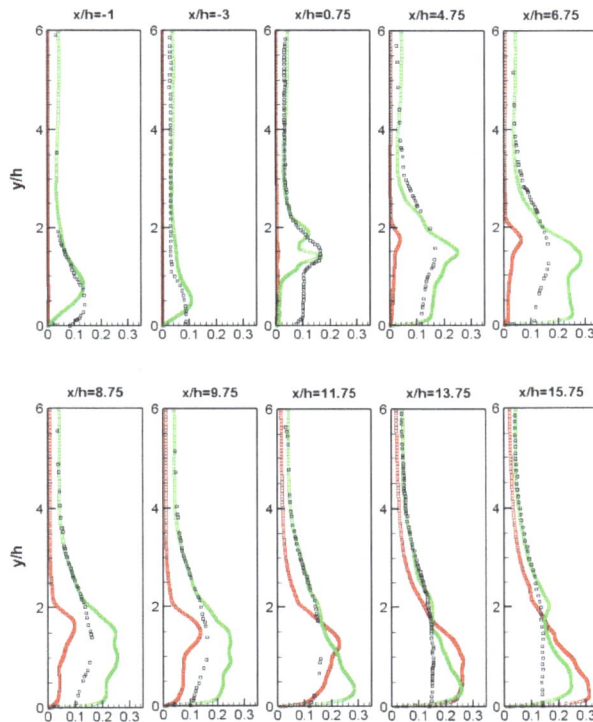

Fig. (6). Profiles of Turbulence Intensity (□:Liu *et al.* [5] □:3D Case, Re=2500, Uniform Flow Inlet □:3D Case, Re=2500, Turbulence Inlet).

Fig. (**6**) shows the profiles of streamwise turbulence intensities in two 3D cases. The reference data from Liu *et al.* [5] is also presented in the graphs. The results and the reference data had the same trend. The difference between the presented data and the reference data is concentrated near the wall, which can be attributed to various reasons. Firstly, the difference may be due to the different Reynolds numbers between the current computations and the experiment data in Liu *et al.* [5]. Secondly, the reference data was obtained from experiments, the number of points near the wall might be the issue which caused mismatch with the current DNS study.

Fig. (7) gives the comparison of the wall-normal location for the maximum stream-wise turbulence intensity in present work and reference data from Liu *et al.* [5]. The curve increases immediately in front of the rib. After the rib, the curve still increases smoothly and it gets to the maximum point at about 5h after the rib, then the curve begins to decrease. The presented prediction and the reference data essentially match with each other.

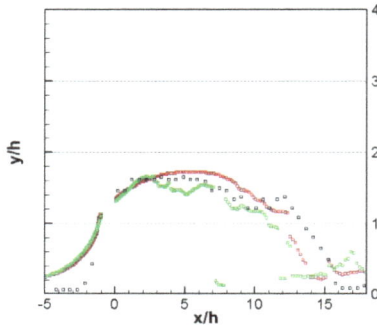

Fig. (7). Location of Maximum Turbulence Intensity.

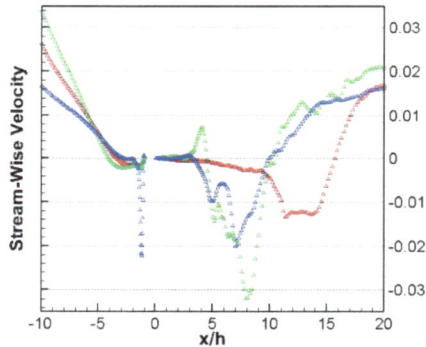

Fig. (8). Stream-Wise Velocity Near Wall (□: 2D Case, Re=2500, Turbulence Inlet; □: 2D Case, Re=5000, Turbulence Inlet; □: 2D Case, Re=1000, Uniform Flow Inlet; □: 2D Case, Re=2500, Uniform Flow Inlet; □: 2D Case, Re=5000, Uniform Flow Inlet).

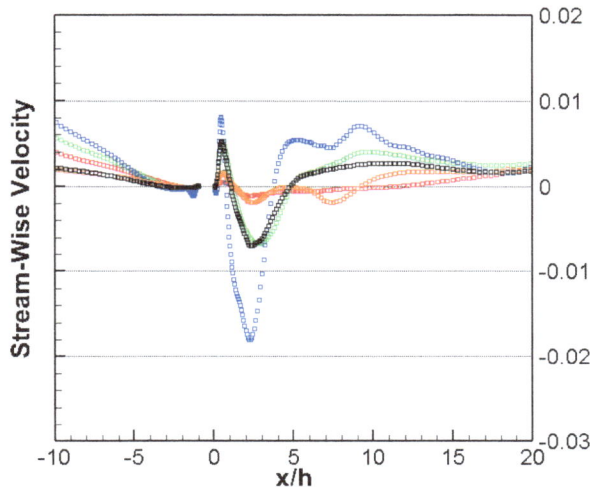

Fig. (9). Stream-Wise Velocity Near Wall (△: 3D Case, Re=2500, Uniform Flow Inlet; △: 3D Case, Re=5000,

niform Flow Inlet; △: 3D Case, Re=2500, Turbulence Inlet).

Flow Structures and Heat Transfer

Figs. (**8** and **9**) show the streamwise velocity near the bottom wall. Fig. (**8**) gives five 2D cases and Fig. (**9**) presents three 3D cases. In these two graphs, $x = -1h$ is the leading edge of the rib and $x = 0h$ is the back edge of the rib. In all 2D and 3D cases, the length of the separate zone in front of the rib increases with the increasing Reynolds number, the separate point is about 2h to 4h in front of the rib. Turbulence inlet condition has significant effects on the flow structure. In the cases with uniform inlet condition, the length of recirculation zone shows a decrease trend with the increasing Reynolds number. On the other hand, the turbulence inlet condition has tangible effects on the flow structure. The turbulence inlet condition decreases the length of separate zone and recirculation zone, which makes the data agreeing better with the reference data in Figs. (**3** and **4**). The observation needs to be further confirmed and validated by more reference data. In 3D cases with Reynold number 2500, the reattachment point with uniform flow inlet condition is about $x = 15h$, but it changes to $x = 10h$ with turbulence inlet condition. This can match the result from Liu *et al.* [3], [5], in their experiment, the reattachment point is about $x = 9.75h$. At the same time, results in 2D and 3D cases show big differences. With the same condition, the fluctuation of near wall stream-wise velocity is more drastic in 3D case than in 2D case. The absolute value of near wall streamwise velocity in 3D case is also bigger than in 2D case. Specially, the back edge of the rib is an important area. After the rib, the near wall streamwise velocity increases immediately in 2D case. However in 3D case, the velocity has a buffer zone after the rib. The near-wall streamwise velocity changes smoothly within the zone and the variation accelerates rapidly beyond the zone.

The maximum streamwise turbulence intensity in different cases are presented by Figs. (**10** and **11**). In both 2D and 3D cases, the turbulence intensity increases with increasing the Reynold number. In the 2D cases, the turbulence intensity increases immediately after the rib and reaches the local peak in the recirculation zone. In front of the rib, the turbulence intensity gets high value around the separation zone in 2D cases with the uniform inlet condition. In the 3D cases, the turbulence intensity increases smoothly after the rib and reaches the local maximum around the reattachment point. In the 2D cases with the fully- developed turbulent inlet condition, the turbulence intensities both in front of the rib and after the rib become lower, which makes the curves smoother. But in the 3D cases, the turbulence inlet condition makes the turbulence intensity more enhanced.

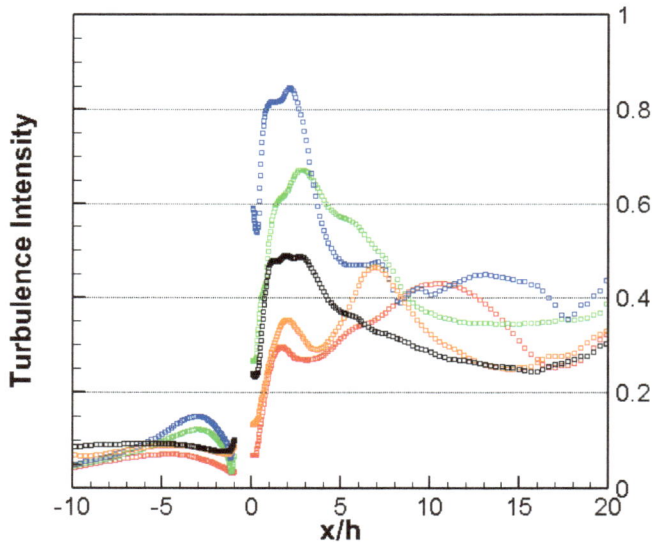

Fig. (10). Turbulence Intensity in 2D Cases (□: 2D Case, Re=2500, Turbulence Inlet; □: 2D Case, Re=5000, Turbulence Inlet; □: 2D Case, Re=1000, Uniform Flow Inlet; □: 2D Case, Re=2500, Uniform Flow Inlet; □: 2D Case, Re=5000, Uniform Flow Inlet).

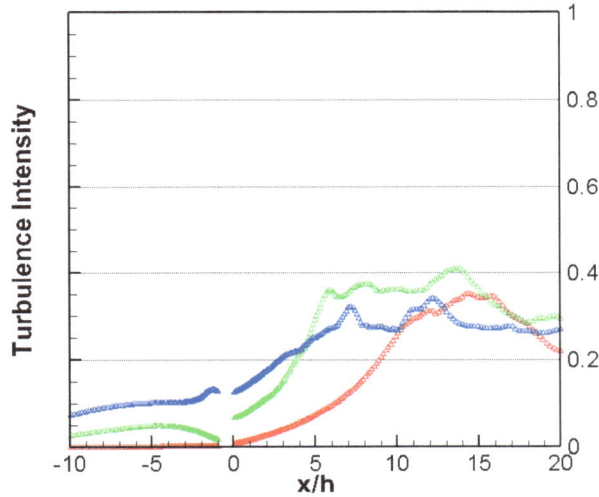

Fig. (11). Turbulence Intensity in 3D Cases (△: 3D Case, Re=2500, Uniform Flow Inlet; △: 3D Case, Re=5000, Uniform Flow Inlet; △: 3D Case, Re=2500, Turbulence Inlet).

Figs. (**12** and **13**) show the location of maximum turbulence intensity in span-wise. In front of the rib, the curves in all cases have the same trend with the intensities being enhanced and the reaching the peaks at the leading edge of the rib. After the rib, the curves in 2D cases go down immediately and get closer to the boundary wall. Behind the reattachment point, the curves begin to increase smoothly. In the 3D cases, the curves still go up after the rib and the location of maximum turbulence intensity occurs above the rib. And further these curves get to the local minimums around the reattachment point.

The time-averaging Nu number in the cases are shown in Figs. (**14** and **15**). In all cases, the Nu number reaches the global maximum at the leading edge of the rib and at the back edge of the rib, the Nu reaches a local maximum point and the local maximum value increases with increasing the Re number. The fully- developed turbulent inlet condition also influences the Nu number, which enlarges the Nu number. The trend of Nu on the upper part and downstream side of the rib changes from decrease to increase with increasing the Re number in the 2D cases. The different area between the 2D cases and 3D cases is concentrated around the rear edge of the rib.

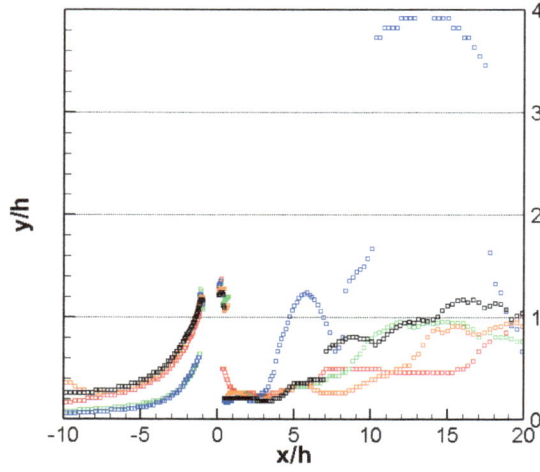

Fig. (12). Location of Maximum Turbulence Intensity in 2D Cases (□: 2D Case, Re=2500, Turbulence Inlet;

□: 2D Case, Re=5000, Turbulence Inlet; □: 2D Case, Re=1000, Uniform Flow Inlet; □: 2D Case, Re=2500,

Uniform Flow Inlet; □: 2D Case, Re=5000, Uniform Flow Inlet).

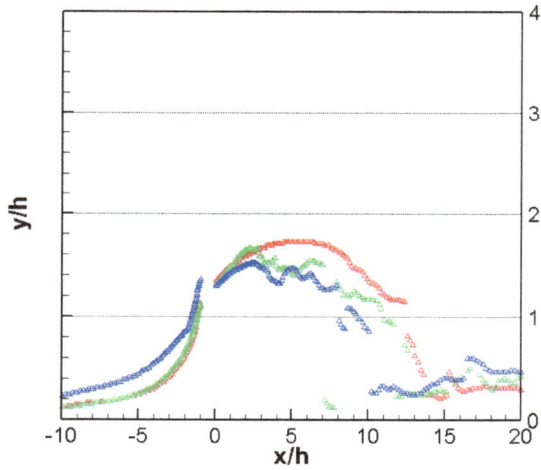

Fig. (13). Location of Maximum Turbulence Intensity in 3D Cases (△: 3D Case, Re=2500, Uniform Flow Inlet;

△: 3D Case, Re=5000, Uniform Flow Inlet; △: 3D Case, Re=2500, Turbulence Inlet).

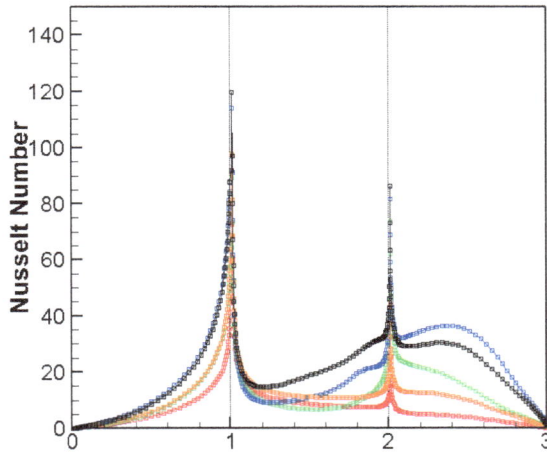

Fig. (14). Nu Number on Rib Surface in 2D (□: 2D Case, Re=2500, Turbulence Inlet; □: 2D Case, Re=5000, Turbulence Inlet; □: 2D Case, Re=1000, Uniform Flow Inlet; □: 2D Case, Re=2500, Uniform Flow Inlet; □: 2D Case, Re=5000, Uniform Flow Inlet).

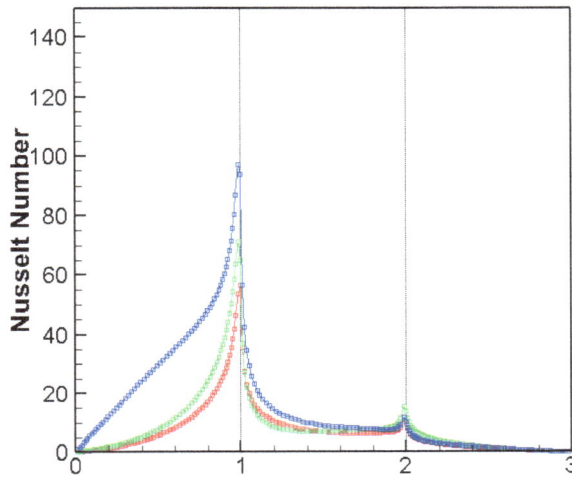

Fig. (15). Nu Number on Rib Surface in 3D (△: 3D Case, Re=2500, Uniform Flow Inlet ; △: 3D Case, Re=5000, Uniform Flow Inlet; △: 3D Case, Re=2500, Turbulence Inlet).

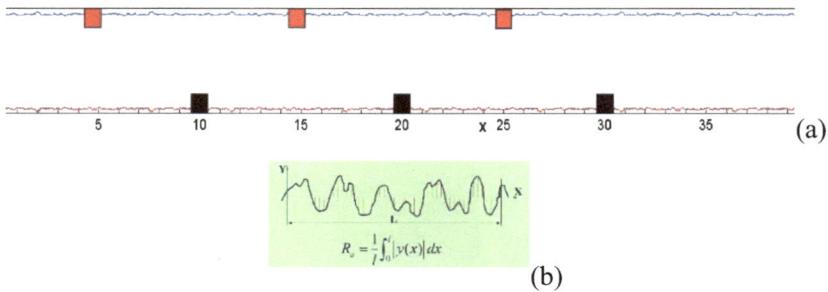

(a)

(b)

Fig. (16). Structure with Roughened Surface (**a**) and Ra Definition (**b**).

RESULTS OF CASE WITH ROUGHENED SURFACE

Structure of the Computed Case with Surface Roughness

Fig. (**16**) shows the structure of case with roughened boundary surface. There were six ribs in cases with the staggering arrangement on the upper and lower surfaces. The flow Reynolds number in the channel was set at Re=10,000 as referenced by [14]. The channel surfaces roughness was obtained by X-ray scanning the metal surface, which gave a roughness-averaging number of Ra at Ra=1.6 as shown in Fig. (**16**).

The simulations were performed for the channel with both smooth and rough surfaces under a fully-developed turbulent inflow conditions. The non-dimensional wall and rib temperatures were set at 1.0 as the boundary conditions, the non-dimensional inlet temperature was isothermal condition fixed at 0.

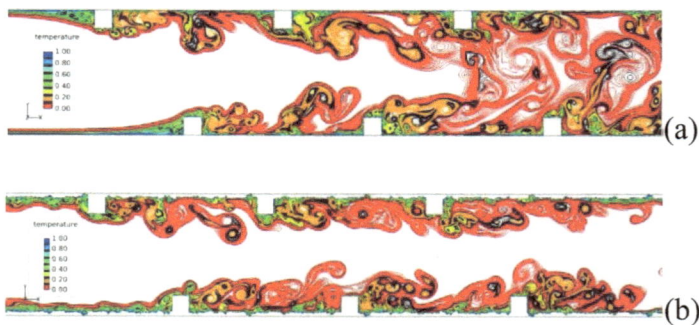

Fig. (17). Instantaneous temperature fields in the ribbed channel with (**a**) smoothed surface; (**b**) roughened surface.

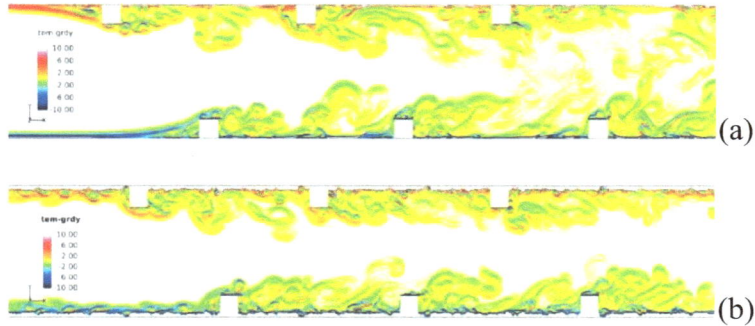

Fig. (18). Instantaneous temperature gradient of $\partial\theta/\partial y$ in the ribbed channel with (**a**) smoothed and (**b**) roughened surface.

Results of Thermal Fields

Fig. (**17a**) demonstrates the instantaneous temperature distributions as the strongly-coupled flow and temperature fields developing in the cooling vane with the smoothed channel surface and the staggered-rib arrangement. Fig. (**17b**) exhibits its counterpart with the roughened surface. The striking features of the strongly-coupled flow and thermal interaction phenomena are seen in the cooling channels with the rib structures, including the thermal boundary layers development in the front of leading ribs(R1), the typical backward-facing step thermal flow patterns near the leeward surface of the ribs(R2) and the flow-heat convections along the horizontal surfaces at the top of the ribs(R3). The comparison between Figs. (**17a** and **b**) indicate that the wall-roughness significantly affects the flow and heat transfer, particularly in the near wall regions.

Figs. (**18a** and **b**) present the temperature gradient component of $\partial\theta/\partial y$ which physically stands for the heat conduction intensity in the y direction. Figs. (**18a** and **b**) exhibits the characteristic and striking streaky structures in a variety of zones, such as R1, R2 and R3, and these structures are very similar to the well-known coherent structures of turbulence, indicating that turbulence has strong effects in the formations of these heat-transfer structures and patterns.

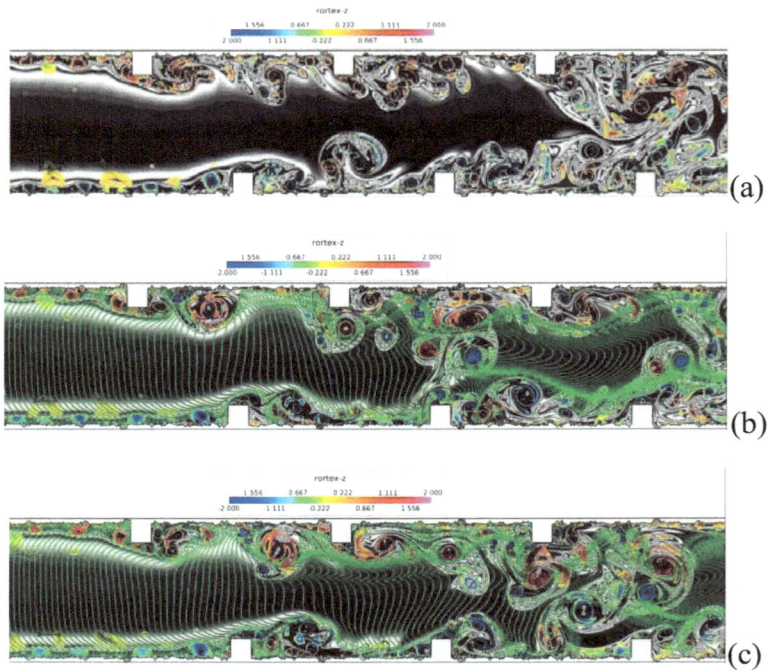

Fig. (19). Instantaneous Rortex and Shear in the ribbed channel.

Results of Rortex

Fig. (**19**) presents Rortex and Shear in the cases with the roughened surfaces, where the (a), (b) and (c) stand for the solutions at different time steps. The color graph presents the magnitude of Rortex, the black and white bands show the distribution of Shear. In the (b) and (c), the green points are the flow tracking points emanated from the channel inlet. In the flow evolution, Rortex is always surrounded by Shear. The shear forms the parclose enwrapping the Rortex, which prevents the Rortex concentrated zone from being penetrated by the flow trajectory points (the green points). On the other hand, the Rortex are generated by the zone with high Shear concentrations. Although the Rortex were found not quite closely-relating to the heat-transfer process, the shear motions demonstrated a very strong correlation with the heat transfer by affecting the Nusselt number. Therefore, it is important to apply the Rortex and Shear decomposition to analyze the strongly-coupled flow and thermal turbulence so that the heat-transfer mechanisms in the near-wall region can be better understood. Recently, Li *et al.* [15] demonstrated a very promising research in this regard, which clearly justified the importance to use Rortex and

Shear method in the future research for the strongly-coupled flow and heat-transfer phenomena.

CONCLUSIONS

The chapter investigated the flow structure and heat transfer of cases with rib wall in both 2D and 3D, and moreover, successfully simulated the dissymmetric rib structure with roughened surface in 2D.

In both 2D cases and 3D cases, as Reynolds number increased, the length of the recirculation zone decreased and the enhancement of heat transfer on ribbed boundary decreased. Nu number got the maximum value at the leading edge of the rib. The turbulence inlet condition generated a bigger the maximum stream-wise turbulence intensity. In front of the rib, the stream-wise turbulence intensity had a small increasing near the separate point and then decreased immediately.

The results also showed the big difference between 2D and 3D cases. Firstly, the length of recirculation zone in 3D cases was shorter than in 2D cases with same Reynolds number. Secondly, the location of maximum stream-wise turbulence intensity after the rib was near wall after the rib in 2D cases, but in 3D cases, it got upper from the height of the rib behind the rib. Thirdly, the effect of the turbulence inlet condition was more remarkable in 3D cases than in 2D cases, especially obviously in Nu and Cf number on the front face of the rib.

The wall roughness was found to significantly affect the heat transfer process in the ribbed channel. The temperature gradient fields were discovered bearing similar streaky characteristics comparing to the well-known turbulence coherent structures.

The new vortex identification method of Rortex was applied in the study. According to the results, the Rortex and Shear decomposition showed that the shear motions presented a strong correlation with the heat-transfer process. Therefore, the Rortex and Shear decomposition and shear-heat transfer correlation are expected to play an important role in the future study of the strongly-coupled flow and thermal phenomena.

CONSENT FOR PUBLICATION

Not applicable.

CONFLICT OF INTEREST

The authors confirm that this chapter contents have no conflict of interest.

ACKNOWLEDGEMENTS

Declared none.

REFERENCE

[1]   J.K. Eaton, and J.P. Johnston, "A review of research on subsonic turbulent flow reattachment", *AIAA J.,* vol. 19, pp. 1093-1100, 1981.
        [http://dx.doi.org/10.2514/3.60048]

[2]   S. Chun, Y.Z. Liu, and H.J. Sung, "Wall pressure fluctuations of a turbulent separated and reattaching flow affected by an unsteady wake", *Exp. Fluids,* vol. 37, no. 4, pp. 531-546, 2004. [J].
        [http://dx.doi.org/10.1007/s00348-004-0839-6]

[3]   Y.Z. Liu, W. Kang, and H.J. Sung, "Assessment of the organization of a turbulent separated and reattaching flow by measuring wall pressure fluctuations", *Exp. Fluids,* vol. 38, no. 4, pp. 485-493, 2005. [J].
        [http://dx.doi.org/10.1007/s00348-005-0929-0]

[4]   G. Bergeles, and N. Athanassiadis, "The flow past a surface-mounted obstacle", *J. Fluids Eng.,* vol. 105, no. 4, pp. 461-463, 1983. [J].
        [http://dx.doi.org/10.1115/1.3241030]

[5]   Y.Z. Liu, F. Ke, and H.J. Sung, "Unsteady separated and reattaching turbulent flow over a two-dimensional square rib", *J. Fluids Structures,* vol. 24, no. 3, pp. 366-381, 2008. [J].
        [http://dx.doi.org/10.1016/j.jfluidstructs.2007.08.009]

[6]   S. Leonardi, P. Orlandi, L. Djenidi, and R.A. Antonia, "Structure of turbulent channel flow with   square bars   on   one   wall",   *Int.   J.   Heat   Fluid   Flow,*   vol.   25,   pp.   384-392,   2004.
        [http://dx.doi.org/10.1016/j.ijheatfluidflow.2004.02.022]

[7]   J. Cui, V.C. Patel, and C-L. Lin, "Large-eddy simulation of turbulent flow in a channel with rib roughness", *Int. J. Heat Fluid Flow,* vol. 24, pp. 372-388, 2003.
        [http://dx.doi.org/10.1016/S0142-727X(03)00002-X]

[8]   K. Matsubara, T. Miura, and H. Ohta, "Transport dissimilarity in turbulent channel flow disturbed by rib protrusion with aspect ratio up to 64", *Int. J. Heat Mass Transf.,* vol. 86, pp. 113-123, 2015. [J].
        [http://dx.doi.org/10.1016/j.ijheatmasstransfer.2015.02.018]

[9]   T. Miura, K. Matsubara, and A. Sakurai, "Heat transfer characteristics and Reynolds stress budgets in single-rib mounting channel", *Journal of Thermal Science and Technology,* vol. 5, no. 1, pp. 135-150, 2010. [J].
        [http://dx.doi.org/10.1299/jtst.5.135]

[10]   H. Xu, "Developing LES/DNS Simulation Capability based on Immersed Boundary Method coupled with FCAC Multigrid and AMR Techniques", In: *The 18th International Conference on Finite Elements in Flow Problem*, Taipei, 2015.

[11]   C. Liu, Y. Gao, S. Tian, and X. Dong, "Rortex a new vortex vector definition and vorticity tensor and vector decompositions", *Phys. Fluids,* vol. 30, 2018.035103
        [http://dx.doi.org/10.1063/1.5023001]

[12]   J. Kim, and P. Moin, "Application of a fractional-step method to incompressible navier-stokes equations", *J. Comput. Phys.,* vol. 59, pp. 308-323, 1985.
[http://dx.doi.org/10.1016/0021-9991(85)90148-2]

[13]   H. Xu, W. Yuan, and M. Khalid, "Design of a high-performance unsteady naiver–stokes solver using a flexible-cycle additive-correction multigrid technique", *J. Comput. Phys.,* vol. 209, pp. 504-540, 2005.
[http://dx.doi.org/10.1016/j.jcp.2005.03.029]

[14]   L. Xi, J. Gao, L. Xu, Z. Zhao, and Y. Li, "Study on heat transfer performance of steam-cooled ribbed channel using neural networks and genetic algorithms", *Int. J. Heat Mass Transf.,* vol. 127, pp. 1110-1123, 2018.
[http://dx.doi.org/10.1016/j.ijheatmasstransfer.2018.08.115]

[15]   H. Li, T. Yu, D. Wang, and H. Xu, "Heat-transfer enhancing mechanisms induced by the coherent structures of wall-bounded turbulence in channel with rib", *Int. J. Heat Mass Transf.,* vol. 137, pp. 446-460, 2019.
[http://dx.doi.org/10.1016/j.ijheatmasstransfer.2019.03.122]

Vortex and Flow Structure inside Hydroturbines

Yuning Zhang [1*], Yuning Zhang [2, 3,]**

[1]*Key Laboratory of Power Station Energy Transfer Conversion and System (Ministry of Education), School of Energy, Power and Mechanical Engineering, North China Electric Power University, Beijing, China*
[2]*College of Mechanical and Transportation Engineering, China University of Petroleum-Beijing, Beijing 102249, China*
[3]*Beijing Key Laboratory of Process Fluid Filtration and Separation, China University of Petroleum-Beijing, Beijing 102249, China*

Abstract: In this chapter, various kinds of vortex in the hydroturbines are briefly introduced with a focus on the swirling vortex rope in Francis turbine and the vortex in the vaneless space of the reversible pump turbine. The vortex induced pressure fluctuation and vibrations are initially demonstrated based on the on-site measurement in the prototype power stations. Then, influences of the vortex in the upstream on the flow status in the downstream are discussed. Finally, detailed characteristics of the swirling vortex in the draft tube section of the hydroturbines are demonstrated based on the plenty of examples together with the aid of a quantitative swirl number analysis.

Keywords: Hydroturbines, Pressure fluctuations, Vibrations, Vortex, Vortex rope.

A SUMMARY OF TYPES OF VORTEX IN HYDROTURBINES

There are various kinds of hydroturbines including Francis turbine, reversible pump turbine, Kaplan turbine, Pelton turbine etc. For a complete review of the flow-induced vortex in hydroturbines together with the vortex identification methods and their applications in hydroturbines, readers are referred to Zhang *et al*. [1]. For reference books relating with the associated pressure fluctuations and vibrations in hydroturbines, readers are referred to Wu et al. [2] and Dörfler *et al*. [3]. Fig. (**2**) of Chen *et al*. [4] shows the Francis turbine of Three Gorges hydro power station, which was the largest hydroturbines (in terms of the electricity generation capacity)

*Corresponding author Yuning Zhang: Key Laboratory of Power Station Energy Transfer Conversion and System (Ministry of Education), School of Energy, Power and Mechanical Engineering, North China Electric Power University, Beijing, China; Tel: 86 (0)1061773958; E-mail: y.zhang@ncepu.edu.cn

of the given type in the world when it was commissioned. In the above figure, the detailed components of the aforementioned hydroturbines are also shown. Basically speaking, the fluid passing components of a typical Francis turbine include the spiral casing, the stationary guide vane, the wicket gate (also named as the adjustable guide vane, with the function of the water flux control), runner (also named as impeller), draft tube (including cone and elbow sections respectively). For other types of the hydroturbines, the basic structures and principles are quite similar. In the present chapter, two paramount types of the hydroturbines are discussed: the aforementioned Francis turbine and the reversible (Francis type) pump turbines of a pumped hydro energy storage power plant.

For different kinds of hydroturbines, the dominant types of vortex are quite different. For example, for the Francis turbine, the dominant vortex is usually the swirling vortex rope in the draft tube (referring to the figure 13 of Chen *et al.* [4]). When the turbine is operated in the partial loads (off-design conditions), this kind of vortex is quite strong with significant rotating momentum. Meanwhile, prominent pressure fluctuation will be also generated by the swirling vortex rope with the propagation to the upstream or the downstream also possibly leading to the vibrations of the whole unit. Generally speaking, the rotational speed of this kind of vortex is rather slow and is far less than the rotational speed of the runner. Other prominent vortex also exists in the channel of runner (termed as the channel vortex).

For the reversible pump turbine, as shown in figure 16 of Zhang *et al.* [5], the dominant vortex is the vortex shown in the vaneless space between the wicket gate and the runner (also named as the impeller). For more details about the reversible pump turbine, readers are referred to Zhang *et al.* [5]. In the vaneless space, as shown in figure 12 of Hasmatuchi *et al.* [6], there are prominent backflows during the low discharge mode. Comparing with the best design point (BEP), the flow status is much distorted in the low discharge working conditions, leading to the strong channel blockage of the flow. For the reversible pump turbine, the vortex rope in the draft tube also exists but is no longer the primary source of the vortex as those shown in the Francis turbine.

For the Kaplan turbine, the tip vortex between the runner and the hub is very significant, leading to serious damage on the fluid components. Generally speaking, because the tip is quite small according to the design (for the enhancement of the efficiency), cavitation usually occurs near the tip. Hence, as shown in figure 14 of Motycak *et al.* [7], the tip leakage vortex is often accompanied by the cavitation bubbles. Because the force and micro-jet generated by the bubble final collapse are quite prominent, serious damage could be observed on the runner edges.

Other types of vortex also include the inter-blade vortex in the runner, the Kármán vortex in the wake flow of the vanes and the cavitating vortex.

The relationships between the vortex and the turbulence could be illustrated as follows. On the one hand, the vortex phenomenon could induce the generations of the turbulence together with associated structures. For example, during the rotating stall status in the reversible pump turbine, the vortex generated in the impeller channels could block the fluid passing through the component. With the increment of the fluid distortion, a strong turbulence flow will be finally demonstrated with prominent pressure fluctuations. On the other hand, the existing turbulence will lead to the intensive generations of the vortex. For example, in the low load condition, the turbulence will be induced inside the guide vane channels due to the large incidence angle. Then, many small vortex will appear especially near the pressure side of the vanes.

THE EFFECTS OF VORTEX ON PRESSURE FLUCTUATION

In this section, various kinds of negative effects of the vortex on the hydroturbine performances will be introduced including the pressure fluctuation together with its characteristic frequency.

Fig. (**2**) by Zhang *et al*. [8] shows the non-dimensional peak-to-peak values of pressure fluctuation versus load variations (from 25.41% to 96.82% of the full load) at four monitoring points (referring to the ref. [8] for the positions). The identified three zones could be summarized as follows:

Zone 1: This zone corresponds to the conditions of the low partial load. The pressure fluctuation is mainly generated by the vortex flow in the vaneless space. Fig. (**4**) of Zhang *et al*. [8] further shows the cascade plot of frequency spectrums measured at the vaneless space. The dominant frequency is $9f_n$ (also termed as blade passing frequency, with f_n representing the impeller rotational frequency), which is generated by vortex induced by the rotating impeller.

Zone 2: This zone corresponds to the conditions of the medium partial load. The pressure fluctuation is mainly generated by the swirling vortex rope in the draft tube cone section. Figure 5 of Zhang et al. [8] further shows the cascade plot of frequency spectrums measured at the draft tube cone section. The dominant frequency of this kind of vortex is less than f_n, which is generated by the vortex induced by the swirling vortex rope.

Zone 3: This zone corresponds to the conditions of the high partial load and conditions near the design point. As shown in Fig. (**4**) of Zhang *et al.* [8], the pressure fluctuation is still generated by the vortex flow in the vaneless space but with a quite different frequency ($18f_n$). Furthermore, because this zone is quite close to the design point, the vortex-induced pressure fluctuation is quite limited.

Except for the aforementioned parameters (*e.g.* load conditions, monitoring positions), the water head of the power plant could also affect the primary characteristics of the pressure fluctuation in the hydroturbines (referring to Fig. (**5**) of Li *et al.* [9]). For more details, readers are referred to our recent work [9].

THE EFFECTS OF VORTEX ON VIBRATIONS OF HYDROTURBINES

In this section, effects of the vortex on the vibrations of the hydroturbine will be introduced based on our recent on-site measurement [10]. Fig. (**1**) shows a schematic view of the components of the whole pumped hydro energy storage power plant. The pressure fluctuations generated by the fluid flow inside the turbine could possibly propagate to the whole unit even the entire power plant (*e.g.* the main control house and related buildings). In China, sometimes, this kind of vortex-induced vibration is quite serious, leading to large vibration of the whole power plant and the great danger to the daily safe operations of the unit. In this section, we take a recent measurement performed in a typical large-scale energy storage power plant to illustrate the characteristics of the vortex-induced phenomenon.

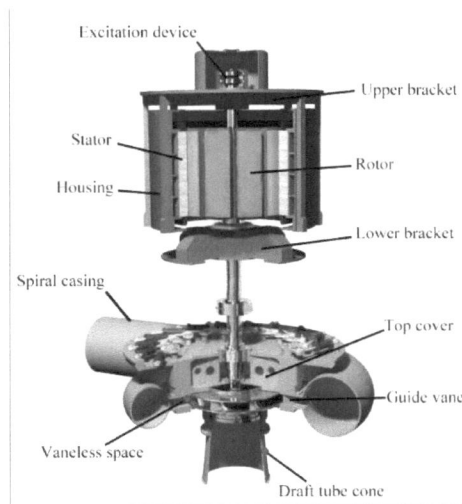

Fig. (1). A schematic view of the components of the whole pumped hydro energy storage power plant. This figure was adapted from the figure 1 of Zhang *et al.* [10] with the permissions from the copyright owner Elsevier.

Fig. (**2**) shows the effects of load variations on the X/Y/Z directions of the vibrations of the top cover. One can find that for the low partial load, the vibrational levels of the unit are quite serious. Based on the analysis of the signals, as shown in Fig. (**3**), three regions were proposed by us in our previous work [10] together with their characteristic frequencies and vibrational levels revealed. We found that the primary source of the observed vibrations origins from the pressure fluctuations in the fluid flow of the turbines. For example, in the region 1, the dominant frequency component is the blade passing frequency, which is induced by the vortex and the unstable flow in the vaneless space. And, in the region 2, the dominant frequency component is a low-frequency component, which is induced by the swirling vortex rope in the draft tube cone section. Additionally, for some vibrations, the mechanical factor (*e.g.* imbalance of the impeller) could also show significant contributions. For example, in region 3, the impeller rotational frequency (f_n=1) also plays an important role on the phenomenon.

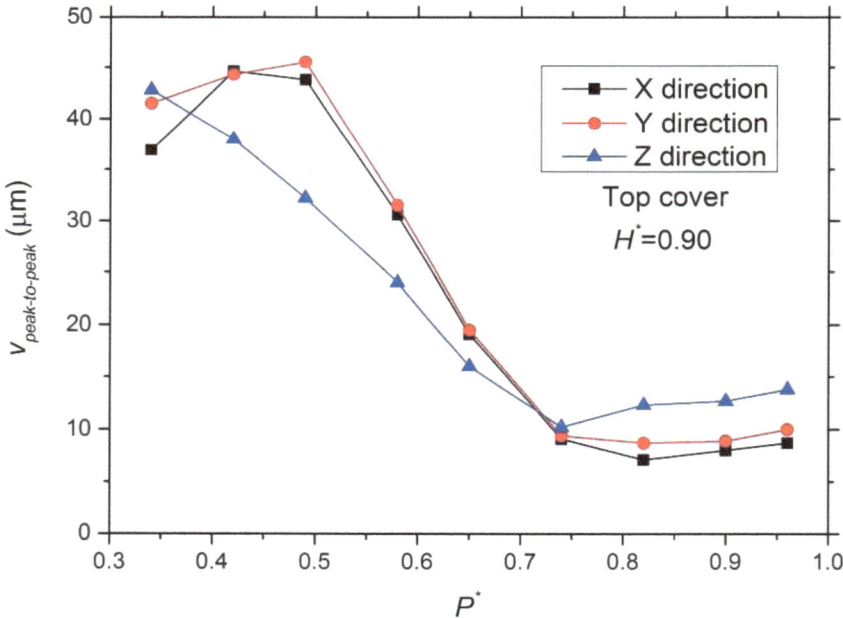

Fig. (2). Effects of load variations on the vibrations of the X/Y/Z directions of the top cover. This figure was adapted from the figure 6 of Zhang *et al.* [10] with the permissions from the copyright owner Elsevier.

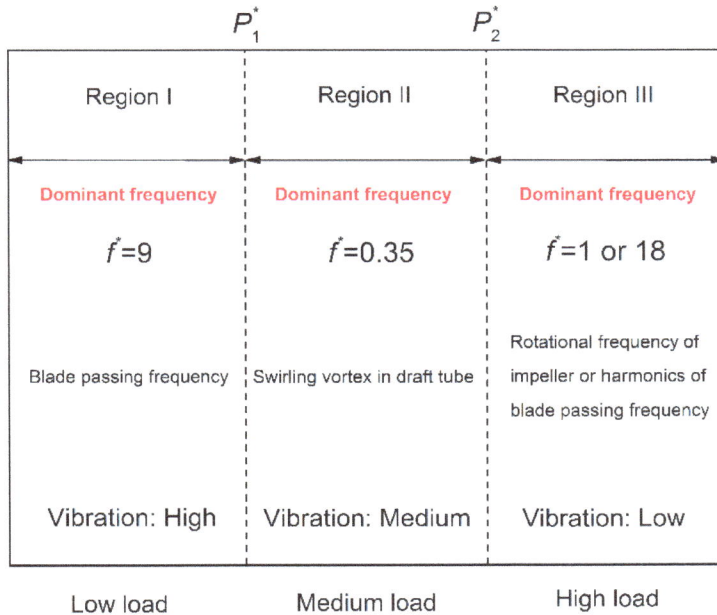

Fig. (3). Three proposed regions together with their characteristic frequencies and vibrational levels. This figure was adapted from the figure 7 of Zhang *et al.* [10] with the permissions from the copyright owner Elsevier.

VORTEX AROUND THE GUIDE PLATE IN THE FRANCIS TURBINE

During the initial design of the many large-scale power stations, the guide plate is usually involved in order to shrink the size of the unit with the electricity generation capacity maintained. Fig. (**1**) of Chen *et al.* [4] shows a demonstration of the effects of the guide plate for the size reduction of the unit. However, such kind of design could possibly lead to the many negative effects on the turbine operations. In this section, the latest findings on this phenomenon are summarized with a focus on the vortex dynamics involved.

Figs. (**7** and **8**) by Chen *et al.* [4] show the vortex generated around the guide plate during the off-design and the design conditions respectively. Different slices refer to the different positions inside the spiral casing of the turbine. For the detailed definitions, readers are referred to Fig. (**6**) of the Chen *et al.* [4].

For the design condition, the generated vortex is quite weak. For the slices shown in Fig. (**8**) of Chen *et al*. [4], the flow is quite smooth in the turbine. And, there are even no obvious vortex around the guide plate in terms of the streamlines.

However, as shown in Fig. (**7**) of Chen *et al*. [4], the vortex is quite obvious for the off-design condition. When the flow passing through the guide plate, two prominent vortices are formed with one above the guide plate and another one below the guide plate. For more detailed analysis of the vortex dynamics, readers are referred to Fig. (**5**) of Chen *et al*. [11]. Such kind of vortex could lead to the unbalanced forces on the guide plate with serious damage as observed finally in the power station.

SWIRLING VORTEX ROPE IN FRANCIS TURBINE

The aforementioned guide plate not only affects the fluid flow in the upstream (*e.g.* the spiral casing) but also could significantly lead to the modification of the vortex rope in the downstream of the turbine. In this section, a detailed analysis is performed with a focus on the vortex rope dynamics.

Firstly, the dominant frequency component will be greatly affected by the addition of the guide plate. Figs. (**10 and 11**) of Chen *et al*. [4] show the time-domain and the frequency-domain spectrums of the simulated pressure fluctuation at several typical positions inside the hydroturbine without and with the guide plate respectively. As shown in the Fig. (**10**) of Chen *et al*. [4], without the guide plate, the primary frequency of the swirling vortex rope is 0.67 Hz as shown in the draft tube monitoring point. With the addition of the guide plate, the dominant frequency in the whole turbine has been modified to be 0.336 Hz (referring to Fig. (**11**) of Chen *et al*. [4]).

Secondly, the vortex patterns of the swirling vortex rope are also strongly affected by the addition of the guide plate. Fig. (**3**) of Chen *et al*. [4] shows the comparisons between the vortex identification using the iso-surface for cases without and with the guide plate respectively. One can find that obvious difference could be found. For more detailed comparisons (*e.g.* in terms of the whole vortex swirling period), readers are referred to Figs. (**12 and 13**) of Chen *et al*. [4].

Swirl Number Analysis

In order to quantify the swirl flow in the draft tube of the Francis turbine, swirl number (S) is usually introduced with the following definitions [11]:

$$S = \frac{\Omega D}{\rho Q_v^{\ 2}} \tag{1}$$

Here, Ω is the angular momentum through a given section of the draft tube; D is the inlet diameter of the draft tube cone section; ρ is the water density; Q_v is the (volumetric) fluid flow rate passing the tested section of the draft tube. Hence, one can find that the swirl number is defined as the ratio between the angular and the axial momentums of the fluid flow. Practically, the calculation of the swirl number of a given flow is usually performed along a series of planes in the fluid flow directions.

In practice, the swirl number could be also calculated based on the computational results of the turbine with the following definitions [11]:

$$S = \frac{\displaystyle\int_0^R (\rho V_a)(rV_t)r\mathrm{d}r}{R\displaystyle\int_0^R (\rho V_a)(rV_a)\mathrm{d}r} \tag{2}$$

Here, r is the radial coordinate defined with the origins at the center of the given section; R is the radius of the tested section in the draft tube; V_a is the axial velocity; V_t is the tangential velocity.

Fig. (**10**) of Chen *et al.* [11] shows the comparisons of the swirl number along the draft tube cone sections for cases without and with the guide plate with different guide vane openings. For the comparisons, six sections along the draft tube were selected for the calculations of the swirl number. As shown in the figure, for the off-design condition (*e.g.* 16 degree), the difference of the swirl number between the two cases are quite remarkable along the whole draft tube. Specifically, the addition of the guide plate leads to higher values of the swirl number, indicating that the ratio of the angular and the axial momentums are also higher. Hence, for the case with the guide plate, more momentum of the fluid flow is transferred from the axial one to the angular one (*e.g.* the swirling pattern). Hence, this explains the previously observed large pressure fluctuations shown in the draft tube section.

VORTEX IN VANELESS SPACE OF REVERSIBLE PUMP TURBINE

The operations of the reversible pump turbine are quite complex with various kinds of operational mode involved. During its daily operations, frequent shifts between different modes are generally involved. For example, as shown in Fig. (**2**) of Zhang *et al.* [5], for the shutdown of the turbine mode, the reversible pump turbine will pass through the turbine mode, runaway, turbine brake mode, zero-flow-rate mode and even reverse pump mode. The vortex during those working modes are quite complex and also very important for the turbine performances.

Fig. (**3**) of Zhang *et al.* [12] shows the vortex identified using the newly proposed Omega method [13, 14] for five typical operational modes of the reversible pump turbine. For more details of the Omega method, readers are referred our recent review [13]. Based on the figure, one can find that the vortex patterns of the reversible pump turbine are strongly related with the given operational conditions of pump turbines. Specifically, strong vortex structure could be observed in the impeller and the vaneless space. Fig. (**4**) of Zhang *et al.* [12] further shows the statistics of the identified vortex iso-surfaces for five typical operational modes of the reversible pump turbine. As shown in the figure, the vortex for the zero-flow-rate mode is the most significant one. And, the vortex for the reverse pump mode is the second largest one. For other modes, the difference is marginal.

CONCLUSIONS

In this chapter, the vortex induced pressure fluctuation and vibrations of hydroturbines is introduced and discussed based on several engineering practical examples. For the Francis turbine, one of the primary interests of the vortex phenomenon is the swirling vortex rope during the draft tube surging. For the reversible pump turbine, the vortex in the vaneless space plays an important role on the turbine performances. A quantitative analysis based on the swirl number is also briefly discussed with an example. Some concluding remarks are:

1. For the reversible pump turbine, during the low-load and the full-load conditions, the vortex induced by the impeller rotation is the dominant one. During the middle-load condition, the low-frequency vortex rope is the dominant one.

2. During the part-load condition, the vortex inside the spiral casing could strongly affect the vortex characteristics inside the draft tube, leading to the modifications of the primary frequencies of the generated pressure signals.

3. Swirl number is an effective non-dimensional parameter for the analysis of the rotating vortex flow, especially those inside the draft tube.

In fact, the cavitation is a very important phenomenon relating with the vortex in hydroturbines through the bubble-silt interactions, the resonances and the bubble-pressure wave interactions. Due to the limited spaces of the present chapter, we do not introduce much details relating with the cavitation research and the readers could refer to the given references for the details.

CONSENT FOR PUBLICATION

Not applicable.

CONFLICT OF INTEREST

The authors confirm that this chapter contents have no conflict of interest.

ACKNOWLEDGEMENT

This work was financially supported by the National Natural Science Foundation of China (Project Nos.: 51976056 and 51606221).

REFERENCES

[1]   Y. Zhang, K. Liu, H. Xian, and X. Du, "A review of methods for vortex identification in hydroturbines", *Renew. Sustain. Energy Rev.,* vol. 81, no. 1, pp. 1269-1285, 2018.
        [http://dx.doi.org/10.1016/j.rser.2017.05.058]
[2]   Y. Wu, S. Li, S. Liu, H.S. Dou, and Z. Qian, *Vibration of hydraulic machinery*, 2013.
        [http://dx.doi.org/10.1007/978-94-007-6422-4]
[3]   P. Dörfler, M. Sick, and A. Coutu, *Flow-induced pulsation and vibration in hydroelectric machinery: engineer's guidebook for planning, design and troubleshooting.* Springer Science & Business Media, 2012.
[4]   T. Chen, Y. Zhang, and S. Li, "Instability of large-scale prototype Francis turbines of Three Gorges power station at part load", *Proc. Inst. Mech. Eng., A J. Power Energy,* vol. 230, no. 7, pp. 619-632, 2016.
        [http://dx.doi.org/10.1177/0957650916661638]
[5]   Y. Zhang, Y. Zhang, and Y.L. Wu, "A review of rotating stall in reversible pump turbine", *Proc. Inst. Mech. Eng., C J. Mech. Eng. Sci.,* vol. 231, no. 5, pp. 1181-1204, 2017.
        [http://dx.doi.org/10.1177/0954406216640579]
[6]   V. Hasmatuchi, M. Farhat, and S. Roth, "Experimental evidence of rotating stall in a pump-turbine at off-design conditions in generating mode", *J. Fluids Eng.,* vol. 133, no. 5, 2011.
        [http://dx.doi.org/10.1115/1.4004088]
[7]   L. Motycak, A. Skotak, and R. Kupcik, "Kaplan turbine tip vortex cavitation–analysis and prevention", *IOP Conference Series: Earth and Environmental Science,* vol. vol. 15, 2012
        [http://dx.doi.org/10.1088/1755-1315/15/3/032060]
[8]   Y. Zhang, T. Chen, J. Li, and J. Yu, "Experimental study of load variations on pressure fluctuations in a prototype reversible pump turbine in generating mode", *J. Fluids Eng.,* vol. 139, no. 7, 2017.
        [http://dx.doi.org/10.1115/1.4036161]
[9]   J. Li, Y. Zhang, and J. Yu, "Experimental investigations of a prototype reversible pump turbine in generating mode with water head variations", *Sci. China Technol. Sci.,* vol. 61, no. 4, pp. 604-611, 2018.
        [http://dx.doi.org/10.1007/s11431-017-9169-7]
[10]   Y. Zhang, X. Zheng, J. Li, and X. Du, "Experimental study on the vibrational performance and its physical origins of a prototype reversible pump turbine in the pumped hydro energy storage power

station", *Renew. Energy,* vol. 130, pp. 667-676, 2019.
[http://dx.doi.org/10.1016/j.renene.2018.06.057]

[11]   T. Chen, X. Zheng, Y.N. Zhang, and S. Li, "Influence of upstream disturbance on the draft-tube flow of Francis turbine under part-load conditions", *J. Hydrodynam.,* vol. 30, no. 1, pp. 131-139, 2018.
[http://dx.doi.org/10.1007/s42241-018-0014-9]

[12]   Y.N. Zhang, K.H. Liu, J.W. Li, H.Z. Xian, and X.Z. Du, "Analysis of the vortices in the inner flow of reversible pump turbine with the new omega vortex identification method", *J. Hydrodynam.,* vol. 30, no. 3, pp. 463-469, 2018.
[http://dx.doi.org/10.1007/s42241-018-0046-1]

[13]   Y.N. Zhang, X. Qiu, F.P. Chen, K.H. Liu, X.R. Dong, and C. Liu, "A selected review of vortex identification methods with applications", *J. Hydrodynam.,* vol. 30, no. 5, pp. 767-779, 2018.
[http://dx.doi.org/10.1007/s42241-018-0112-8]

[14]   C. Liu, Y. Gao, X. Dong, Y. Wang, J. Liu, Y. Zhang, X. Cai, and N. Gui, "Third generation of vortex identification methods-Omega and Liutex/Rortex based systems", *J. Hydrodynam.,* vol. 31, no. 2, pp. 205-223, 2019.
[http://dx.doi.org/10.1007/s42241-019-0022-4]

A Comparative Study of Compressible Turbulent Flows Between Thermally and Calorically Perfect Gases

Xiaoping Chen*

National-Provincial Joint Engineering Laboratory for Fluid Transmission System Technology, Zhejiang Sci-Tech University, Hangzhou, Zhejiang 310018, China

Abstract: In this chapter, direct numerical simulations (DNSs) of compressible turbulent flows for thermally perfect gas (TPG) and calorically perfect gas (CPG), including two wall temperature of 298.15K (low temperature condition) and 596.30K (high temperature condition), are performed to investigate the influence of a gas model on the turbulent statistics and flow structures. The results show that the influence of TPG is negligible and remarkable for low and high-temperature conditions, respectively. Many of the statistical characteristics used to express low-temperature conditions for CPG still can be applied to high-temperature conditions for TPG. The smaller the influence of the gas model on the mean and fluctuating velocity, the stronger the Reynolds analogy. The static temperature for TPG is smaller than that for CPG, whereas an inverse trend is found for turbulent and root square mean Mach numbers. Omega could capture both strong and weak vortices simultaneously for compressible flow, even TPG, which is difficult from Q. Compared to the results of CPG, the vortex structure becomes smaller, sharper and more chaotic considering TPG.

Keywords: Calorically perfect gas, Compressible flow, Direct numerical simulation, Thermally perfect gas, Vortex structure.

INTRODUCTION

Compressible flow has critical importance in gas dynamics and engineering application. The heat flux and frictional resistance in turbulent flow are obviously higher than that for laminar flow. For compressible turbulent flows, the near wall

---

*Corresponding author Xiaoping Chen: National-Provincial Joint Engineering Laboratory for Fluid Transmission System Technology, Zhejiang Sci-Tech University, Hangzhou, Zhejiang, 310018, China; Tel+86 18626878196; E-mail: chenxp@zstu.edu.cn

temperature may be higher than 500K. Here, it will be changed the thermodynamic environment, which means that the calorically perfect gas (CPG) no longer appropriate and the thermally perfect gas (TPG) [1] should be considered. Therefore, assumed to be TPG, the behavior of turbulent statistical characteristics and flow structures in compressible turbulent flow urgently needs to be investigated.

Direct numerical simulation (DNS) does not involve any modeling errors and solves the Navier-Stokes equations directly [2], which is a powerful tool to simulate the turbulent flows, including channel flows [3-10], boundary layers [11-15], compression ramps [16, 17], and blunt cones [18-20]. A large number of DNS study is reported to investigate the statistical characteristics in the compressible turbulent flows, such as strong Reynolds analogy (SRA) and Morkovin's hypothesis. The Morkovin hypothesis [21] denotes that, when the Mach number isn't very large, the relationship of turbulent statistical characteristics between incompressible and compressible flows can be connected by mean variations of fluid properties. Huang *et al.* [5] investigated compressibility effects based on the DNS data of compressible turbulent channel flow performed by Coleman *et al.* [4]. Lechner *et al.* [6] and Foysi *et al.* [7] studied compressible effects and turbulence scaling in the compressible turbulent channel flow. The differences in turbulence statistics near both the adiabatic and isothermal walls were reported by Mamano et al [8] and Morinishi et al [9]. In addition, to some extent, the energy equation can be explained by SRA-a relationship between velocity and temperature fluctuations. Morkovin [21] firstly proposed SRA in 1963. Then, researches proposed several modified SRA, such as ESRA was introduced by Cebeci *et al.* [22], GSRA was introduced by Gaviglio [23], RSRA was introduced by Rubesin [24], HSRA was introduced by Huang *et al.* [5] and GHSRA was introduced by Duan *et al.* [25]. So far, many studies of supersonic turbulent boundary layer flows have been performed to check the validity of SRA and Morkovin's hypothesis by the DNS data. For example, Duan and Martin [13, 14] assessed the influence of Mach number and wall temperatures on the SRA and Morkovin's hypothesis. Liang and Li [15] investigated many turbulent characteristics, such as Walz equation, mean and fluctuating velocity, compressibility effect and SRA

 For the compressible turbulent flow, the behaviour of instantaneous vortex structures is also very important. Many criterions for identify turbulent structures have been introduced in many literatures, such as $\grave{\Delta}$-criterion [26-27], ϱ-criterion [28], λ_2-criterion [29], λ_{ci}-criterion [30], the Ω criterion [31, 32], and Rortex [33]. Based on the DNS results of boundary layer, these criterions were evaluated by Sayadi *et al.* [34] and Pierce *et al.* [35]. They found that these criterion produce the same images as chosen the threshold of $10^{-3}\left(\partial u/\partial y\right)^{2}\big|_{w}$. Coleman *et al.* [4] illustrated that near-wall streak in the stream-wise direction become more coherent as the Mach number increased. They also argued that the weakly compressible hypothesis modifies the near-wall structures little. Morinishi *et al.* [9] showed that the thermal wall boundary condition has very little effect on the near-wall streaks in semi-local units. For the supersonic boundary layers, Lagha *et al.* [36, 37] studied the influence of Mach number on the near-wall structure. Guo and Adams [38] illustrated that near-wall streak structures are larger than that of incompressible flow. Most previous studies have been carried out on calorically perfect gas (CPG), a good understanding is gained due to these works.

So far, many DNS results for TPG were performed based on DNS. Marxen *et al.* [39, 40] studied the effects of gas model on the stability of hypersonic boundary layer. In the hypersonic boundary layer, Jia and Cao [41] investigated the behavior of stability of flat plate under different variable specific heat. Recently, taking temporally evolving compressible turbulent flows as a research object, Chen *et al.* [42-46] not only performed several DNS, but also investigated the similarities and differences between TPG and CPG. However, the turbulent statistics and flow structures in compressible turbulent flows for TPG have not been studied clearly, especially for vortex structures, SRA and Morkovin's hypothesis.

In the present study, based on the DNS database, we focus on the behavior of turbulent statistical characteristics and flow structures in the compressible turbulent channel flow for TPG, and discuss how they depend on the wall temperature.

GOVERNING EQUATIONS

The governing equations are the time-dependent three-dimensional Navier-Stokes equations in non-dimensional form, which can be described as follows:

$$\frac{\partial \rho}{\partial t}+\frac{\partial}{\partial x_{j}}\left(\rho u_{j}\right)=0 \tag{1}$$

$$\frac{\partial\left(\rho u_{i}\right)}{\partial t}+\frac{\partial}{\partial x_{j}}\left(\rho u_{i}u_{j}+p\delta_{ij}-\frac{1}{\mathrm{Re}}\sigma_{ij}\right)=\rho f_{i} \tag{2}$$

$$\frac{\partial E}{\partial t}+\frac{\partial}{\partial x_{j}}\left[\left(E+p\right)u_{j}-\frac{1}{\mathrm{Re}}\left(u_{i}\sigma_{ij}+q_{j}\right)\right]=\rho f_{i}u_{i} \tag{3}$$

where ρ is density, u_j is velocity vector, T is temperature.   Re is Reynolds number, and f_i is body force vector. The pressure p, shear stress tensor σ_{ij}, total energy E and conductive heat flux q_j are given by

$$p = \frac{\rho T}{\gamma Ma^2} \tag{4}$$

$$E = \rho \left(C_V T + \frac{1}{2} u_i u_i \right) \tag{5}$$

$$\sigma_{ij} = \mu \left(\frac{\partial u_i}{\partial x_j} + \frac{\partial u_j}{\partial x_i} \right) - \frac{2}{3} \mu \frac{\partial u_k}{\partial x_k} \delta_{ij} \tag{6}$$

$$q_j = \frac{C_p \mu}{\mathrm{Pr}} \frac{\partial T}{\partial x_j} \tag{7}$$

where δ_{ij} is Kronecker tensor. Pr is Prandtl number, and *Ma* is Mach number. The viscosity μ is calculated by Sutherland's law.

$$\mu = T^{3/2} \frac{1 + 110.4/T_w^{**}}{T + 110.4/T_w^{**}} \tag{8}$$

The specific heat at constant volume C_V, specific heat at constant pressure C_p, and specific heat ratio γ are computed as follows

$$C_p = C_V + R, \; \gamma = C_p / C_V \tag{9}$$

where R is the gas constant.

Thermally Perfect Gas (TPG)

For the TPG, C_V is consisted of three parts [1]

$$C_V = C_{V,tr} + C_{V,r} + C_{V,v} \tag{10}$$

$$C_{V,tr} = \frac{3}{2} \frac{T_w^{**}}{u_b^{**2}} R^{**} \tag{11}$$

$$C_{V,r} = \frac{T_w^{**}}{u_b^{**2}} R^{**} \tag{12}$$

$$C_{V,v} = \frac{\exp\left(\theta^{**}/T^{**}\right)}{\left[\exp\left(\theta^{**}/T^{**}\right)-1\right]^2} \left(\frac{\theta^{**}}{T^{**}}\right)^2 \frac{T_w^{**}}{u_b^{**2}} \tag{13}$$

where $C_{V,tr}$, $C_{V,r}$ and $C_{V,v}$ are translational, rotational and vibrational specific heat at constant volume, respectively. The characteristic vibrational temperature θ equals to 3371K and 2256K for N_2 and O_2, respectively. The subscript "b" denotes bulk condition and subscript "w" denotes wall condition. The dimensional flow variable is denoted by the corresponding variable with superscript "**".

Calorically Perfect Gas (CPG)

Compared with TPG, the difference of specific heat for CPG concentrates on its vibrational part, whose value is equal to zero. Therefore, the equations (10-13) can be used here. Moreover, the specific heat is constant, so they can be computed simply as follows

$$C_V = \frac{1}{\gamma(\gamma-1)Ma^2}, \; C_p = \frac{1}{(\gamma-1)Ma^2}, \; \gamma=1.4 \tag{14}$$

DESCRIPTION OF DNS

The code developed by Chen *et al.* [42-46] and Li *et al.* [10, 15, 17-20, 47-48] is used to generate the DNS data for TPG and CPG, respectively. The description of DNS can be found in Ref. [42-46] in detail. For example, the 7th WENO scheme is used for convection terms [49]. An 8th central difference is used for the viscous terms. The 3rd Runge-Kutta method is used for the time integration.

Table **1** gives the flow and computational parameters for the two gas models. Two constant dimensional wall temperature is 298.15K (low-temperature condition) to 596.30K (high-temperature condition), which take account of negligible and remarkable vibrational energy, respectively. The same Mach number Ma, Reynolds number Re, Prandtl number Pr, wall temperature T_w and viscosity μ are used. For example, the Mach number ($Ma = u_b/c_w$) is based on the speed of sound at the isothermal wall and the bulk velocity. The Reynolds number ($Re = \rho_b u_b H / \mu_w$) is based on viscosity at the isothermal, channel half-width, bulk velocity, and wall and the bulk density. In addition, several existing DNS results for CPG are also performed.

The instantaneous fully developed turbulent flow [43] is selected as the initial condition for all cases. The steady-state is reached after 450 dimensionless times computed, whose time step is 0.0005 dimensionless time. Then, 180 dimensionless times (400 statistical samples) are continuously average. Table **2** gives grid parameters, grid spacing and computational domain. Note that, this study shares the same computational domain size ($L_x \times L_y \times L_z = 4\pi H \times 2H \times 4\pi H/3$) and the number of grid points ($n_x \times n_y \times n_z = 571 \times 261 \times 251$), whose subscripts "*x*", "*y*" and "*z*" denote stream-wise, wall-normal and span-wise directions, respectively. In addition, Table **3** gives several time-averaged parameters for the two gas models. The friction Mach number, $Ma_\tau = u_\tau / c_w$, is based on friction velocity and the wall sound speed ($u_\tau = \sqrt{\tau_w / \rho_w}$, $\tau_w = \sqrt{\mu_w (\partial u / \partial y)|_w}$). The friction Reynolds number, $Re_\tau = \rho_w u_\tau H / \mu_w$, is based on the wall friction velocity and the channel half-width. Bradshaw [50] introduced a non-dimensional "inner layer" parameter ($B_q = q_w / \rho_w C_p u_\tau$) to reveal the heat flux.

Table 1. Flow and computational parameters for the two gas models.

Cases	T_w^* (K)	Ma	Re	Pr	T_w	γ	Gas Models
TL	298.15	3.0	4880	0.7	1.0	------	Thermally perfect gas
TH	596.30	3.0	4880	0.7	1.0	------	Thermally perfect gas
CL	298.15	3.0	4880	0.7	1.0	1.4	Calorically perfect gas
CH	596.30	3.0	4880	0.7	1.0	1.4	Calorically perfect gas
Coleman *et al.* [4]	------	3.0	4880	0.7	1.0	1.4	Calorically perfect gas
Foysi *et al.* [7]	500.00	3.0	6000	0.7	1.0	1.4	Calorically perfect gas

Table 2. Grid resolution and domain size for the two gas models.

Cases	L_x/H	L_y/H	L_z/H	n_x	n_y	n_z	Δx^+	Δy_w^+	Δy_{max}^+	Δz^+
TL	4π	2	$4\pi/3$	571	261	251	9.248	0.222	8.191	7.014
TH	4π	2	$4\pi/3$	571	261	251	9.427	0.233	8.349	7.148
CL	4π	2	$4\pi/3$	571	261	251	10.196	0.245	9.028	7.731
CH	4π	2	$4\pi/3$	571	261	251	10.132	0.224	8.898	7.683
Coleman *et al.* [4]	4π	2	$4\pi/3$	144	119	80	39	0.2	------	24
Foysi *et al.* [7]	4π	2	$4\pi/3$	512	221	256	13.65	0.89	9.38	8.91

Table 3. Time-averaged parameters for the two gas models.

Cases	Ma_τ	Re_τ	$-B_q$	$\langle \rho_w \rangle$	$\langle \rho_c \rangle$	$\langle T_c \rangle$	$\langle \mu_c \rangle$	$\langle \gamma_w \rangle$	$\langle \gamma_c \rangle$
TL	0.127	420.24	0.1193	2.039	0.957	2.069	1.503	1.325	1.296
TH	0.118	443.59	0.1344	2.317	0.943	2.380	1.838	1.399	1.366
CL	0.115	463.28	0.1402	2.483	0.951	2.593	1.699	1.400	1.400
CH	0.114	460.38	0.1395	2.481	0.948	2.597	1.942	1.400	1.400
Coleman *et al.* [4]	0.116	451	0.137	2.388	0.952	2.490	1.894	1.400	1.400

DNS RESULTS AND DISCUSSION

Turbulent Statistics

To represent the degree of vibrational energy excitation, Fan [51] introduced a temperature-dependent parameter, named vibrational energy excited degree.

$$\phi_{C_V} = \frac{\langle C_{V,v} \rangle}{\langle C_{V,tr} \rangle + \langle C_{V,r} \rangle} \qquad (16)$$

Fig. (**1a**) shows the vibrational energy excited degree ϕ_{C_V} for TPG. Note that, ϕ_{C_V} significantly increases with the increase in wall temperature. The peak value of

ϕ_{C_V} is 0.24 for TH, which indicates that the vibrational energy is important, whose flow flied is significantly different from that for CH. ϕ_{C_V} is smaller than 0.1 among the channel for TL, whose flow flied is closed to that for CL. ϕ_{C_V} is dependent on the temperature, which is obviously affected by the gas models. For example, the mean temperature for TRG is obviously smaller than that for CRG, as shown in Fig. (**1b**). Moreover, the mean temperature decreases with the increase of wall temperature.

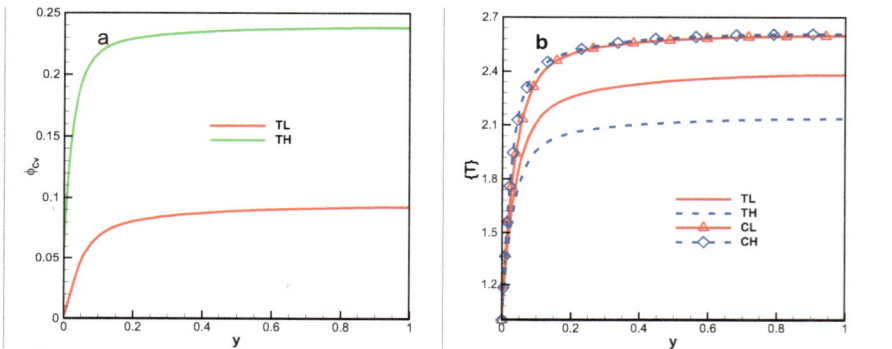

Fig. (1). Distributions of (**a**) vibrational energy excited degree and (**b**) mean temperature.

The Van Driest transformed velocity $\langle u \rangle_{vd}^{+}$ is an important parameter to indicate the turbulent statistical characteristics.

$$\langle u \rangle_{vd}^{+} = \int_{0}^{\langle u \rangle^{+}} \sqrt{\frac{\langle \rho \rangle}{\langle \rho_w \rangle}} d\langle u \rangle^{+} \,, \quad \langle u \rangle^{+} = \frac{\langle u \rangle}{u_{\tau}} \tag{17}$$

Fig. (**2**) shows that the mean velocity versus wall scaling y^{+}, in the log-law layer, collapses the profiles for different gas model and wall temperature conditions.

$$y^{+} = y \frac{\rho_w u_{\tau}}{\mu_w} \tag{18}$$

Moreover, these profiles also show good agreement with incompressible log law.

$$\langle u \rangle_{vd}^{+} = \frac{1}{\kappa} \ln y^{+} + C \tag{19}$$

where κ is Kármán constant and equals to 0.41. C is the logarithmic coefficient, whose value is larger than that of incompressible flow [49].

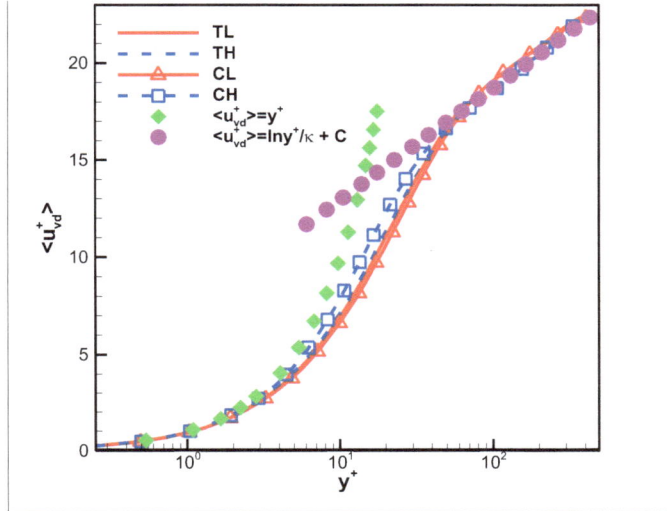

Fig. (2). Distributions of van Driest transformed velocity.

The turbulent intensity could be often quantified by the fluctuating velocity, such as root mean square (RMS) velocity fluctuation, whose definition is $f''_{rms} = \left\langle f''f'' \right\rangle^{1/2}$ (f can be the velocity in stream-wise, wall-normal and span-wise directions). When the RMS velocity fluctuation is normalized by conventional wall variables, these profiles are distinct in the different cases, as shown in Fig. (**3a**). However, Fig. (**3a**) shows that a much better collapse of the data is achieved as used semi-local scaling. The semi-local scaling was proposed by Huang *et al.* [5], and its effectiveness had been demonstrated by several researches [13-15]. y^* and u_τ^* are defined as:

$$y^* = y\frac{\langle\rho\rangle u_\tau^*}{\langle\mu\rangle}, \ u_\tau^* = \sqrt{\frac{\tau_w}{\langle\rho\rangle}} \tag{21}$$

Figs. (**4a** and **4b**) show that the slope of y^* to y^+ ratio increases with the increase in wall temperature, and the profiles of u_τ^* are also different.

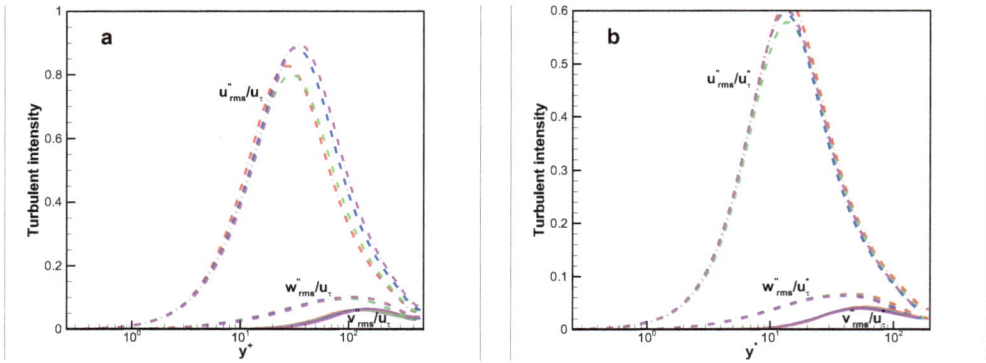

Fig. (3). Distributions of RMS velocity fluctuations (*a*) *versus* y^+ and normalized by u_τ and (*b*) *versus* y^* and normalized by u_τ^*. Red curves denote TL, green curves denote TH, blue curves denote CL, and pink curves denote CH.

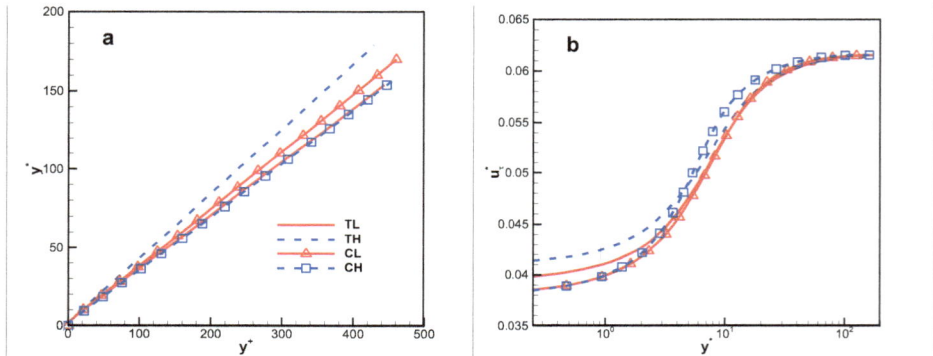

Fig. (4). Variations of (*a*) y^* *versus* y^+ and (*b*) u_τ^* *versus* y^*.

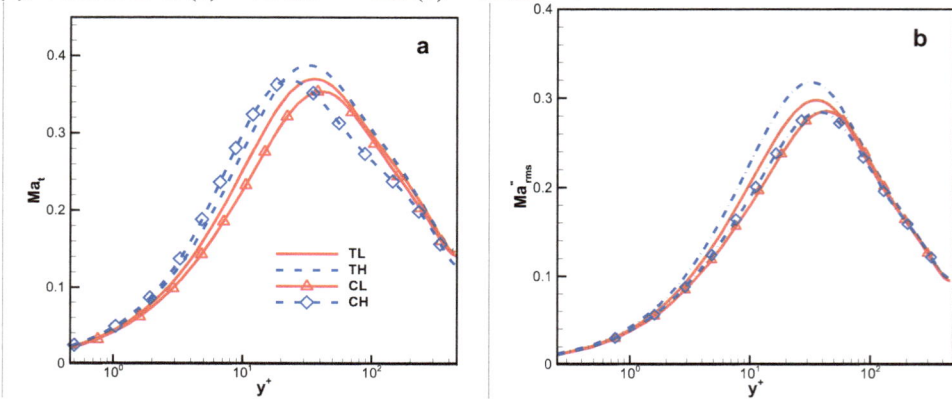

Fig. (5). Distributions of (**a**) turbulent Mach number and (**b**) fluctuating Mach number.

An indicator for compressibility effects is the turbulent Mach number ($Ma_t = \sqrt{u_j''^2}/\{c\}$). Smits and Dussauge [52] believed that when the value of Ma_t is larger than 0.3, compressibility effects become important. Fig. (**5a**) shows some excited regions where Ma_t is larger than 0.3 as observed in all cases. The different mean sound speed results that the magnitude of Ma_t for TPG is larger than that for CPG. Although the RMS Mach number fluctuation Ma'_{rms} exhibits a similar trend as Ma_t, the magnitude of Ma'_{rms} is smaller than Ma_t, as shown in Figs. (**5a and b**).

Strong Reynolds Analogy

Four of SRA relations proposed by Morkovin [21] are given below

$$\frac{T'_{rms}/\{T\}}{(\gamma-1)Ma^2\left(u'_{rms}/\{u\}\right)} \approx 1 \tag{20}$$

$$R_{u''T''} = \frac{\langle u''T''\rangle}{u''_{rms}T''_{rms}} \approx 1 \tag{21}$$

$$R_{u''v''} \approx -R_{v''T''}\left[1-\frac{\langle v''T'_T\rangle}{\langle v''T''\rangle}\right] \tag{22}$$

$$\Pr_t = \frac{\langle \rho u''v''\rangle\left(\partial\{T\}/\partial y\right)}{\langle \rho v''T''\rangle\left(\partial\{u\}/\partial y\right)} \approx 1 \tag{23}$$

where $\Pr_t$ is called the turbulent Prandtl number and T_T is the total temperature. Fig. (**6a**) shows that the SRA, as expressed in Eq. (20), is significantly deviated from 1.0. Morinishi *et al.* [9] explained that the condition $\langle T'^2\rangle/\{T\}^2 \ll \left(\langle T_T'^2\rangle - 2\langle T'T'_T\rangle/\{T\}^2\right)$ is not satisfied. Based on Eq. (20), extend SRA (ESRA) is derived by Cebeci and Smith [21]

$$\frac{T'_{rms}/\{T\}}{(\gamma-1)Ma^2\left(u'_{rms}/\{u\}\right)} \approx \left[1+C_p\frac{\{T_w\}-\{T_{T,c}\}}{\{u\}\{u_c\}}\right] \tag{24}$$

However, Fig. (**6b**) shows that ESRA is not adequate. Several modified SRAs have been proposed, such as GSRA (Gaviglio [23]), RSRA (Rubesin [24]) and HSRA (Huang *et al.* [5]), which correspond to $c = 1.0$, $c = 1.34$ and $c = \mathrm{Pr}_t$, respectively.

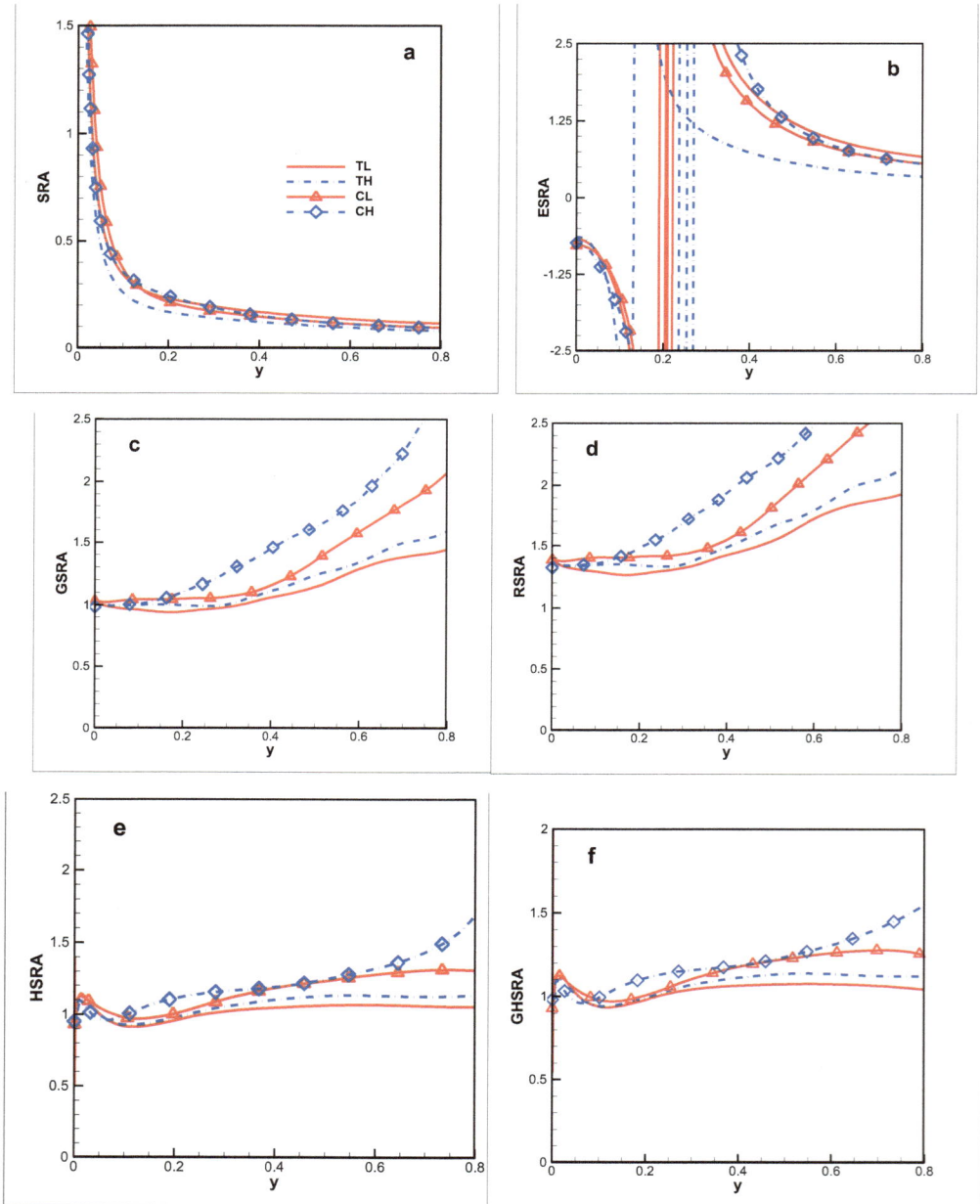

Fig. (6). Distributions of Distributions of (**a**) SRA, (**b**) ESRA, (**c**) GSRA, (**d**) RSRA, (**e**) HSRA and (**f**) GHSRA.

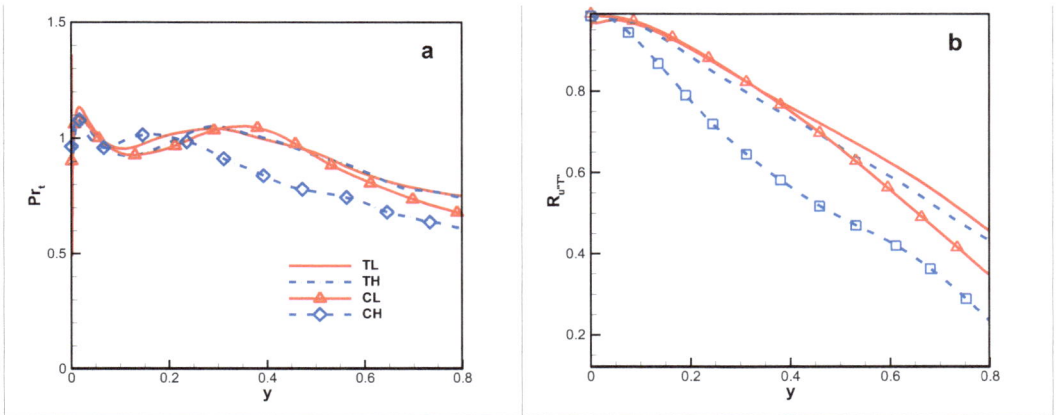

Fig. (7). Distributions of (**a**) turbulent Prandtl number and (**b**) velocity-temperature correlation.

$$\frac{T''_{rms}/\{T\}}{(\gamma-1)Ma^2\left(u''_{rms}/\{u\}\right)} \approx \frac{1}{c\left[1-\partial\{T_T\}/\partial\{T\}\right]} \tag{25}$$

More recently, by removing the assumption of CPG, Duan *et al.* [25] extended an SRA form (GHSRA)

$$T''_{rms} = -\frac{1}{\mathrm{Pr}_t}\frac{\partial\{T\}}{\partial\{u\}}u''_{rms} \tag{26}$$

The distributions of the HSRA Fig. (**6e**) and GHSRA Fig. (**6f**) are closed and have better agreement than that of GSRA Fig. (**6c**) and RSRA Fig. (**6d**). Note that the SRA and modified SRA have little relationship with the gas model and wall temperature. Similar results can be found in Pr_t, as shown in Fig. (**7a**).

The cross-correlation coefficient $R_{u''T'}$ is expressed in Eq. (21) and shown in Fig. (7b). The positive value of $R_{u''T'}$ indicates a positive correlation between u'' and T'. $R_{u''T'}$ is near to unity close to the wall, and its value quickly degrades away from the wall.

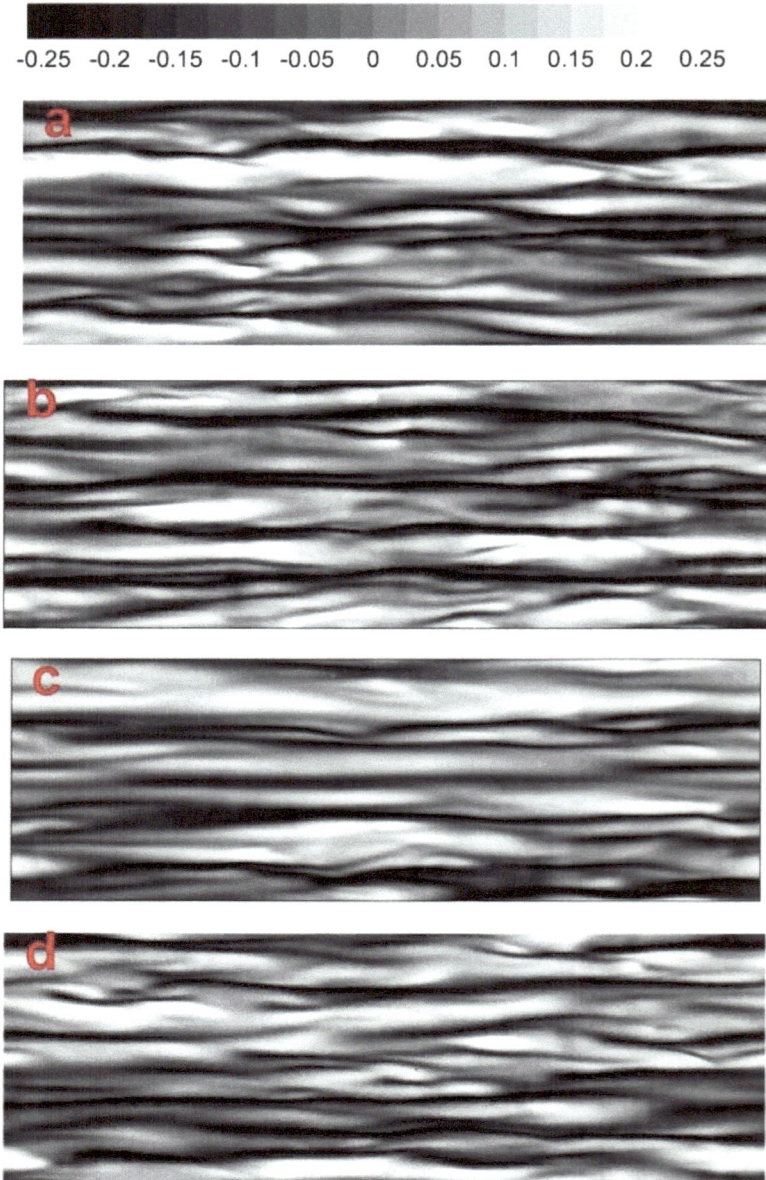

Fig. (8). Instantaneous fluctuating stream-wise velocity for (**a**) TL, (**b**) TH, (**c**) CL and (**d**) CH to visualize near–wall streaks on (x, z) –planes near the wall, $1 - |y| = 0.04$.

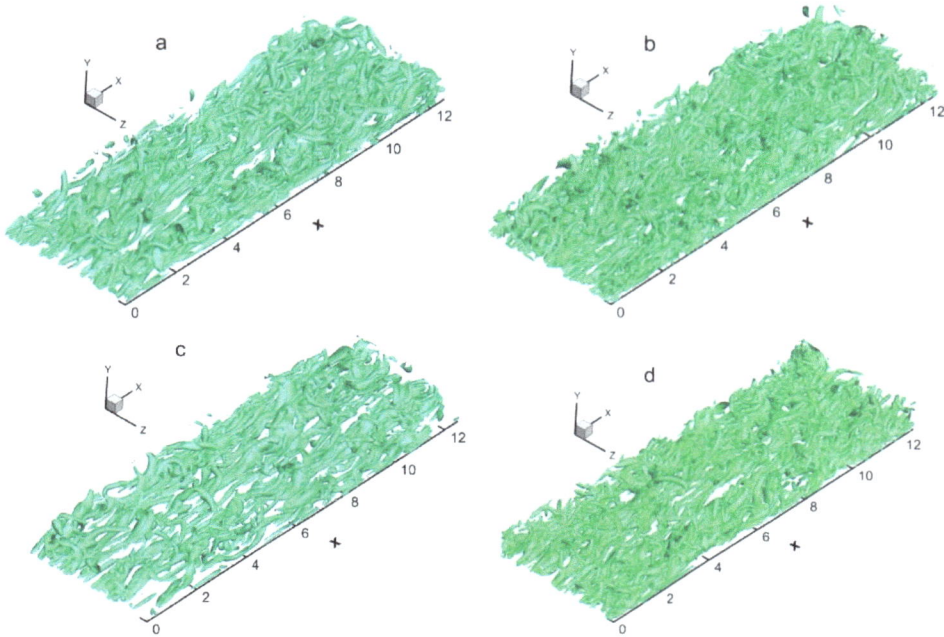

Fig. (9). Iso-surfaces of Q=0.5 for (**a**) TL, (**b**) TH, (**c**) CL and (**d**) CH.

Flow Structures

Fig. (**8**) shows the instantaneous fluctuating stream-wise velocity near the wall. The near-wall streaks in the span-wise direction for all cases have been observed. The average span-wise spacing is much larger than the traditional value for incompressible flow and weak compressible flow. For compressible channel flow, its spacing is approximately 100 semi-local wall scaling as suggested by Morinishi *et al.* [9]. This finding is also valid in this study.

A common vortex identification method, Q criterion is used, to investigate the flow structures in more detail.

$$Q = \frac{1}{2}\left[\left(\frac{\partial u_i}{\partial x_i}\right)^2 - \frac{\partial u_i}{\partial x_j}\frac{\partial u_j}{\partial x_i}\right] \qquad (27)$$

The instantaneous iso-surfaces of Q are shown in Fig. (**9**). Note that the vortex structures in the stream-wise direction are few changed. Much chaotic, random, small-scale, and no obvious direction coherent structures are filled in the near-wall region. However, hairpin vortex structures can be found away from the wall.

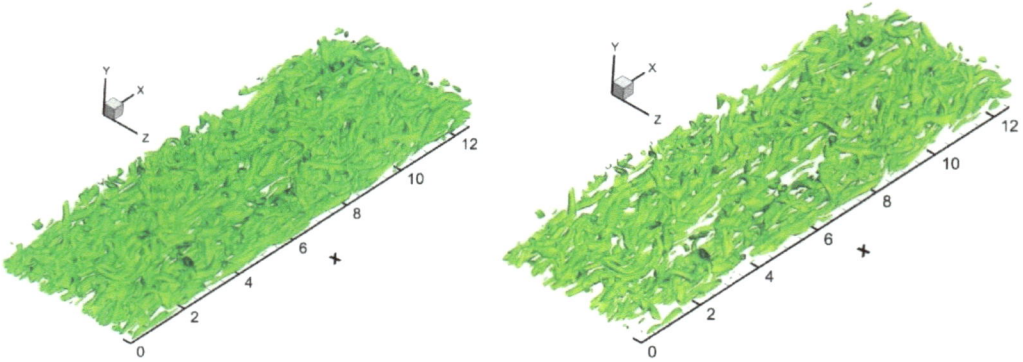

Fig. (10). Iso-surfaces of (left) Ω= 0.52 and (right) Ω= 0.6 for TL.

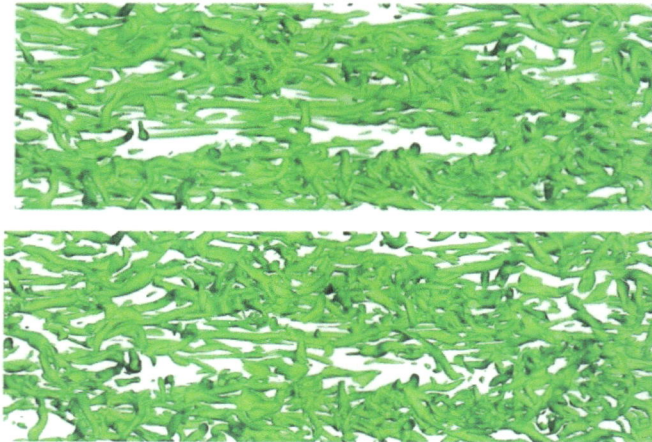

Fig. (11). Top view of iso-surfaces of (up) Q=0.5 and (down) Ω=0.6 for TL.

Recently, Liu *et al.* [31, 32] proposed Ω criterion, which is defined as a ratio of vorticity square over the sum of vorticity square and deformation square:

$$\Omega = \frac{b}{a+b+\varepsilon} \tag{29}$$

$$a = \frac{1}{2}\left|\frac{\partial u_i}{\partial x_j} + \frac{\partial u_j}{\partial x_i}\right|, \ b = \frac{1}{2}\left|\frac{\partial u_i}{\partial x_j} - \frac{\partial u_j}{\partial x_i}\right| \tag{30}$$

where ε is introduced in the denominator of Ω for removing noise case by case.

$$\varepsilon = 0.001(b-a)_{\max} \tag{31}$$

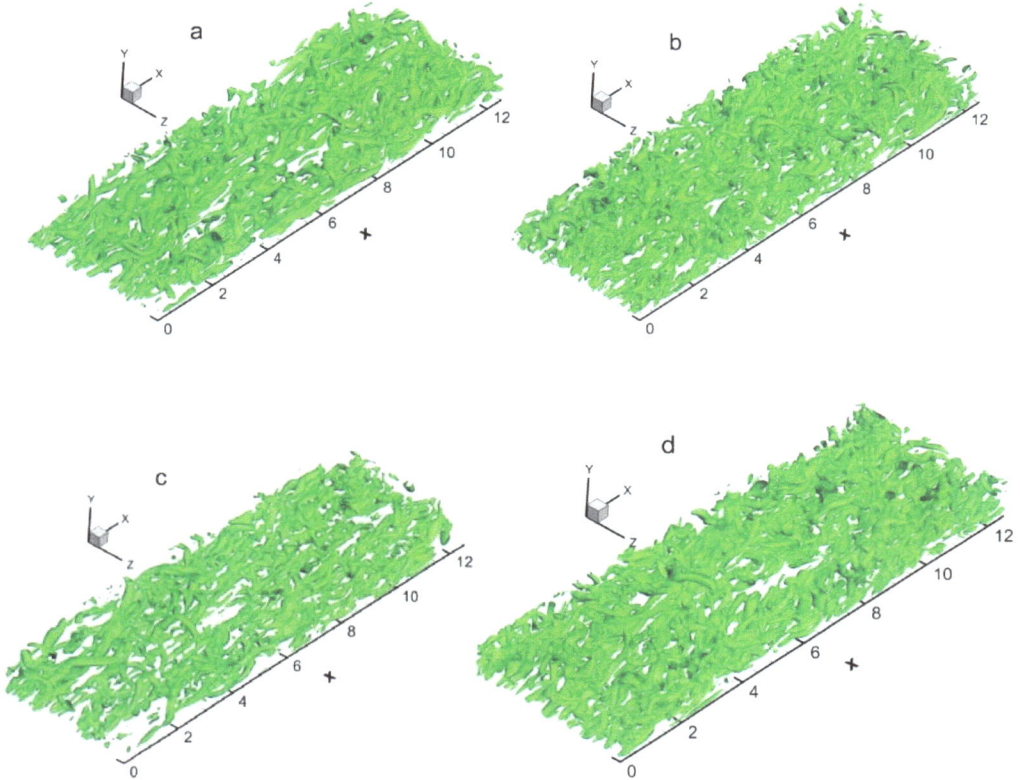

Fig. (12). Iso-surfaces of $\Omega=0.6$ for (*a*) TL, (*b*) TH, (*c*) CL and (*d*) CH.

Fig. (**10**) shows the instantaneous iso-surfaces of Ω for TL. Note that both strong and weak vortices are well captured in both conditions. Moreover, the effects of coupling conditions on the vortex structures are similar to Q criterion, as shown in Fig. (**11**). In addition, the instantaneous iso-surfaces of Ω for different cases are shown in Fig. (**12**). Note that, compared with calorically perfect gas, the large–scale hairpin vortices are smaller, and the coherent vortices are arranged in a span-wised manner. Moreover, the vertical structures become sharper and more chaotic.

CONCLUSION AND OUTLOOK

DNSs of temporally evolving compressible turbulent channel flows for TPG and CPG, share the same Mach number of 3.0 and Reynolds number of 4880, are performed to investigate the influence of gas model on the statistical characteristics and flow structures. Two wall temperatures, whose values are 298.15K (low-temperature condition) and 596.30K (high-temperature condition), respectively, are considered. The reliability of DNS results for CPG is validated by the results in previous studies. The vibrational energy for TPG is negligible and remarkable for low and high-temperature conditions, respectively.

Many characteristics, derived for low-temperature condition, still can be applied to high-temperature condition, even TPG. The statistical velocity is relatively insensitive to the gas model, such as mean stream-wise velocity and RMS velocity fluctuation. The mean temperature for TPG is lower than that for CPG, whereas an inversed trend is found for turbulent and RMS Mach numbers.

In the terms of SRA, although the original SRA relation does not work for both the gas models, reasonably good results are provided by modifying SRA, particularly for the versions of Huang *et al.* and Zhang *et al.* A perfect anti-correlation is observed between the stream-wise velocity and temperature fluctuations. SRA has little relationship with the gas model, including the turbulent Prandtl number.

In terms of flow structure, streaks occur in the two gas models. Computations with considering TPG can increase the stream-wise coherency of near-wall streaks, which is much larger than that with CPG. Moreover, the stream-wise vortices structure for TPG is longer than that for CPG. In addition, the results show that Ω could capture both strong and weak vortices simultaneously for supersonic flow, even under thermally perfect gas, which is difficult to obtain by Q. The vortex structure becomes smaller, sharper and more chaotic with increased wall temperature.

CONSENT FOR PUBLICATION

Not applicable.

CONFLICT OF INTEREST

The authors confirm that this chapter contents have no conflict of interest.

ACKNOWLEDGEMENT

This work was supported by the National Natural Science Foundation of China (Grant No. 51976198).

REFERENCES

[1]   J.D. Anderson, *Hypersonic and High-Temperature Gas Dynamic.* McGraw–Hill: New York, 2006. [http://dx.doi.org/10.2514/4.861956]

[2]   S.B. Pope, *Turbulent Flows.* Cambridge University Press: Cambridge, 2000. [http://dx.doi.org/10.1017/CBO9780511840531]

[3]   G. N. Coleman, "Direct simulation of compressible wall-bounded turbulence", *Center for Turbulence Research, Annual Research Briefs,* pp. 139-144, 1992.

[4]   G.N. Coleman, J. Kim, and R.D. Moser, "A numerical study of turbulent supersonic isothermal–wall channel flow", *J. Fluid Mech.,* vol. 305, pp. 159-183, 1995. [http://dx.doi.org/10.1017/S0022112095004587]

[5]   P.G. Huang, G.N. Coleman, and P. Bradshaw, "Compressible turbulent channel flows: DNS results and modeling", *J. Fluid Mech.,* vol. 305, pp. 185-218, 1995. [http://dx.doi.org/10.1017/S0022112095004599]

[6]   R. Lechner, J. Sesterhenn, and R. Friedrich, "Turbulent supersonic channel flow", *J. Turbul.,* vol. 2, pp. 1-25, 2001. [http://dx.doi.org/10.1088/1468-5248/2/1/001]

[7]   H. Foysi, S. Sarkar, and R. Friedrich, "Compressibility effects and turbulence scaling in supersonic channel flow", *J. Fluid Mech.,* vol. 509, pp. 207-216, 2004. [http://dx.doi.org/10.1017/S0022112004009371]

[8]   S. Tamano, and Y. Morinishi, "Effect of different thermal wall boundary conditions on compressible turbulent channel flow at M =1.5", *J. Fluid Mech.,* vol. 548, pp. 361-373, 2006. [http://dx.doi.org/10.1017/S0022112005007639]

[9]   Y. Morinishi, S. Tamano, and K. Nakabayashi, "Direct numerical simulation of compressible turbulent channel flow between adiabatic and isothermal walls", *J. Fluid Mech.,* vol. 502, pp. 273-308, 2004. [http://dx.doi.org/10.1017/S0022112003007705]

[10]   X.L. Li, D.X. Fu, and Y.W. Ma, "DNS and scaling law analysis of compressible turbulent channel flow", *Sci. China A,* vol. 44, no. 5, pp. 645-654, 2001. [http://dx.doi.org/10.1007/BF02876712]

[11]   S. Pirozzoli, F. Grasso, and T.B. Gatski, "Direct numerical simulation and analysis of a spatially evolving supersonic turbulent boundary layer at M=2.25", *Phys. Fluids,* vol. 16, pp. 530-545, 2004. [http://dx.doi.org/10.1063/1.1637604]

[12]   T. Maeder, N.A. Adams, and L. Kleiser, "Direct simulation of turbulent supersonic boundary layers by an extended temporal approach", *J. Fluid Mech.,* vol. 429, pp. 187-216, 2001. [http://dx.doi.org/10.1017/S0022112000002718]

[13]   L. Duan, and M.P. Martín, "Direct numerical simulation of hypersonic turbulent boundary layers. Part 2: Effect of wall temperature", *J. Fluid Mech.,* vol. 655, pp. 419-445, 2010. [http://dx.doi.org/10.1017/S0022112010000959]

[14]   L. Duan, and M.P. Martín, "Direct numerical simulation of hypersonic turbulent boundary layers. Part 3: Effect of Mach number", *J. Fluid Mech.,* vol. 672, pp. 245-267, 2011. [http://dx.doi.org/10.1017/S0022112010005902]

[15]   X. Liang, and X.L. Li, "DNS of a spatially evolving hypersonic turbulent boundary layer at Mach 8", *Sci. China Phys. Mech. Astron.,* vol. 56, pp. 1408-1418, 2013. [http://dx.doi.org/10.1007/s11433-013-5102-9]

[16]   N.A. Adams, "Direct numerical simulation of turbulent compression ramp flow", *Theor. Comput. Fluid Dyn.,* vol. 12, pp. 109-129, 1998. [http://dx.doi.org/10.1007/s001620050102]

[17] X.L. Li, D.X. Fu, Y.W. Ma, and X. Liang, "Direct numerical simulation of shock/turbulent boundary layer interaction in a supersonic compression ramp", *Sci. China Phys. Mech. Astron.,* vol. 53, no. 9, pp. 1651-1658, 2010.
      [http://dx.doi.org/10.1007/s11433-010-4034-x]

[18] X.L. Li, D.X. Fu, and Y.W. Ma, "Direct numerical simulation of hypersonic boundary-layer transition over a blunt cone", *AIAA J.,* vol. 46, no. 11, pp. 2899-2913, 2008.
      [http://dx.doi.org/10.2514/1.37305]

[19] X.L. Li, D.X. Fu, and Y.W. Ma, "Direct numerical simulation of hypersonic boundary layer transition over a blunt cone with a small angle of attack", *Phys. Fluids,* vol. 22, no. 2, 2010.025105 [http://dx.doi.org/10.1063/1.3313933]

[20] X. Liang, X.L. Li, D.X. Fu, and Y.W. Ma, "Effect of wall temperature on boundary layer stability over a blunt cone at Mach 7.99", *Comput. Fluids,* vol. 39, no. 2, pp. 259-271, 2010.
      [http://dx.doi.org/10.1016/j.compfluid.2009.09.015]

[21] M.V. Morkovin, Effects of Compressibility on Turbulent Flows.*Mechanique de la Turbulence.,* A. Favre, Ed., , 1964, pp. 367-380.

[22] T. Cebeci, and A. M. O. Smith, *Analysis of turbulent boundary layers.*
      [http://dx.doi.org/10.1115/1.3423784]

[23] J. Gaviglio, ""Reynolds analogies and experimental study of heat transfer in the supersonic boundary," Inter", *J. Heat Mass Trans.,* vol. 30, no. 5, pp. 911-926, 1987.
      [http://dx.doi.org/10.1016/0017-9310(87)90010-X]

[24] M. W. Rubesin, *Extra compressibility terms for Favre-averaged two-equation models of inhomogeneous turbulent flows,* 1990.

[25] L. Duan, and M.P. Martin, "Direct numerical simulation of hypersonic turbulent boundary layers. Part 4: Effect of high enthalpy", *J. Fluid Mech.,* vol. 684, pp. 25-59, 2011.
      [http://dx.doi.org/10.1017/jfm.2011.252]

[26] A.M. Perry, and M.S. Chong, "A description of eddying motions and flow patterns using critical-point concepts", Annu. Rev. Fluid Mech., vol. 19, pp. 125-155, 1987.
      [http://dx.doi.org/10.1146/annurev.fl.19.010187.001013]

[27] M.S. Chong, and A.E. Perry, "A general classification of three-dimensional flow fields", *Phys. Fluids A Fluid Dyn.,* vol. 2, pp. 765-777, 1990
      [http://dx.doi.org/10.1063/1.857730]

[28] J.C.R. Hunt, A.A. Wray, and P. Moin, *Eddies, stream, and convergence zones in turbulent flows,*1988.

[29] J. Jeong, and F. Hussain, "On the identification of a vortex", *J. Fluid Mech.,* vol. 285, pp. 69-94, 1995. [http://dx.doi.org/10.1017/S0022112095000462]

[30] P. Chakraborty, S. Balachandar, and R.J. Adrian, "On the relationships between local vortex identification schemes", *J. Fluid Mech.,* vol. 535, pp. 189-214, 2005.
      [http://dx.doi.org/10.1017/S0022112005004726]

[31] C.Q. Liu, Y.Q. Wang, Y. Yang, and Z.W. Duan, "New omega vortex identification method", *Sci. China Phys. Mech. Astron.,* vol. 59, no. 8, 2016.684711.
      [http://dx.doi.org/10.1007/s11433-016-0022-6]

[32] X.R. Dong, Y.Q. Wang, X.P. Chen, Y.L. Dong, Y.N. Zhang, and C.Q. Liu, "Determination of Epsilon for Omega Vortex Identification Method", *J. Hydrodynam.,* vol. 30, pp. 541-548, 2018.
      [http://dx.doi.org/10.1007/s42241-018-0066-x]

[33] C.Q. Liu, Y.S. Gao, S.L. Tian, and X.R. Dong, "Rortex-A new vortex vector definition and vorticity tensor and vector decompositions", *Phys. Fluids,* vol. 30, no. 3, 2018.035103.
      [http://dx.doi.org/10.1063/1.5023001]

[34] T. Sayadi, C. Hamman, and P. Moin, *Direct numerical simulation of complete transition to turbulence via h-type and k-type secondary instabilities,* 2011.

[35] B. Pierce, P. Moin, and T. Sayadi, "Application of vortex identification schemes to direct numerical simulation data of a transitional boundary layer", *Phys. Fluids,* vol. 25, 2013.015102.
[http://dx.doi.org/10.1063/1.4774340]

[36] M. Lagha, J. Kim, and J.D. Eldradge, "A numerical study of compressible turbulent layers", *Phys. Fluids,* vol. 23, 2011.015106.
[http://dx.doi.org/10.1063/1.3541841]

[37] M. Lagha, J. Kim, and J.D. Eldradge, "Near-wall dynamics of compressible boundary layers", *Phys. Fluids,* vol. 23, 2011.065109.
[http://dx.doi.org/10.1063/1.3600659]

[38] Y. Guo, N.A. Adams, and L. Kleiser, "Modeling of nonparallel effects in temporal direct numerical simulations of compressible boundary-layer transition", *Theor. Comput. Fluid Dyn.,* vol. 7, no. 2, pp. 141-157, 1995.
[http://dx.doi.org/10.1007/BF00311810]

[39] O Marxen, G Iaccarino, and ESG Shaqfeh, *Numerical simulation of hypersonic boundary-layer instability using different gas models,* 2007.

[40] O. Marxen, T. Magin, G. Iaccarino, and E.S.G. Shaqfeh, *Hypersonic boundary-layer instability with chemical reactions.* AIAA P, 2010, pp. 2010-0707.

[41] W.L. Jia, and W. Cao, "Effects of variable specific heat on the stability of hypersonic boundary layer on a flat plate", *Appl. Math. Mech.,* vol. 31, pp. 979-986, 2010.
[http://dx.doi.org/10.1007/s10483-010-1333-7]

[42] X.P. Chen, X.L. Li, and J. Fan, "Hypersonic turbulent channel flow in thermally perfect gas", *Sci. China Phys. Mech. Astron.,* vol. 41, pp. 969-979, 2011.
[http://dx.doi.org/10.1360/132011-97]

[43] X.P. Chen, X.P. Li, H-S. Dou, and Z.C. Zhu, "Effects of variable specific heat on energy transfer in a high–temperature supersonic channel flow", *J. Turbul.,* vol. 19, no. 5, pp. 365-389, 2018. [http://dx.doi.org/10.1080/14685248.2018.1441532]

[44] X.P. Chen, H-S. Dou, Q. Liu, Z.C. Zhu, and W. Zhang, "Comparative study of Reynolds stress budgets of thermally and calorically perfect gases for high-temperature supersonic turbulent channel flow", *Proc. Inst. Mech. Eng. Part G J. Aerosp. Eng.,* vol. 32, pp. 315-325, 2018.

[45] X.P. Chen, and F. Fei, "Effects of Dimensional Wall Temperature on Reynolds Stress Budgets in a Supersonic Turbulent Channel Flow with Thermally Perfect Gas", *Int. J. Comput. Fluid Dyn.,* vol. 233, no. 11, pp. 4222-4234, 2019.

[46] X.P. Chen, X.L. Li, and Z.C. Zhu, "Effects of dimensional wall temperature on velocity–temperature correlations in supersonic turbulent channel flow of thermally perfect gas", *Sci. China Phys. Mech. Astron.,* vol. 62, no. 6, 2019.064711.
[http://dx.doi.org/10.1007/s11433-018-9318-4]

[47] Y.S. Zhang, W.T. Bi, F. Hussain, and Z.S. She, "A generalized Reynolds analogy for compressible wall-bounded turbulent flows", *J. Fluid Mech.,* vol. 739, pp. 392-420, 2014.
[http://dx.doi.org/10.1017/jfm.2013.620]

[48] Y.S. Zhang, W.T. Bi, F. Hussain, X.L. Li, and Z.S. She, "Mach-number-invariant mean-velocity profile of compressible turbulent boundary layers", *Phys. Rev. Lett.,* vol. 109, no. 5, 2012.054502.
[http://dx.doi.org/10.1103/PhysRevLett.109.054502] [PMID: 23006178]

[49] G.S. Jiang, and C.W. Shu, "Efficient implementation of weighted ENO schemes", *J.*

Comput. Phys., vol. 126, pp. 202-222, 1996.
[http://dx.doi.org/10.1006/jcph.1996.0130]

[50]   P. Bradshaw, "Compressible turbulent shear layers", *Annu. Rev. Fluid Mech.,* vol. 9, pp. 33-54, 1977. [http://dx.doi.org/10.1146/annurev.fl.09.010177.000341]

[51]   J. Fan, "Criteria on high-temperature gas effects around hypersonic vehicles", *Chin. J. Theor. Appl. Mech.,* vol. 42, no. 4, pp. 591-596, 2010.

[52]   A.J. Smits, and J.P. Dussauge, *Turbulent boundary layer structure in supersonic flow.* 2nd ed. American Institute of Physics, 2006.

Current Developments in Mathematical Sciences, 2020, Vol. 2, 265-279   265

The Experimental Study on Vortex Structures in Turbulent Boundary Layer at Low Reynolds Number

Yanang Guo, Xiaoshu Cai*, Wu Zhou, Lei Zhou and Xiangrui Dong

Institute of Particle and Two-phase Flow Measurement, College of Energy and Power Engineering, University of Shanghai for Science and Technology, Shanghai 200093, China

Abstract: Experiments with a moving single-frame and long-exposure (MSFLE) imaging method, which is a Lagrangian-type measurement, is carried out to study the vortex structures in a fully developed turbulent boundary layer at low Reynolds number on a flat plate. In order to give the process of the vortex generation and evolution, on the one hand, the measurement system moves at the substantially same velocity as the vortex structure; on the other hand, a long exposure time is selected for recording the paths of the particles. In the experiment, the vortex structure characteristics as well as the temporal-spatial development can be shown by the streamwise-normal (x-y)-plane and streamwise-spanwise (x-z)-plane images which are extracted from a fully developed turbulent boundary layer. The result shows that the interaction between high- and low-speed streaks induces the generation, deformation and 'breakdown' of the vortex structures, and badly influences the vortex evolution.

Keywords: Boundary layer, High- and low-speed streaks, Low Reynolds number, Moving single-frame and long-exposure time, Vortex structures, Vortex generation, Vortex evolution, Vortex breakdown, Turbulent.

INTRODUCTION

The coherent structures play a significant role in the friction drag, heat and mass transfer, and the turbulence kinetic energy of turbulent boundary layer. However, the vortex structure is dominant and badly affects the generation and evolution of other coherent structures. Thus, the study on vortex structures is the starting point of the turbulence research.

In 1952, Theodorsen [1] proposed the horseshoe vortex as the basic structures in

*Corresponding author Xiaoshu Cai: Institute of Particle and Two-phase Flow Measurement, College of Energy and Power Engineering, University of Shanghai for Science and Technology, Shanghai 200093, China; Tel: +86-21-55275059; E-mail: usst_caixs@163.com

wall-bounded turbulent flow. Kline *et al.* [2] analyzed a turbulent boundary layer flow field by hydrogen bubble. The high- and low-speed spots and the long streamwise streaks of hydrogen bubble were found in the near-wall region. In addition, they used dye to be the tracer for flow visualization and suggested that bursting is an important factor for energy generation of turbulence. A visual study of a turbulent boundary layer flow was conducted by photographing the motions of small tracer particles ($d = 62 - 74\mu$m) using a stereoscopic medium-speed camera system moving with the flow by Brodkey *et al.* [3]. They found that the ejections could be a consequence of low-speed fluid being trapped between fingers of high-speed fluid. Head *et al.* [4] conducted an experiment by a smoke tunnel with a 45 degree light sheet to visualize the hairpin vortex in turbulent boundary layer flow. Adrian *et al.* [5-9] studied the structure of energy-containing turbulence in the outer region of a zero-pressure gradient boundary layer by using particle image velocimetry (PIV) to measure the instantaneous velocity fields in a streamwise-normal plane. They found that the hairpin vortices in the outer layer occur in streamwise-aligned structures and could form large scale packets. The experimental investigations by the hydrogen bubble technique were performed by Lian [10] to show the coherent structures of turbulent boundary layer. In their experiment, the streamwise- and normal-vortices were observed along the interface regions between high- and low-speed streaks, while, the transverse (spanwise) vortices were observed at the front of the high-speed regions. Lozanoduran *et al.* [11] and Zandonade *et al.* [12] pointed out that the coherent structures in turbulent boundary layer could be illustrated by the high- and low-speed streaks and the vortices. The generation of these high- and low-speed streaks is related to the ejection and sweep, while, the generation of the vortices is related to the shear layer. Gao *et al.* [13] proposed and implemented a moving tomographic particle image velocimetry method to measure temporal evolution of velocity fields in three-dimensional volumes and to track coherent structures within a turbulent boundary layer with $Re_\tau \approx 2410$.

Although the vortex structure in turbulent boundary layer has been widely studied, and people already have had some understandings on its structural characteristics, its mechanism is still not clear. To further study the essence of vortex structure in turbulent boundary layer, the dynamic evolution of vortex structure with time and space must be obtained. In this paper, the moving single-frame and long-exposure (MSFLE) imaging method is utilized to study the vortex structure in a fully developed turbulent boundary layer on the flat plate at low Reynolds number. The evolution process of the vortex structure as well as the interaction between streaks and vortices have been discussed.

EXPERIMENTAL METHODS

Moving Single-Frame and Long-Exposure (MSFLE)

For most imaging measurements, the camera is usually fixed without moving, such as PIV and single-frame and long-exposure (SFLE). Fig. (**1**) gives the schematic diagram of the SFLE imaging measurement. The principle of SFLE is introduced in the following.

Firstly, the tracer particles with good tracking property are interspersed in the flow field; then, these tracer particles are illuminated by a light sheet from the laser; finally, the scattered light of the particles can be received by the camera. It means that the trajectory of the particles can be recorded in a single frame image by setting a proper exposure time of the camera. The length of the path line represents the movement distance of the tracer particle during the exposure time, thus the velocity of the tracer particle V, can be obtained as:

$$V = \frac{S}{M\Delta t} \tag{1}$$

Where S is the total length of trajectory, M is magnification factor of lens and Δt is exposure time. Moreover, the direction of the particle velocity can be determined by two consecutive frames.

However, it would be failed to capture the process of the fast-moving vortex structure evolution if the camera is fixed. Here, the MSFLE imaging method, which is developed from the SFLE method, is utilized to show both of the temporal and spatial development of the vortex structure in a single frame image, without using vortex identification criteria or Galilean velocity decomposition. Compared to SFLE method, the advantage of MSFLE is that the camera can move in the measuring system so that vortex structures with the same speed as the camera can be captured. MSFLE is a Lagrangian-type measurement and it is easy to observe the evolution process of vortex structure intuitively.

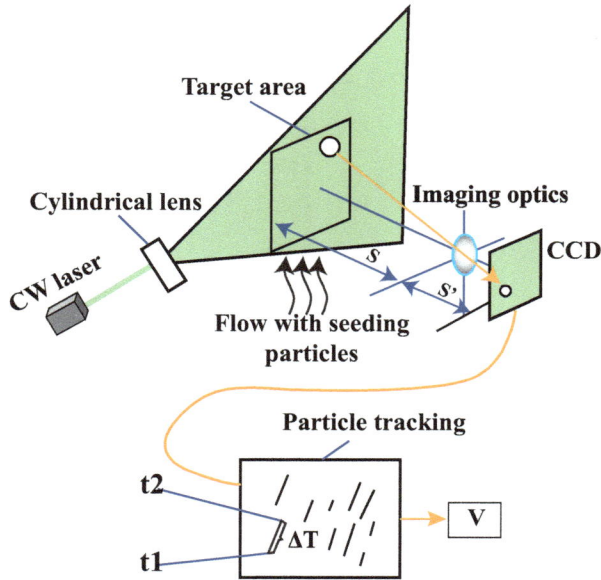

Fig. (1). The schematic diagram of single-frame and long-exposure measurement.

Experiment Apparatus

The schematic diagram of the experimental measurement system is shown by Fig. (2) and the experimental equipment is shown in Fig. (3). The experiment is conducted in a low-speed circulating water tunnel. The water pump, with a rated head and a rated flow rate of $5m$ and $40\ L/min$ separately, can transport water from the water tank to the inlet of the tunnel and give a constant inlet pressure. Then, water can be recycled after flowing through the ball valves and the floater flowmeter and finally back to the water tank. The length of the water tunnel is $2500mm$, and the cross section area of the tunnel is $80 \times 80mm^2$. A plexiglass plate with the size of $1500mm \times 78mm \times 4.5mm$ is settled in the bottom of the water tunnel and locates from $x = 700mm$ to $x = 2200mm$. The leading edge of the plate is set as an ellipse, which has a ratio of 4:1 for the long axis to the short axis, to alleviate the disturbance of the incoming flow. The tracer particles are polyamide resin (PSP) with an average diameter of $5\ \mu m$, the density of $1.04\ g/ml$ and the refractive index of 1.584. The light source is $450nm$ wavelength continuous laser diode, which generates a laser sheet with 1mm thickness. Images are captured by a CMOS camera which has a resolution of 1280×1024 pixels, a pixel size of $4.8\ \mu m$ and the lens with a magnification factor of 0.14 (or 0.3). The entire measurement system is installed on a horizontal guide rail which can move along the streamwise direction. The traveling speed of the measurement system is controlled by a servo motor based on the velocity of the flow in water tunnel.

Since the main flow locates above the boundary layer, the velocity of the main flow can be calculated by formula(1) and is about $98\ mm/s$ when the flow flux is $1.6 m^3/h$, so the speed of vortex structures in the boundary layer is lower than $98\ mm/s$. In addition, the evolution process of vortex structures at different heights in the boundary layer can be observed by setting different movement speed of camera. Therefore, in order to obtain abundant evolution process of vortex structures, different movement speeds of the camera can be adopted for different measurements. With extensive tests, it is found that the vortex structures is most abundant when the moving speed of camera is $90\ mm/s$ and $75\ mm/s$ respectively for the streamwise-normal measurement and the streamwise-spanwise measurement.

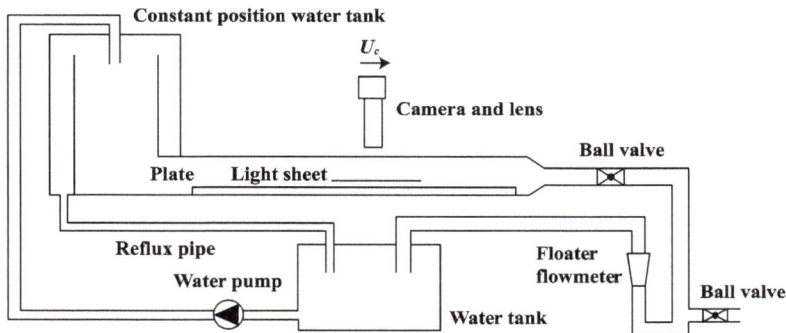

Fig. (2). Schematic diagram of the experimental measurement system.

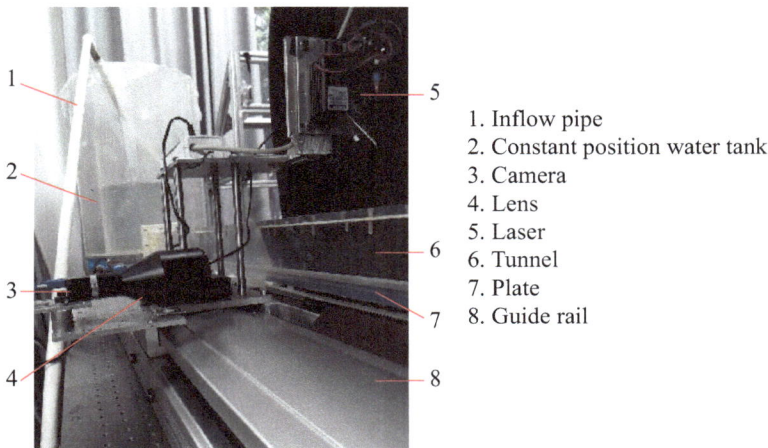

1. Inflow pipe
2. Constant position water tank
3. Camera
4. Lens
5. Laser
6. Tunnel
7. Plate
8. Guide rail

Fig. (3). Diagram of experimental equipment.

Experimental Validation

The velocity profile of the boundary layer is measured by SFLE method at $x = 400mm$ and $x = 1300mm$ separately. The average velocity and the normal distance from the plate are normalized by the friction velocity u_τ, where the u_τ is obtained by the wall shear stress τ_ω, which is obtained by the velocity gradient of the viscous sublayer.

$$u_\tau = \sqrt{\frac{\tau_\omega}{\rho}} \;,\; \tau_\omega = \mu \frac{\partial u}{\partial y} \tag{2}$$

where τ_ω denotes the wall shear stress, ρ is the density, $\partial u/\partial y$ is the velocity gradient of the viscous sublayer and μ is the dynamic viscosity coefficient. Thus, u_τ is equal to $6mm/s$ at $x = 400mm$ and equals to $5.48\ mm/s$ at $x = 1300mm$.

u^+ and y^+ are calculated by $y^+ = yu_\tau/v$, $u^+ = u/u_\tau$, and the solid line is calculated by the Spalding [14] velocity profile equation,

$$y^+ = U^+ + e^{-KB}\left[e^{KU^+} - 1 - KU^+ - \frac{(KU^+)^2}{2!} - \frac{(KU^+)^3}{3!}\right] \tag{3}$$

$K = 0.4$, $B = 5.5$. The results are compared in Fig. (**4**). As can be seen, the experimental result coincides with the Spalding theory which means our experiment method is quite accurate. The results of two different locations are similar to each other which demonstrates that the turbulent boundary layer is fully developed in the test area.

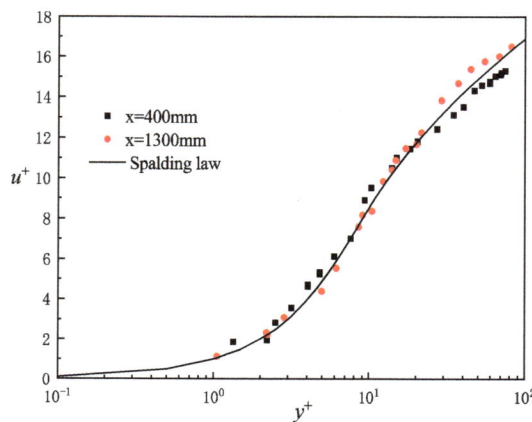

Fig. (4). The measured velocity profiles at *x*=400mm and *x*=1300mm and Spalding law.

EXPERIMENTAL RESULTS AND ANALYSIS

Measurements in the Streamwise-Normal (*x-y*)-Plane

Fig. (**5**) shows the sketch of measurement system in the (*x-y*)-plane. The axes x, y, z are coordinates in streamwise, normal and spanwise direction, respectively. The light sheet is fixed vertically in the middle of the plate along the streamwise direction, for better visualizing the vortex structure generation as well as evolution in the (*x-y*)-plane. The movement speed of the measurement system is determined as $U_C = 90mm/s$. The magnification factor of the lens is 0.14, and the field of view is $44 \times 35mm^2$. Measuring range in flow direction is from $x = 540mm$ to $x = 1336mm$. The corresponding Reynolds number Re at $x = 540mm$ and $x = 1336mm$ are 52920 and 130928 respectively. $Re = U_\infty x/v$, where v is the kinematic viscosity. The single frame exposure time is 100ms and the frame rate is $9.99fps$. The time interval between two frames is around $50\mu s$.

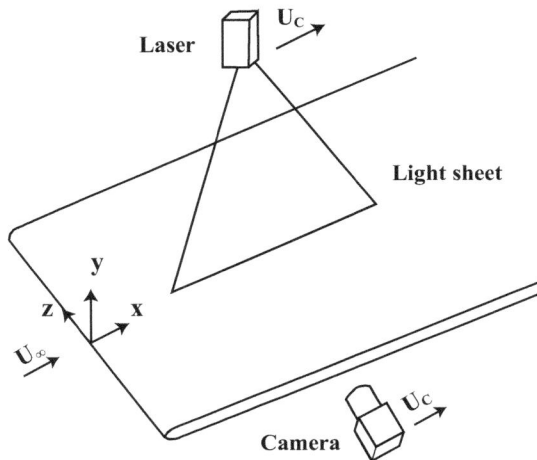

Fig. (5). Sketch of measurement system in the (*x-y*)-plane.

Fig. (**6a-i**) show the evolution of the vortex structure in the (*x-y*)-plane. The fluid flows along the x direction, and the lower edge of the figures has the same horizontal position with the upper surface of the flat plate. The vertical coordinate of the images is the normal direction. Fig. (**6j**) gives the streamwise distribution of each frame shown by Fig. (**6a-i**). Since the camera is moving and has a moving velocity of U_C, the light spots appeared at the same locations on two consecutive frames mean they have same velocity with U_C. In this way, if the velocity of tracer particles is higher than U_C, the tracers move downstream; however, the tracer particles moving upstream mean their velocity is lower than U_C. The path direction of the particles can be determined by two consecutive frames, and is shown as arrows in Fig. (**6**) for better recognizing the movement of the fluid.

Fig. (**6a**) is assumed to be the frame at $t = 0$, and the height of the dotted line from the bottom which is around $2.5mm$, is recognized as the viscous sublayer. On the one hand, from the side view of the fluid along the flow direction, the circled area marked as A, B, and C perform like vortex structures that rotates clockwise in an approximate circle. Although two-dimensional measuring method is used here, these vortex structures are three dimensional and may be the heads of the hairpin vortices captured by the light sheet at different heights. Focus on the vortex in regions A, B and C, the fluid above these vortices has a velocity larger than U_C thus can be regarded as high-speed streaks; while, the fluid below these vortices moves slower than U_C and thus is treated as low-speed streaks. Actually the generation of a rotational vortex attributed to a twist motion (or called 'shear') between these high- and low-speed streaks can be clearly shown. On the other hand, the light spots inside the squared area in Fig. (**6a**) indicates a laminar flow which has the same speed $U_C = 90mm/s$ as the camera moving speed. However, this laminar structure quickly decays and is merged into upper high-speed streaks due to the entrainment, (see Fig. **6b**). At $t = 1.3$s in Fig. (**6c**), the previous vortex marked B (hereinafter referred to as 'Vortex B') in Fig. (**6a**) is disappeared due to the impact effect of the high-speed streaks. Unlike Vortex B, the Vortex A in Fig. (**6c**) is less affected by high-speed streaks and does not change much, since it locates in much lower area and is far from high-speed streaks or upwelling zone. Although the Vortex C is under a heavy impact effect of upwelling or high-speed streaks, it becomes dominant and entrains the surrounding fluids to rotate with an increasing size, due to its larger size and higher stability than others. Moreover, compared with Fig. (**6a**), the core of Vortex C generally does not change its location, which means that the Vortex C remains moving by $U_C = 90mm/s$ from $t = 0$ to $t = 1.3$s. At $t = 2.6$s in Fig. (**6d**), the Vortex A is hindered by the down sweep of the high-speed streaks thus acts like moving upstream due to its lower velocity than U_C; however, the Vortex C moves downstream with an increasing velocity. The area in a rectangle indicates a strong shear layer which is generated by the interaction between high- and low-speed streaks. At $t = 2.9$s, the Vortex A and C are all disappeared, while a new flow rotation appears in a rectangle region in Fig. (**6e**), due to the strong shear effect of high- and low-speed streaks. Finally, a vortex marked as Vortex D forms at $t = 3.2$s in Fig. (**6f**). At $t = 3.6$s, the Vortex D moves upstream with a lower velocity than U_C, and another new vortex generates at the location where the Vortex D generated, marked as Vortex E in Fig. (**6g**). At $t = 4.4$s in Fig. (**6h**), the size of the Vortex D and E decreases due to the compressing effect of the high- and low-speed streaks. Both of these vortices move upstream and become closer since the Vortex E has a lower velocity than the Vortex D. At $t = 4.9$s, the vortices D and E finally merge into one vortex with a larger size but the same rotating direction, see the Vortex F in Fig. (**6i**).

It can be observed from the evolution process of the vortex structures that the vortices can increase their size by entraining the surrounding fluid rotating along with them, and also can be deformed or even broken by fluid impact, which means that vortices are fluid, but not rigid bodies. Most vortex structures observed in the experiment locate in the logarithmic layer with elliptic shapes, and they are clockwise rotating, with the dimensions of its major and minor axes around $7mm \times 5mm$. In addition, as shown by Fig. (**6**), the size of vortex structure increases with distance from the plate.

The Fig. (**6**) suggests that the high- and low-speed streaks badly affect the decay or growth of the vortex structures with different scales and different rotation strength. For instance, the high- and low-speed streaks can cause a 'breakdown' or a decay of a vortex with a small size and low stability, and also can cause a deformation or movement of a vortex with large size and high stability. Also, the streaks can form a vortex through shear or merge two vortices into one.

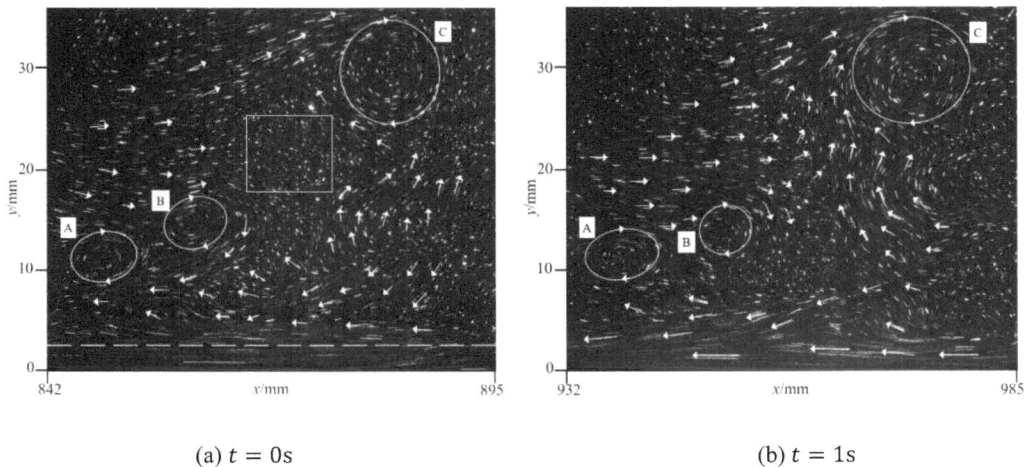

(a) $t = 0$s

(b) $t = 1$s

Fig. 6 cont.....

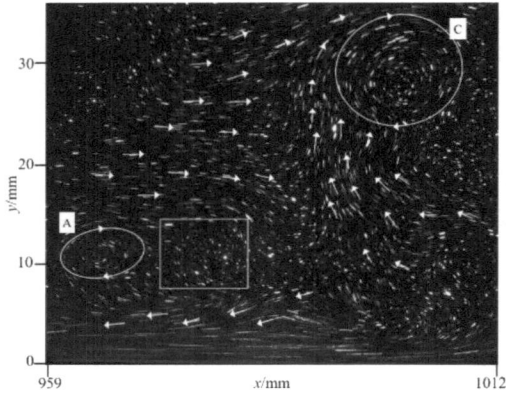

(c) $t = 1.3$s

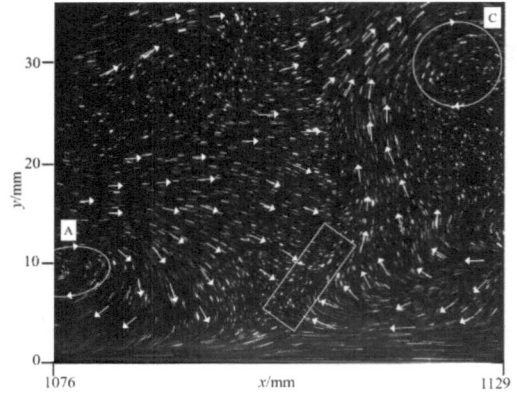

(d) $t = 2.6$s

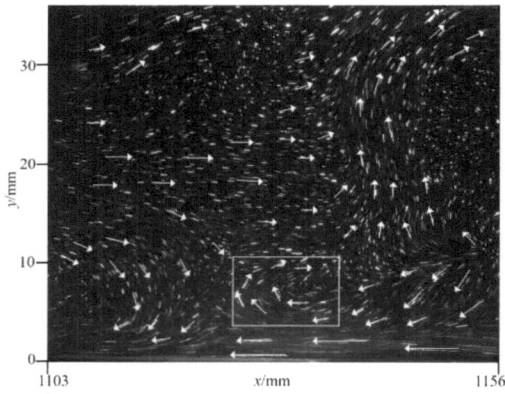

(e) $t = 2.9$s

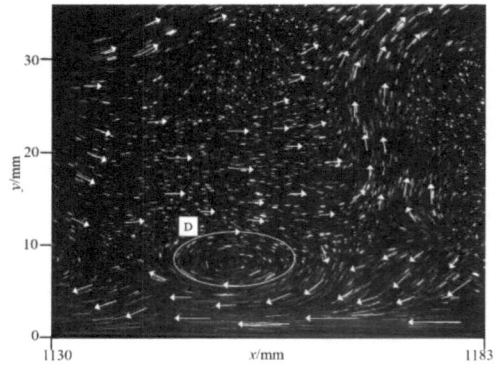

(f) $t = 3.2$s

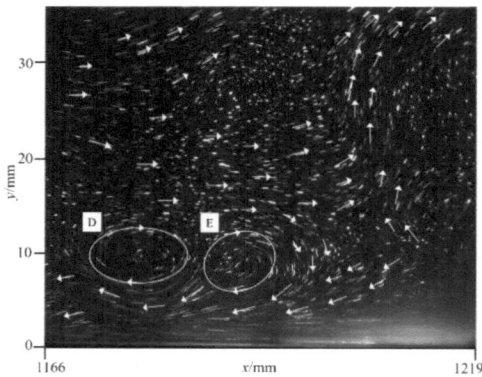

(g) $t = 3.6$s

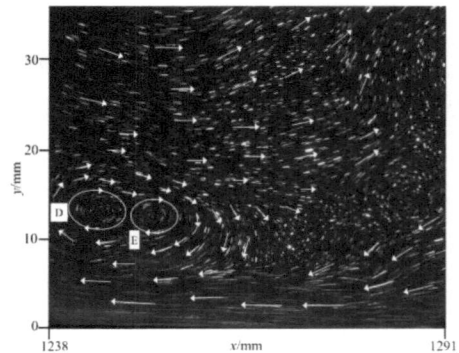

(h) $t = 4.4$s

Fig. 6 cont.....

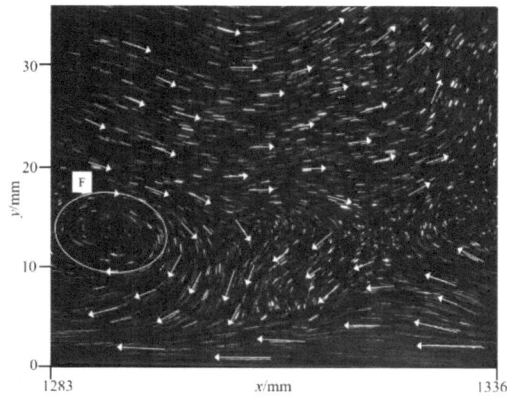

(i) $t = 4.9s$

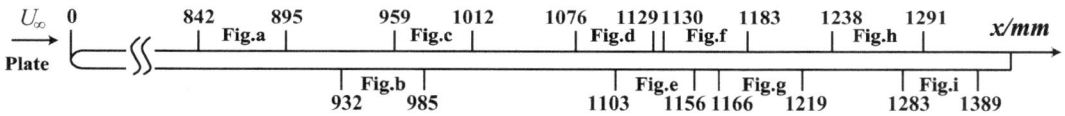

(j) Sketch of the locations in streamwise direction for each image above

Fig. (6). Evolution of the vortex structure shown in the $(x\text{-}y)$-plane.

Measurements in the Streamwise-Spanwise $(x\text{-}z)$-Plane

The measurement system in the $(x\text{-}z)$-plane is also given in Fig. (**7**). The light sheet is parallel to and $7mm$ away from the plate. The lens of the camera is normal to the flat plate, for the purpose of visualizing the vortex structure generation and evolution in the $(x\text{-}z)$-plane. The movement speed of the camera is determined as $U_C = 75mm/s$. The magnification factor of the lens is 0.3, and the field of view is $20 \times 16mm^2$. The measuring range in flow direction is from $x = 440mm$ to $x = 1200mm$, and the corresponding Reynolds number at $x = 440mm$ and $x = 1200mm$ are 43120 and 117600, respectively. The single frame exposure time, the frame rate and the time interval between two frames have the same configuration as the above case.

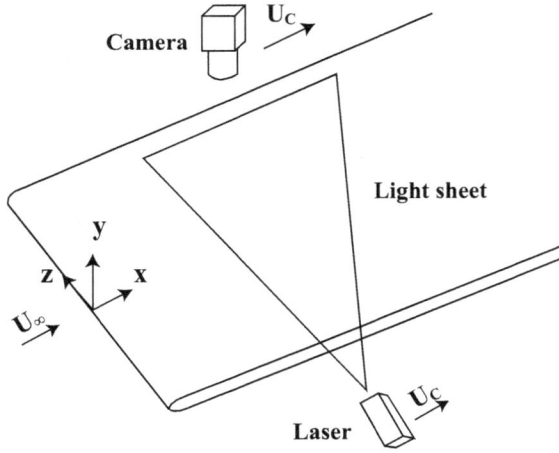

Fig. (7). Sketch of measurement system in the (*x-z*)-plane.

Fig. (**8a-f**) show the evolution of the vortex structures in the (*x-z*)-plane. The vertical coordinate of the images is the spanwise direction. Fig. (**8g**) gives the streamwise distribution of each frame shown by Fig. (**8a-f**).

Fig. (**8a**) is assumed to be the frame at $t = 0$. It can be observed that there is a pair of counter-rotating vortices marked as A and B in the figure, which could be recognized as a pair of the legs of a hairpin vortex. Its oval shape is the result of the intersecting between the horizontal light sheet and the tilted hairpin vortex legs. In addition, the high-speed streaks in the center of vortex pair move downstream at a velocity of $92mm/s$. At $t = 0.3s$, in Fig. (**8b**), the low-speed streaks near the Vortex A move along the positive spanwise direction, and thus cause that the Vortex B is out of the sight and the Vortex A is deformed by the impaction of the high-speed streaks in center area. At $t = 0.6s$, the Vortex A disappears and is merged into the high-speed streaks, (see the rectangular area in Fig. (**8c**)). However, although the Vortex A disappears, a new strong shear forms at the same location due to the upper high-speed streaks and lower low-speed streaks from the image. This strong shear then generates a clockwise rotating vortex at $t = 0.8s$, which is marked as Vortex C in Fig. (**8d**). At $t = 1.2s$, in Fig. (**8e**), the Vortex C enlarges its size by entraining the surrounding fluids and stretches along the streamwise direction. It means that the legs of the hairpin vortex coarsen and stretch during the evolution. At $t = 1.5s$, in Fig. (**8f**), the vortex leg is captured to lift in normal direction and stretch in streamwise direction, according to the shape as well as the rotation direction of the Vortex C.

Fig. (**8**) indicates that the high- and low-speed streaks can cause a 'breakdown' or decay of a vortex, and also can cause generation or development of a vortex. It can be concluded that the evolution of the vortex in turbulent boundary layer is closely related to the high- and low-speed streaks.

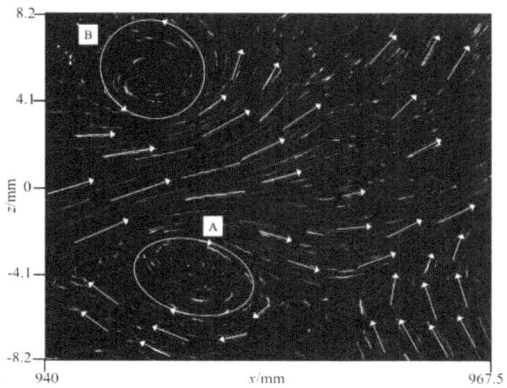

(a) $t = 0$s

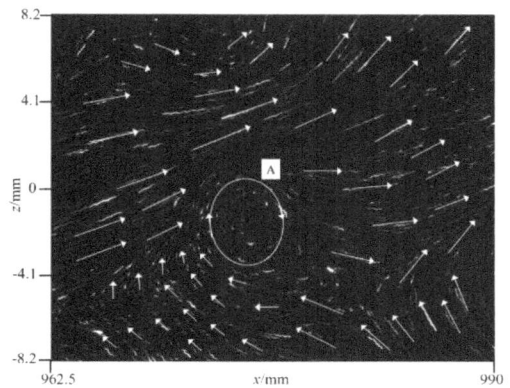

(b) $t = 0.3$s

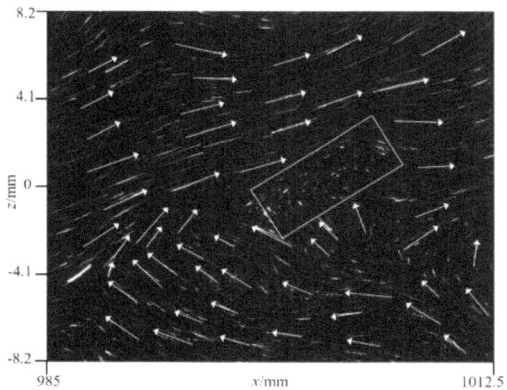

(c) $t = 0.6$s

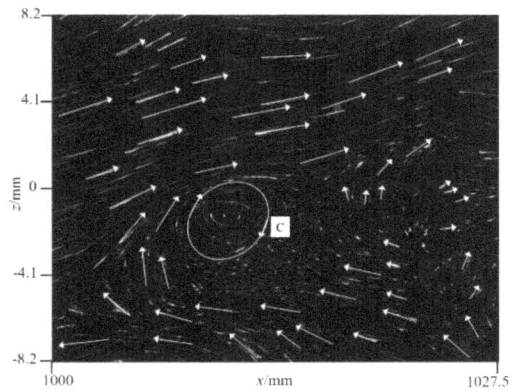

(d) $t = 0.8$s

Fig. 8 cont.....

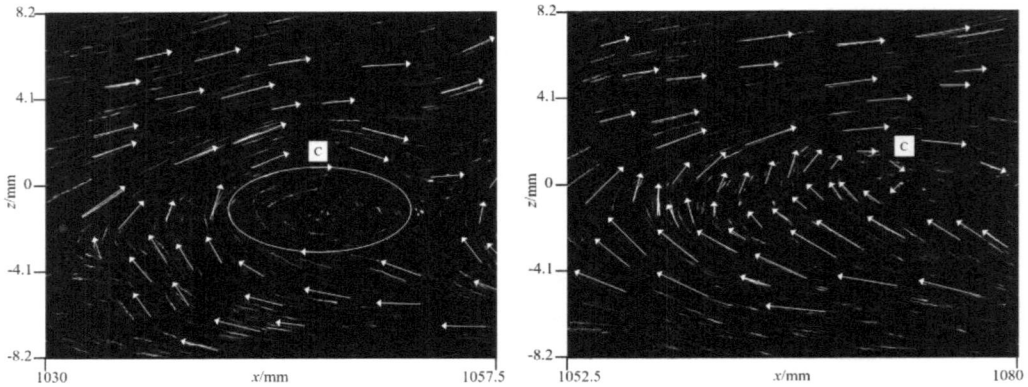

(e) $t = 1.2$s   (f) $t = 1.5$s

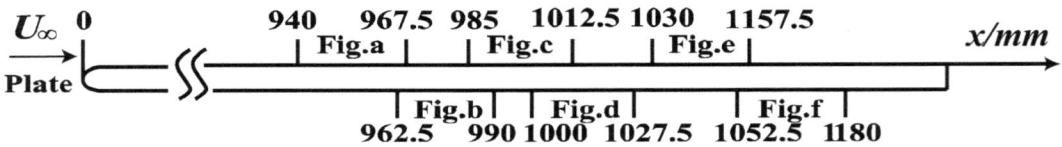

(g) Sketch of the locations in streamwise direction for each above image

Fig. (8). Evolution of the vortex structure shown in the $(x\text{-}z)$-plane.

CONCLUSIONS

Moving single-frame and long-exposure (MSFLE) imaging method, which is a Lagrangian-type measurement, is experimentally conducted to study a fully developed turbulent boundary layer at low Reynolds number on a flat plate. The vortex structure characteristics and their temporal-spatial development are obtained from the $(x\text{-}y)$-plane and $(x\text{-}z)$-plane images of the turbulent boundary layer, and the relation between the vortex structure and the high- and low-speed streaks is discussed.

The experimental result indicates that the impact effect of the high-speed streaks can cause a 'breakdown' or decay of a vortex with a small size and low stability, and also can lead to a deformation or movement of a vortex with large size and high stability. The generation of a vortex attributes to the strong shear effect, which is derived from a twist motion between high- and low-speed streaks. In conclusion, the evolution of the vortex in turbulent boundary layer is closely related to the high- and low-speed streaks.

CONSENT FOR PUBLICATION

Not applicable.

CONFLICT OF INTEREST

The authors confirm that this chapter contents have no conflict of interest.

ACKNOWLEDGEMENTS

This work is partly supported by the National Natural Science Foundation of China (Grants No. 51576130, No. 51327803), whose help is gratefully acknowledged.

REFERENCES

[1]   T. Theodorsen, "Mechanism of turbulence", *Proc. 2nd Midwestern Conf. on Fluid Mechanics.,* 1952 Columbus

[2]   S.J. Kline, W.C. Reynolds, F.A. Schraub, and P.W. Runstadler, "The structure of turbulent boundary layers", *J. Fluid Mech.,* vol. 30, no. 4, p. 741, 1967. [http://dx.doi.org/10.1017/S0022112067001740]

[3]   A.K. Praturi, and R.S. Brodkey, "A stereoscopic visual study of coherent structures in turbulent shear flow", *J. Fluid Mech.,* vol. 89, no. 2, pp. 251-272, 1978. [http://dx.doi.org/10.1017/S0022112078002608]

[4]   M.R. Head, and P. Bandyopadhyay, "New aspects of turbulent boundary layer structure", *J. Fluid Mech.,* vol. 107, pp. 297-338, 1981. [http://dx.doi.org/10.1017/S0022112081001791]

[5]   R.J. Adrian, C.D. Meinhart, and C.D. Tomkins, "Vortex organization in the outer region of the turbulent boundary layer", *J. Fluid Mech.,* vol. 422, pp. 1-54, 2000. [http://dx.doi.org/10.1017/S0022112000001580]

[6]   R.J. Adrian, "Hairpin vortex organization in wall turbulence", *Phys. Fluids,* vol. 19, no. 4, pp. 1-16, 2007. [http://dx.doi.org/10.1063/1.2717527]

[7]   J. Zhou, R.J. Adrian, S. Balachandar, and T.M. Kendall, "Mechanisms for generating coherent packets of haripin vortices", *J. Fluid Mech.,* vol. 387, pp. 353-396, 1999. [http://dx.doi.org/10.1017/S002211209900467X]

[8]   B.J. Balakumar, and R.J. Adrian, "Large- and very-large-scale motions in channel and boundary-layer flows", *Philos. Trans.- Royal Soc., Math. Phys. Eng. Sci.,* vol. 365, no. 1852, pp. 665-681, 2007. [http://dx.doi.org/10.1098/rsta.2006.1940] [PMID: 17244580]

[9]   Q. Chen, R.J. Adrian, Q. Zhong, D. Li, and X. Wang, "Experimental study on the role of spanwise vorticity and vortex filaments in the outer region of open-channel flow", *J. Hydraul. Res.,* vol. 52, no. 4, pp. 476-489, 2014. [http://dx.doi.org/10.1080/00221686.2014.919965]

[10]   Q.X. Lian, "A visual study of the coherent structure of the turbulent boundary layer in flow with adverse pressure gradient", *J. Fluid Mech.,* vol. 215, pp. 101-124, 1990. [http://dx.doi.org/10.1017/S0022112090002579]

[11]   A. Lozano-Durán, and J. Jiménez, "Time-resolved evolution of coherent structures in turbulent channels: Characterization of eddies and cascades", *J. Fluid Mech.,* vol. 759, pp. 432-471, 2014. [http://dx.doi.org/10.1017/jfm.2014.575]

[12]   J.C. del Álamo, J. Jiménez, P. Zandonade, and R.D. Moser, "Self-similar vortex clusters in the turbulent logarithmic region", *J. Fluid Mech.,* vol. 561, pp. 329-358, 2006. [http://dx.doi.org/10.1017/S0022112006000814]

[13]   Q. Gao, C. Ortiz-Dueñas, and E.K. Longmire, "Evolution of coherent structures in turbulent boundary layers based on moving tomographic PIV", *Exp. Fluids,* vol. 54, no. 12, 2013. [http://dx.doi.org/10.1007/s00348-013-1625-0]

[14]   D.B. Spalding, "A Single Formula for the 'Law of the Wall'", *J. Appl. Mech.,* vol. 28, no. 3, pp. 455- 458, 1961. [http://dx.doi.org/10.1115/1.3641728]

Experimental Studies on Coherent Structures in Jet Flows Using Single-Frame-Long-Exposure (SFLE) Imaging Method

Lei Zhou[1], Xiaoshu Cai[1,*], Wu Zhou[1] and Yiqian Wang[2]

[1]*Institute of Particle and Two-phase Flow Measurement, University of Shanghai for Science and Technology, Shanghai, China*

[2]*School of Mathematical Science, Soochow University, Suzhou, 215006, China*

Abstract: On the axisymmetric water jet experimental apparatus, the flow field structures in entrainment boundary layers are measured using Single Frame Long Exposure image method, in the range of Reynolds number (*Re*) 1849~2509. It is found that engulfing and nibbling entrainment model occur intermittently with time, in the region of L=2~3.5d streamwise and H=1~1.25d radial direction. It concludes that the occurrence probability of engulfing increases with Reynolds number when Re>1915, the influence of Reynolds number on the occurrence probability of this structures decreases when Re>2311; the occurrence frequency of this coherent structures obtained by fast Fourier transform is between 10 and 19Hz; special vortex structures were observed in the flow field during the occurrence of engulfing. The jet flow field is measured using Moving Single Frame Long Exposure image method in Lagrangian coordinate system, and it is found that the vortex structures generally exist near the interface of turbulent regions and non-turbulent regions.

Keywords: Entrainment layer, Jet, Move single frame long Exposure, Vortex structures, Single frame single exposure.

INTRODUCTION

The observation of alternating vortical structures on the two sides of a gaseous jet using stroboscopic cinematography by Brown [1] provided one of the earliest experimental evidence for the existence of coherent structures in jet flows. Davis, Fisher and Barratt [2] then reported that a chain of vortex rings exists in round turbulent jets and Beavers and Wilson [3] further confirmed that these vortical

*Corresponding author Xiaoshu Cai: Institute of Particle and Two-phase Flow Measurement, University of Shanghai for Science and Technology, Shanghai, China; Tel: +86-21-55275059; E-mail: usst_caixs@163.com

structures are all located inside the shear layer region. Ever since, the investigation of coherent vortical structures in jet flows became a fundamental topic and received attentions from researchers. Westerweel, Hofmann, Fukushima and Hunt [4] studied the characteristics of the TNTI (turbulent/non-turbulent interface) in self-similarity jets by combining the techniques of PIV (Particle Image Velocimetry) and LIF (Laser Induced Fluorescence) and concluded that the profile of TNTI rapidly changes along the axial direction with violent mass, momentum and energy exchange. Gan [5-7] investigated the initial development of gaseous turbulent vortex rings by PIV and found that this coherent pattern forms around $x/d = 2.5$, where x is the axial coordinate and d is the nozzle diameter. Two-dimensional (2D) and three-dimensional (3D) PTV (Particle Tracking Velocimetry) are also applied to study the small-scale coherent structures in the developed region of turbulent jets. Moreover, Silva, Taveira and Borrell [8-12] classified the entrainment at the TNTI into engulfing (big-scale eddy motions) and nibbling small-scale eddy motions).

Undoubtedly, the coherent structures play an important role in the entrainment of turbulent jet flows. However, the generation and the development of these coherent structures, especially their relationship with the TNTI is still unclear. To study the dynamics as well as the physical mechanism of the entrainment in jet flows, a detailed analysis on the evolution of the coherent structures in a water jet flow by utilizing Single Frame Long Exposure (SFLE) and Moving Single Frame Long Exposure (MSFLE) developed in our group is carried out in this chapter.

EXPERIMENTAL METHODS

Single-Frame-Long-Exposure (SFLE) and Moving SFLE (MSFLE)

The trajectories of illuminated tracer particles are obtained by using a relatively long exposure time and a typical trajectory is shown in Fig. (**1a**). The length of the trajectory S can be regarded as the distance covered by the tracer particle during the exposure time Δt plus the diameter of the tracer particle D as shown in Fig. (**1b**). Given that Δt is sufficiently small, the velocity magnitude of the particle V can be estimated as

$$V = \frac{L}{K * \Delta t} = \frac{S-D}{K * \Delta t}$$

where K is the magnification factor of the camera lens, D is the diameter of the tracer particle and L is the distance covered by the tracer particle over Δt.

In the frame work of SFLE, the exposure time can be adjusted to accommodate to the targeting velocity, or Reynolds number based on the nozzle velocity V_j and the nozzle diameter d, *i.e.*, $Re = V_j d/v$, where v is kinematic viscosity. In addition, information of the flow field with various resolution can be obtained by adjusting the magnification factor of the lens. Additional information of the moving direction can be acquired by the correlation between two consecutive snapshots or the SFME (Single-Frame-Multiple-Exposure) [13] technique. A schematic diagram of the SFLE is shown in Fig. (**2**). To study the coherent vortical structures from the instantaneous flow field, moving SFLE (MSFLE) is developed to remove the mean flow velocity by making the camera moving with a constant velocity in the axial direction. SFLE and MSFLE, which have the advantages of providing more intuitively vision of the flow field and more comprehensive information about the moving trajectories during an exposure time, thus are selected in this study to investigate the coherent vortical structures in a water submerged jet flow.

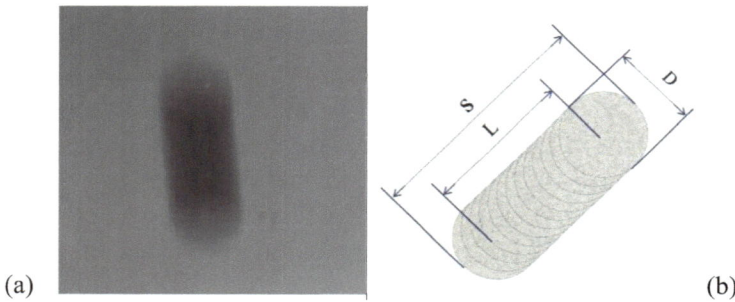

(a)   (b)

Fig. (1). (a) A typical trajectory by SFLE; **(b)** Sketch of a particle trajectory.

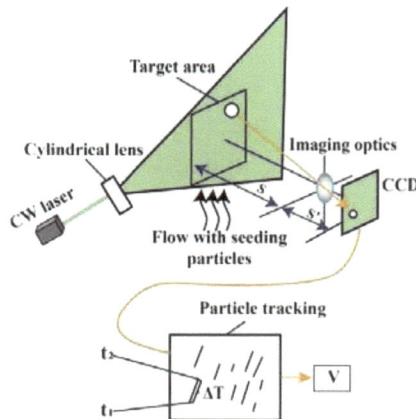

Fig. (2). A schematic diagram of SFLE.

Experiment Apparatus

The experimental apparatus of the water submerged jet used in this study is shown in Fig. (**3**). The fluid is originally stored in a pressure storage tank and has thoroughly mixed with tracer particles. After passing through a flowmeter with range from 0 to 250ml/min, the fluid is ejected into a 100mm × 100mm × 400mm roofless plexiglass container through a nozzle with diameter of 1.6mm. Finally, the fluid goes into the drain tank through a hole on the downstream side of the plexiglass container to keep the water level constant. The tracer particles are SiC (1 − 2μm) which are small enough and have the almost same density as the water. The measuring system consists of a laser diode, optical lenses and a XIMEA camera. The continuous wave laser emits 532nm wavelength light that passes through the optical lenses and forms a light sheet which illuminates the tracer particles. The scattered light is then received by the XIMEA camera perpendicular to the light sheet. Telecentric lens with magnification factor of 1, resolution of 1024 × 1280 and pixel of 4.5μm is selected in this study. A sketch of experimental apparatus can be found in Fig. (**4**).

EXPERIMENTAL RESULTS AND ANALYSIS

In this study, both SFLE and MSFLE are adopted. With the location of the camera fixed, the jet flow field of Reynolds number *Re* from 1849 to 2509 are measured with exposure time of 7ms and frame rate of 141 frames per second in the frame work of SFLE. For the MSFLE, the velocity of the camera along the axial direction is 0.133m/s, the exposure time is 10ms and the frame rate is 98 frames per second. The field of view is 4.608mm × 5.760mm for all cases.

Fig. (3). The experimental apparatus of the water submerged jet flow.

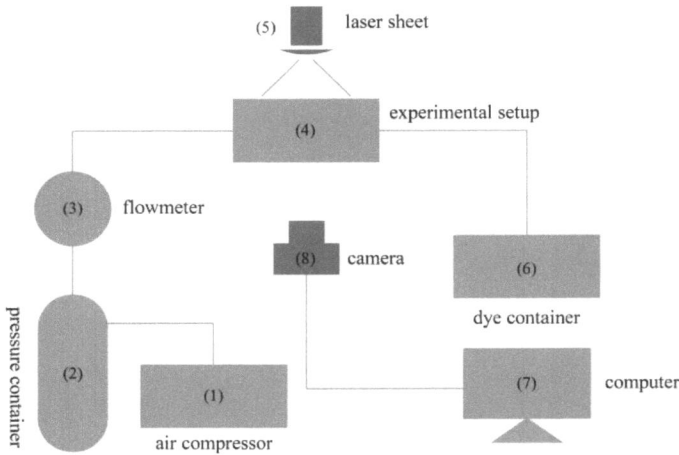

Fig. (4). Sketch of the experimental apparatus.

The Measurements of Coherent Structures in Jet Entrainment Boundary Layer

It is observed that in the region of turbulent/non-turbulent interface (TNTI), a regular entrainment pattern exists. The jet flow is in a mild entrainment mode of 'nibbling' until the pathlines become distorted followed by special pathline structures. And then the fierce entrainment mode of 'engulfing' appear which become weaker afterwards until the flow recovered to the 'nibbling' mode. Fig. (**5**) gives the flow structures development in consecutive snapshots from a selected initial time of t_0.

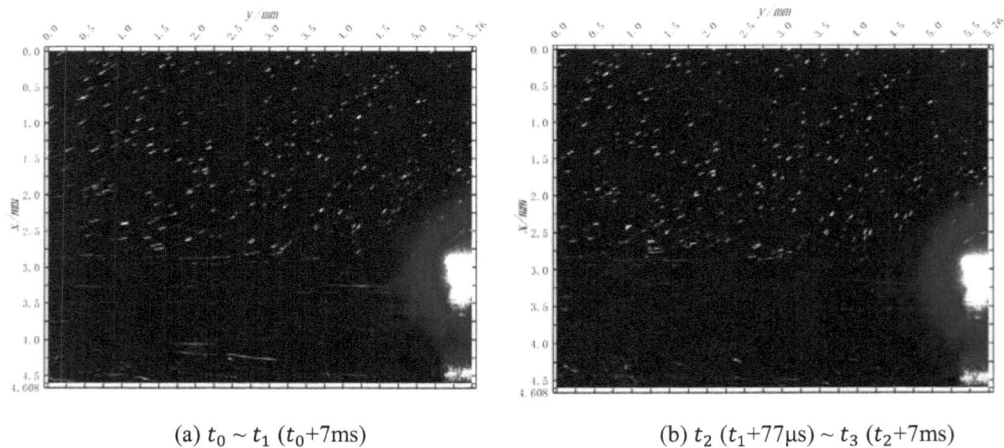

(a) $t_0 \sim t_1$ (t_0+7ms)   (b) t_2 (t_1+77μs) $\sim t_3$ (t_2+7ms)

Fig. 5 cont.....

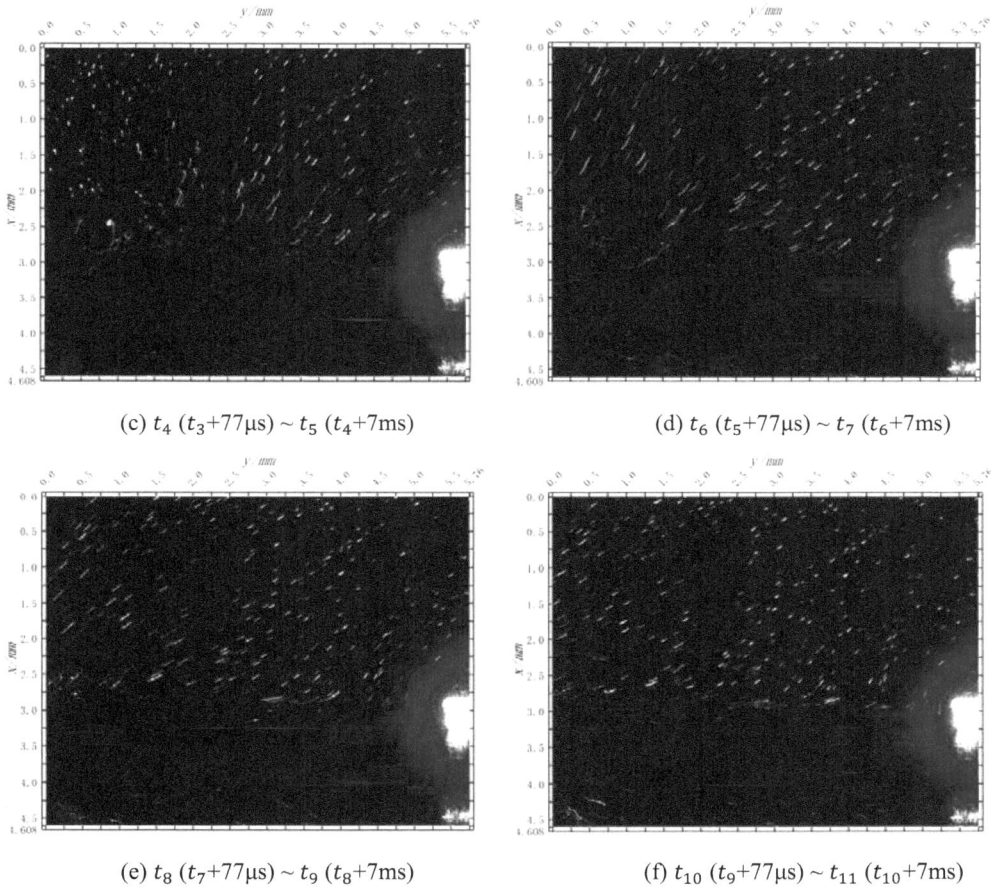

(c) t_4 (t_3+77µs) ~ t_5 (t_4+7ms)

(d) t_6 (t_5+77µs) ~ t_7 (t_6+7ms)

(e) t_8 (t_7+77µs) ~ t_9 (t_8+7ms)

(f) t_{10} (t_9+77µs) ~ t_{11} (t_{10}+7ms)

Fig. (5). Consecutive snapshots of the flow field near the TNTI with $Re = 2047$.

As shown in Fig. (**5a**) with the exposure time of 7ms, the flow is clearly in a mild entrainment mode. After an interval of 77µs between consecutive exposures, some special pathline structures appear as shown in Fig. (**5b**), which indicates a drastic change in the forcing field. After another interval of 77µs, the magnitude and direction of velocity change a lot in Fig. (**5c**) and the entrainment becomes fierce and chaotic which means the turbulent region is 'engulfing' the non-turbulent fluid. During the time of $t_6 \sim t_7$ in Fig. (**5d**), the fierce entrainment structures moves further downstream until the observed structures move out of the view field of the camera, as shown in Fig. (**5e**). The flow restores the mild entrainment mode during $t_{10} \sim t_{11}$ in Fig. (**5f**). This phenomenon keeps happening in the whole process from the observation.

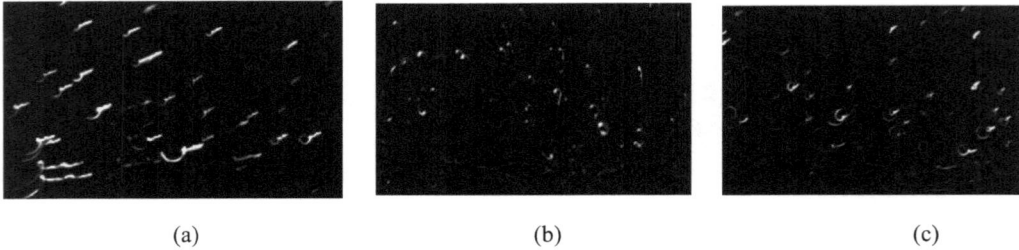

(a)   (b)   (c)

Fig. (6). Special pathline structures with $Re = 2179$.

Special Pathline Structures

In Fig. (**5b**), special pathline structures exist in the flow field. For example, the pathline in Fig. (**6a**) changes fiercely, forming shapes resembling tennis rackets, which can be named as 'tennis racket' vortex. It is found that these structures are formed by the sudden blocking of the downstream fluid. 'Earphone' vortices are also found in the jet flow in Fig. (**6b**) and also named after their shape. It is observed that the non-turbulent fluid is entrained with spiral pathlines into the turbulent region and the 'earphone' vortex is formed by these spiral force along the movement direction. It also should be noted that these special pathline structures are all found in the preceding snapshot of the 'engulfing' moment.

Investigation on the Coherent Structures of Jet Entrainment Boundary Layer

In the region of $x = 2d \sim 3.5d$ in the axial direction and $r = 1d \sim 1.25d$ in the radial direction, the non-dimensional fluctuation velocity is defined as $\tilde{u} = u'/V_J$, where u' is the fluctuation velocity in the axial direction. Thus, the 'engulfing' mode is detected when $\tilde{u} > 0.02$, while the 'nibbling' mode is detected otherwise.

With the method introduced above, consecutive snapshot sequences are obtained and labeled 'engulfing' or 'nibbling' (labeling the 'engulfing' snapshot as 1 and 'nibbling' snapshot as zero). The results are shown by Table **1**. In order to verify the experiments being repeatable, the results from two independent executions (Condition 1 and Condition 2) are compared in Figs. (**7** and **8**), and clearly shows the consistency. The proportion of 'engulfing' snapshots is shown in Fig. (**7**) and it can be concluded that the appearance probability of the fierce 'engulfing' entrainment mode increases with the Reynold number, with a saturation value around Re=2311.

By labeling the first 'engulfing' snapshot in one period as 1 and all other snapshots in this period as zero, this can count the interval frames between 'engulfing' events. And the results are shown in Fig. (**8**). It can be seen that the interval frames are between 7 and 15, and there is a decreasing trend of the period between 'engulfing' events with Reynolds number. In addition, IFFT is applied on this labeled new signal and the main frequency is found between 10 ~ 19 Hz in Reynolds number considered as shown in Fig. (**9**). Moreover, the main frequency clearly increases with Reynolds number.

Table 1. The frame number of 'engulfing' mode with various *Re*.

Re	**Engulfing**	**Total**	**Proportion/ %**
1981	201	1357	14.8
2047	232	1353	17.1
2113	287	1414	20.3
2179	354	1403	25.3
2245	387	1409	27.5
2311	409	1400	29.2
2377	417	1383	30.2

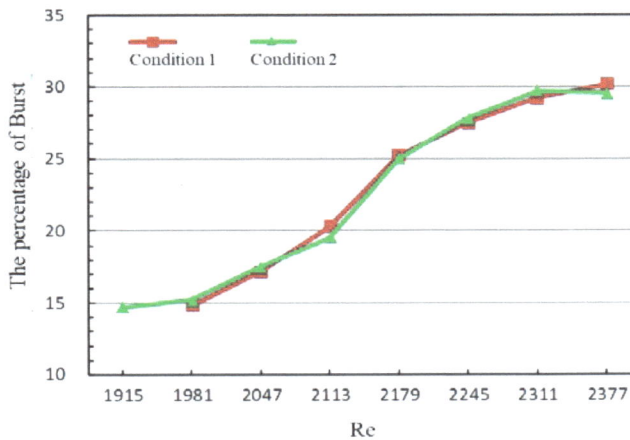

Fig. (7). Proportion of 'engulfing' snapshots.

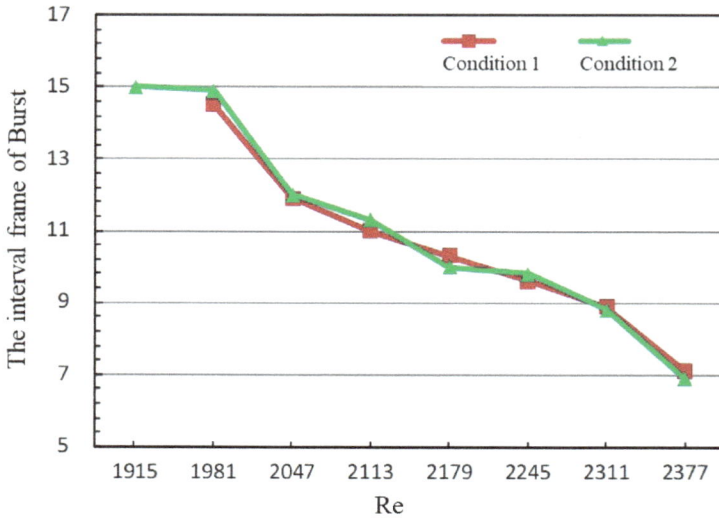

Fig. (8). The interval between 'engulfing' snapshots.

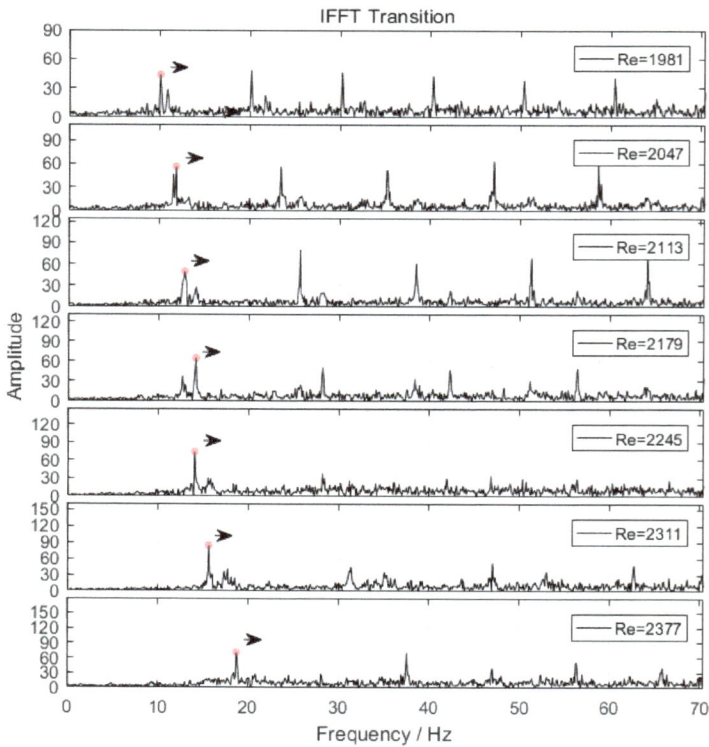

Fig. (9). The frequency of the coherent structures appearance.

Results from MSFLE

In optical measurements of the flow field, compromise has to be made between the view field and the resolution. High resolution results can reveal the detailed flow structures, which requires a large magnification factor of the lens but thus leads to a small view field. To overcome this issue, the MSFLE method is adopted. Clearly, the coherent structures in jet entrainment boundary layer are kind of Lagrangian moving with the main stream. With the traditional single-frame measurement by a fixed camera leading to a limitation of the view field, it is difficult to observe the movement of coherent structures due to the coherent structures shift with main stream velocity. To remedy this, the camera is moving with a constant speed along the axial direction in the frame work of MSFLE. It can provide an intuitive and complete vision of the coherent structures development which has the same speed as the camera. Another advantage of the MSFLE method is that neither vortex identification method nor Galilean velocity decomposition are needed to study the development of coherent structures.

The MSFLE method is applied to the jet flow with $Re = 1909$. The moving speed of the camera is set as 0.133m/s to match the velocity of the coherent structures that we are interested in. The exposure time is $10ms$, the frame rate is 98 frame per second, and interval between exposure is $207\mu s$. The coordinates of the view field can be obtained by the sequence of snapshots and the moving speed of the camera. In Fig. (**10**), the view field is found to be located at around $x/d = 11.03$, and the trajectories in 10ms are shown. It can be seen that the vortical structures are located near the TNTI, and moving along the interface. The Lagrangian coordinate of the snapshot in Fig. (**11a**) is $x/d = 22.06$, while that in Fig. (**11b**) is $x/d = 22.91$. It can be seen in Fig. (**11a**) that a pair of counting rotating vortices exists at the position label 1, while two large vortices are found at the positions 2 and 3. In the following snapshot shown in Fig. (**11b**), the vortices labeled in Fig. (**11a**) can still be identified because the moving speed of the camera matches the mainstream velocity. In addition, the lower vortex of the vortex pair at the position 1 is broken up into two vortices while other labeled vortices haven not broken and remain their shapes.

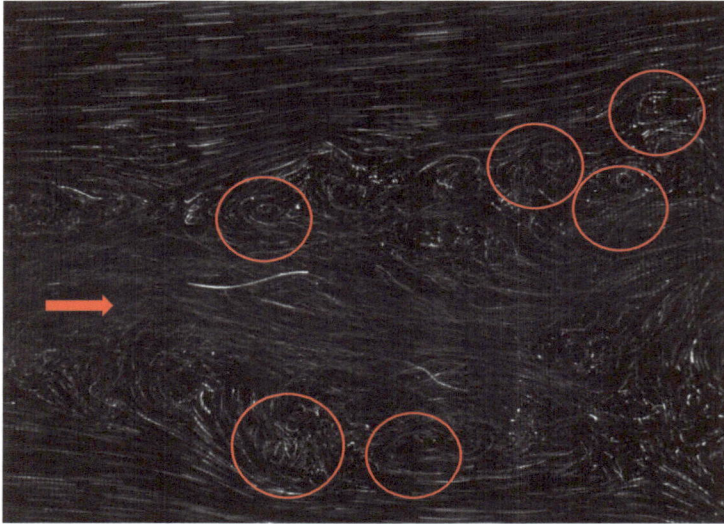

Fig. (10). Experimental result at $x/d = 11.03$ with MSFLE method.

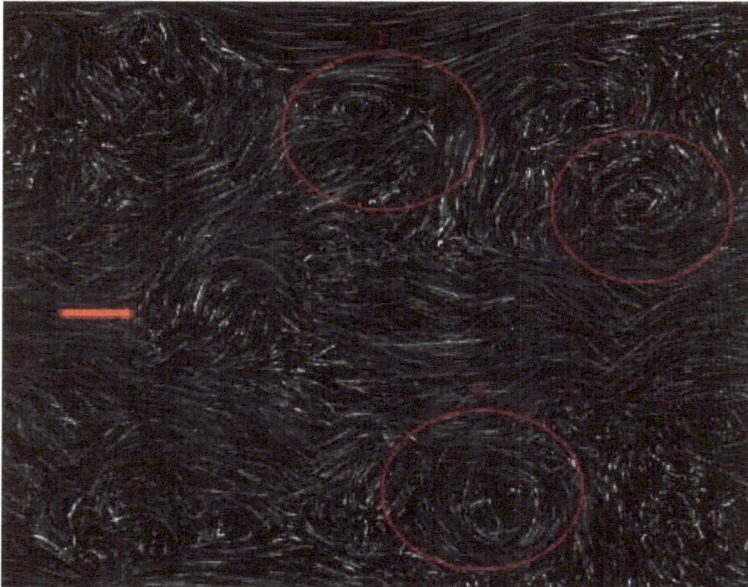

Fig. 11 cont.....

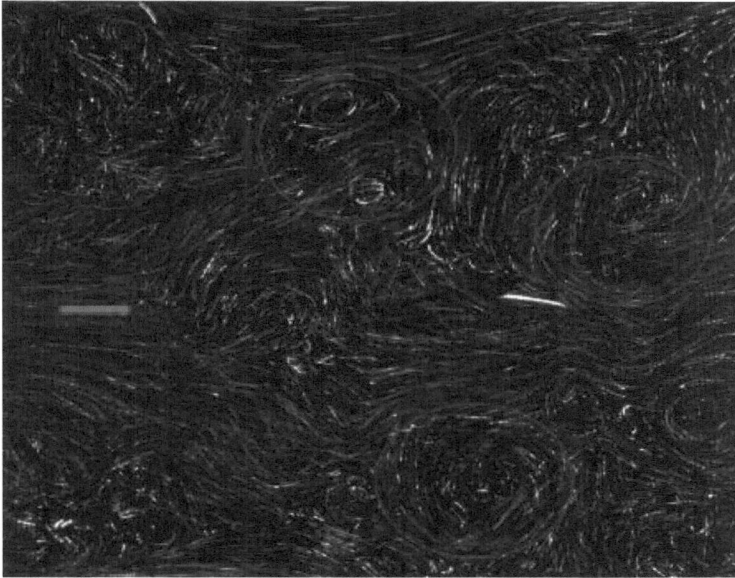

(b)

Fig. (11). Consecutive snapshots obtained by MSFLE.

CONCLUSIONS

The Single-Frame-Long-Exposure (SFLE) and Moving SFLE (MSFLE) methods are applied to investigate the flow field of water submerged jet flows, which clearly shows the development of the coherent structures in the jet entrainment boundary layer. The main conclusions are summarized as follows. (1) In the region of $x = 2d \sim 3.5d$ in the axial direction and $r = 1d \sim 1.25d$ in the radial direction when $Re > 1915$, a coherent alternating of 'engulfing' and dominating 'nibbling' entrainment modes exist. It is also observed that special pathline like 'tennis racket' vortices and 'earphone' vortices appear just before the 'engulfing' entrainment mode. (2) With Reynolds number between 1915 and 2377, the main frequency of the coherent alternating entrainment modes is between 10~19 Hz. In addition, the appearance probability of such a pattern increases with Reynolds number, while the period decreases with Reynolds number. (3) The coherent structures in jet flows are mainly found near the TNTI. Vortices of various scales are found in this region, which are observed to deform, break up and merge. These results show that the SFLE and MSFLE methods are effective in capturing the coherent motions inside jet entrainment boundary layer flows.

CONSENT FOR PUBLICATION

Not applicable.

CONFLICT OF INTEREST

The authors confirm that this chapter contents have no conflict of interest.

ACKNOWLEDGEMENTS

This work is partly supported by the National Natural Science Foundation of China (Grant No. 51576130, No. 51327803), whose help is gratefully acknowledged.

REFERENCES

[1] G.B. Brown, "On vortex motion in gaseous jets and the origin of their sensitivity to sound", *Proc. Phys. Soc.,* vol. 47, no. 4, pp. 703-732, 1935.
[http://dx.doi.org/10.1088/0959-5309/47/4/314]

[2] P.O.A.L. Davies, M.J. Fisher, and M.J. Barratt, "The characteristics of the turbulence in the mixing region of a round jet", *J. Fluid Mech.,* vol. 15, no. 3, pp. 337-367, 1963. [http://dx.doi.org/10.1017/S0022112063000306]

[3] Z.Y. Dong, *Jet Mechanic.* Science Press: Beijing, 2005. (In Chinese)

[4] J. Westerweel, T. Hofmann, C. Fukushima, and J. Hunt, "The turbulent/non-turbulent interface at the outer boundary of a self-similar turbulent jet", *Exp. Fluids,* vol. 33, no. 6, pp. 873-878, 2002. [http://dx.doi.org/10.1007/s00348-002-0489-5]

[5] L. Gan, and T.B. Nickels, "An experimental study of turbulent vortex rings during their early development", *J. Fluid Mech.,* vol. 649, no. 6, pp. 467-496, 2010. [http://dx.doi.org/10.1017/S0022112009993971]

[6] L. Gan, "Detection of passive scalar interface directly from PIV particle images in inhomogeneous turbulent flows", *Flow Turbul. Combus.,* vol. 97, no. 1, pp. 141-170, 2016. [J]. [http://dx.doi.org/10.1007/s10494-015-9691-4]

[7] L. Gan, and A. Maffioli, "The size and lifetime of organized eddies in an on-solid-body rotating turbulence experiment", *Eur. J. Mech. BFluids,* vol. 74, pp. 41-49, 2019.
[http://dx.doi.org/10.1016/j.euromechflu.2018.10.026]

[8] H. Markus, A. Liberzon, N.V. Nikitin, B. Luthi, and W. Kinzelbach, "A Lagrangian investigation of the small-scale features of turbulent entrainment through particle tracking and direct numerical simulation", *J. Fluid Mech.,* vol. 598, no. 7, pp. 465-475, 2008.

[9] M. Wolf, B. Lüthi, M. Holzner, A. Liberzon, D. Krug, and A. Tsinober, "A Lagrangian, small-scale investigation of turbulent entrainment in an axisymmetric jet", *J. Phys. Conf. Ser,* vol. 318, no. SECTION 3, 2011.
[http://dx.doi.org/10.1088/1742-6596/318/3/032053]

[10] C.B. Da Silva, R.R. Taveira, and G. Borrell, "Characteristics of the turbulent/nonturbulent interface in boundary layers, jets and shear-free turbulence", *J. Phys. Conf. Ser.,* vol. 506, no. 1, 2014. [http://dx.doi.org/10.1088/1742-6596/506/1/012015]

[11] H. Yadav, A. Agrawal, and A. Srivastava, "Mixing and entrainment characteristics of a pulse jet", *Int. J. Heat Fluid Flow,* vol. 61, pp. 749-761, 2016. [http://dx.doi.org/10.1016/j.ijheatfluidflow.2016.08.006]

[12] A. Glezer, "An experimental study of a turbulent vortex ring", *J. Fluid Mech.,* vol. 211, pp. 243-283, 1981.
[http://dx.doi.org/10.1017/S0022112090001562]

[13] J.L. Chen, C. Li, and X.S. Cai, "Search on the single frame imaging method for measuring multi- parameter fields in flow field", *Journal of experiments in Fluid Mechanics,* vol. 29, no. 6, pp. 67-72, 2015.

Current Developments in Mathematical Sciences, 2020, Vol. 2, 293-311

CHAPTER 13

Hybrid Compact-WENO Scheme for the Interaction of Shock Wave and Boundary Layer

Jianming Liu[1,2] **and Chaoqun Liu**[2,*]

[1]*School of Mathematics and Statistics, Jiangsu Normal University, Xuzhou 221116, China*

[2]*Department of Mathematics, University of Texas at Arlington, Arlington, Texas 76019, USA*

Abstract: In this chapter, an introduction to hybrid Weighted Essentially non-oscillatory (WENO) method is given. The hybrid techniques including both central and compact finite difference schemes are introduced. The paper review about the driven mechanism of the high order finite scheme required for compressible flow with shock is presented. The detailed constructing processes of the compact and WENO schemes are given and the hybrid detector is introduced. Further, in particular, a series of examples in the field of the compressible flow are designed to illustrate the different methods.

Keywords: Navier-Stokes equation, High order finite difference scheme, Hybrid Compact-WENO, Shock wave and boundary layer interaction turbulence.

A SHORT REVIEW ON STUDY OF HIGH ORDER FINITE DIFFERENCE SCHEME FOR COMPRESSIBLE FLOWS

The compressible flow field is in general governed by the Navier-Stokes equations deduced from physical conservation law. Due to the complexity of the flow problems, if there is no high-order scheme, it is impossible to obtain exact results which can embody complex fluid structures with different scales. To date, there are many effective numerical discretization methods for compressible flow, such as finite difference method (FDM), finite volume method (FVM), discontinuous Galerkin method (DGM) [1], high-order flux reconstruction [2], Spectral volume/difference Method (SVM) [3], *etc*. The finite difference method is one of the main numerical methods of computational fluid dynamics (CFD) because of its simplicity and easy to achieve high precision with the longest history. In addition, it is one of the most mature, widely used, and the most effective method. Hence, in this chapter, we focus on the high-order FDM.

---

*Corresponding author Chaoqun Liu:** Department of Mathematics, University of Texas at Arlington, Arlington, Texas 76019, USA; Tel: +1-8172725151; Fax: +1-8172725802; E-mail: cliu@uta.edu

In the field of CFD, in general, we consider the scheme with at least third-order accuracy is high order. Since the 1990s, the research and application of high-order finite difference methods have begun to make substantial progress. In the present, the multi-scale complex flow such as simulating turbulence requires high-order numerical methods, which has become the consensus of the scientific community. Many scholars have developed a number of high-order numerical methods with advanced algorithms and good computational effects. Emerging advanced numerical methods include the ENO/WENO method [4-7], non-oscillatory containing no free parameter and dissipative (NND) scheme [8], group velocity control (GVC) scheme [9], compact scheme [10-12], *etc*. Among these numerical methods, the compact scheme only requires small number of grid points to get the high-order accuracy. The compact scheme had relatively small dissipation and gained a lot of favor in the field of direct numerical simulation (DNS) of turbulence flow [10-13]. The high-order central finite difference scheme is another kind of more effective method, which is straightforward and easy to implement and doesn't need to calculate the derivative performed by the compact scheme. Although compact scheme and high-order central scheme have obvious advantages in the simulation of multi-scale turbulent flow, it is hard to simulate the compressible flows with shock wave. On the other hand, to date, upwind or bias upwind high-order WENO scheme has achieved great success in capturing the shocks sharply. In the DNS of turbulence flow, the small length scale vortex is very important in the flow transition and turbulence process and thus very sensitive to any artificial numerical dissipation [11]. But the dissipation caused by WENO scheme is still harmful to the simulation of the flow transition of turbulent flow. Hence, in order to capture the shock waves sharply and resolve the small-scale turbulent flow simultaneously, a combination of compact or central and WENO schemes is desirable [11, 13-15].

Governing Equations

We consider the compressible Navier-Stokes equations in Cartesian tensor form as

$$\frac{\partial \rho}{\partial t} + \frac{\partial \rho u_j}{\partial x_j} = 0,$$

$$\frac{\partial \rho u_i}{\partial t} + \frac{\partial \rho u_i u_j}{\partial x_j} = -\frac{\partial p}{\partial x_i} + \frac{\partial \tau_{ij}}{\partial x_j}, \tag{1}$$

$$\frac{\partial E}{\partial t} + \frac{\partial u_i E}{\partial x_i} = -\frac{\partial p u_i}{\partial x_i} + \frac{\partial u_i \tau_{ij}}{\partial x_j} - \frac{\partial q_i}{\partial x_i},$$

where

$$\tau_{ij} = \mu \left(\frac{\partial u_i}{\partial x_j} + \frac{\partial u_j}{\partial x_i} - \frac{2}{3} \frac{\partial u_k}{\partial x_k} \delta_{ij} \right),$$

$$q_i = -k \frac{\partial T}{\partial x_i}.$$

With the reference values of characteristic length L, the free stream speed U_∞, temperature T_∞, viscous coefficient μ_∞, density ρ_∞, and the pressure $\rho_\infty U_\infty^2$, the Navier-Stokes equation (1) can be non-dimensioned and the only changed terms are

$$\tau_{ij} = \frac{\mu}{Re} \left(\frac{\partial u_i}{\partial x_j} + \frac{\partial u_j}{\partial x_i} - \frac{2}{3} \frac{\partial u_k}{\partial x_k} \delta_{ij} \right),$$

$$q_i = -\frac{1}{Re(\gamma-1)PrMa_\infty^2} k \frac{\partial T}{\partial x_i}.$$

In order to illustrate the establishment of the finite difference scheme of Equation (1), we use the following simple scalar model equation as sample.

$$\frac{\partial u}{\partial t} + \frac{\partial f}{\partial x} = \mu \frac{\partial^2 u}{\partial x^2}. \tag{2}$$

The first derivative term in Equation (2) is used to simulate the convective term, and the second derivative term is used to simulate the viscous term in Navier-Stokes equation (1).

High-order Central Finite Difference Scheme

The main difficulty in the calculation of compressible fluid problems lies in the discretization of convective terms. To simulate the complex turbulence phenomena, it is significant to discretize the inviscid term by a high order method. In this study, we only consider the fourth/sixth- order approximation. The convect term $\frac{\partial f}{\partial x}$ at the grid point x_i can be approximated by

$$\left. \frac{\partial f}{\partial x} \right|_{x_i} = \frac{\hat{f}_{i+1/2} - \hat{f}_{i-1/2}}{\Delta x}. \tag{3}$$

The fourth order central difference scheme to approximate the face flux can be formulated as

$$\hat{f}_{i+1/2} = \frac{1}{12} (- f_{i-1} + 7f_i + 7f_{i+1} - f_{i+2}), \tag{4}$$

and

$$\hat{f}_{i-1/2} = \frac{1}{12}(-f_{i-2} + 7f_{i-1} + 7f_i - f_{i+1}). \tag{5}$$

Equations (4) and (5) give the fourth order central difference scheme as [15]

$$\left.\frac{\partial f}{\partial x}\right|_{x_i} = \frac{1}{12\Delta x}(f_{i-2} - 8f_{i-1} + 8f_{i+1} - f_{i+2}) + O(\Delta x^4). \tag{6}$$

For sixth order central approximation, similarly, the inviscid flux can by approximated by

$$\hat{f}_{i+1/2} = \frac{1}{60}(f_{i-2} - 8f_{i-1} + 37f_i + 37f_{i+1} - 8f_{i+2} + f_{i+3}), \tag{7}$$

and

$$\hat{f}_{i-1/2} = \frac{1}{60}(f_{i-3} - 8f_{i-2} + 37f_{i-1} + 37f_i - 8f_{i+1} + f_{i+2}). \tag{8}$$

Then the sixth order central difference scheme can be written as

$$\left.\frac{\partial f}{\partial x}\right|_{x_i} = \frac{1}{60\Delta x}(-f_{i-3} + 9f_{i-2} - 45f_{i-1} + 45f_{i+1} - 9f_{i+2} + f_{i+3}) + O(\Delta x^6). \tag{9}$$

For the compact sixth order finite difference schemes, one can refer to the papers [16,17]. In the governing equations of fluid, the second-order derivative only appeared in the terms of viscosity and heat conductivity. In general, the second-order derivative should also be discretized by high order method. Based on the Taylor Series expansion, we can easily obtain the fourth-order finite difference scheme for second-order derivation as

$$\frac{\partial^2 u}{\partial x^2} = \frac{1}{\Delta x^2}\left(-\frac{1}{12}u_{i-2} + \frac{4}{3}u_{i-1} - \frac{5}{2}u_i + \frac{4}{3}u_{i+1} - \frac{1}{12}u_{i+2}\right) + O(\Delta x^4). \tag{10}$$

Similarly, the sixth-order finite difference scheme can be formulated by

$$\frac{\partial^2 u}{\partial x^2} = \frac{1}{\Delta x^2}\left(\frac{1}{90}u_{i-3} - \frac{3}{20}u_{i-2} + \frac{3}{2}u_{i-1} - \frac{49}{18}u_i + \frac{3}{2}u_{i+1} - \frac{3}{20}u_{i+2} + \frac{1}{90}u_{i+3}\right) + O(\Delta x^6). \tag{11}$$

Compact Finite Difference Scheme

In order to achieve high-order accuracy without adding grid points, people developed compact scheme. The compact scheme is generally constructed by the Hermite interpolation method, in which both function and its derivatives at grid points are involved. So it can get higher precision with few points. The research shows that the compact scheme has also high resolution. Although the compact scheme is not direct and demand extra work, Fourier analysis indicates that the compact scheme has better spectral resolution than a traditional finite difference scheme [11,18].

By the Taylor series, it is very easy to get the following standard sixth order compact scheme for the first order derivative convect term [11]

$$\frac{1}{3}f'_{i-1} + f'_i + \frac{1}{3}f'_{i+1} = \Delta x^{-1}\left[-\frac{1}{36}f_{i-2} - \frac{7}{9}f_{i-1} + \frac{7}{9}f_{i+1} + \frac{1}{36}f_{i+2}\right]. \qquad (12)$$

In general, the approximated compact scheme is not unique. The sixth-order central compact scheme can also be obtained by [12]

$$\frac{1}{5}f'_{i-1} + \frac{3}{5}f'_i + \frac{1}{5}f'_{i+1} = \Delta x^{-1}\left[\frac{7}{15}(f_{i+1} - f_{i-1}) + \frac{1}{60}(f_{i+2} - f_{i-2})\right]. \qquad (13)$$

The high-order central scheme in Equation (6) and (9) and the standard compact scheme in Equation (12) and (13) did not have dissipation and need filters even for smooth area to prohibit dispersion error. In order to tackle this shortage, the upwind scheme is often suggested to keep the high order without the filter. The upwind scheme can be obtained by the monotone flux decomposing. The simplest smooth flux splitting is the Lax-Friedrichs splitting [5, 19], which can be formulated by $f^{\pm}(u) = \frac{1}{2}[f(u) \pm \alpha u]$. Where α is taken as the maximum $f'(u)$ over the relevant range of u.

The Equation (14) gives the upwind traditional finite difference scheme. For the positive flux function f^+, the fifth-order upwind scheme can be obtained by

$$f'^+_i = \frac{1}{60\Delta x}[-3(f^+_{i+2} - f^+_{i+1}) + 27(f^+_{i+1} - f^+_i) + 47(f^+_i - f^+_{i-1}) - 13(f^+_{i-1} - f^+_{i-2}) + 2(f^+_{i-2} - f^+_{i-3})]. \qquad (14)$$

For high order compact scheme, it is also possible to develop the upwind version [9,12].

$$\frac{3}{5}f'^+_i + \frac{2}{5}f'^+_{i-1} = \frac{1}{60\Delta x}[-(f^+_{i+2} - f^+_{i+1}) + 11(f^+_{i+1} - f^+_i) + 47(f^+_i - f^+_{i-1}) +$$
$$3(f^+_{i-1} - f^+_{i-2})]. \tag{15}$$

In the paper [11], the authors give the compact scheme at the half-point with seventh-order accuracy as

$$\frac{1}{2}H'^+_{i-\frac{3}{2}} + H'^+_{i-\frac{1}{2}} + \frac{1}{4}H'^+_{i+\frac{1}{2}} = \frac{1}{\Delta x}\left(\frac{1}{240}H^+_{i-\frac{7}{2}} - \frac{1}{12}H^+_{i-\frac{5}{2}} - \frac{11}{12}H^+_{i-\frac{3}{2}} + \frac{1}{3}H^+_{i-\frac{1}{2}} + \frac{31}{48}H^+_{i+\frac{1}{2}} +$$
$$\frac{1}{60}H^+_{i+\frac{3}{2}}\right). \tag{16}$$

Here H is the primitive function of flux defined by [19]

$$H^+\left(x_{i+\frac{1}{2}}\right) = \int_{-\infty}^{x_i+\frac{\Delta x}{2}} \hat{f}^+(\xi)d\xi = \sum_{j=-\infty}^{j=i} \int_{x_{j-\frac{\Delta x}{2}}}^{x_{j+\frac{\Delta x}{2}}} \hat{f}^+(\xi)d\xi = \Delta x \sum_{j=-\infty}^{i} f^+_j. \tag{17}$$

Then the numerical flux in Equation (3) can be obtained by

$$\hat{f}_{i+\frac{1}{2}} = H'_{i+\frac{1}{2}}. \tag{18}$$

In the above schemes, we only give the discretization of the positive flux. Similarly, we can get the high-order upwind scheme for the negative flux function f^-.

WENO Schemes

In Equation (3), the derivative of flux is written in the form of numerical flux. Based on the weighted method of different numerical flux, the total flux can be weighted by the fluxes on three different stencils for fifth-order WENO scheme (WENO5) [19] as follows

$$\hat{f}^{WENO5}_{i+\frac{1}{2}} = \omega_1 \hat{f}^{(1)}_{i+\frac{1}{2}} + \omega_2 \hat{f}^{(2)}_{i+\frac{1}{2}} + \omega_3 \hat{f}^{(3)}_{i+\frac{1}{2}}. \tag{19}$$

Here $\hat{f}^{(1)}_{i+\frac{1}{2}}$, $\hat{f}^{(2)}_{i+\frac{1}{2}}$, $\hat{f}^{(3)}_{i+\frac{1}{2}}$ are the numerical flux on small stencil $\{i-2, i-1, i\}$, $\{i-1, i, i+1\}$ and $\{i, i+1, i+2\}$, respectively. Using the Taylor Series expansion, it is feasible to get the third-order approximation. The final expressions for positive flux f^+ are given as

$$\hat{f}^{(1)}_{i+\frac{1}{2}} = \frac{1}{3}f_{i-2} - \frac{7}{6}f_{i-1} + \frac{11}{6}f_i,$$

$$\hat{f}^{(2)}_{i+\frac{1}{2}} = -\frac{1}{6}f_{i-1} + \frac{5}{6}f_i + \frac{1}{3}f_{i+1}, \tag{20}$$

$$\hat{f}^{(3)}_{i+\frac{1}{2}} = \frac{1}{3}f_i + \frac{5}{6}f_{i+1} - \frac{1}{6}f_{i+2}.$$

The ideal weight factor for Equation (18) is $C_1 = \frac{1}{10}, C_2 = \frac{6}{10}, C_3 = \frac{3}{10}$. In order to suppress the shock oscillation and improve the numerical stability, Jiang and Shu (1996) [7] give a group of weighted coefficients

$$\omega_k = \frac{\alpha_k}{\alpha_1 + \alpha_2 + \alpha_3}, \quad \alpha_k = \frac{C_k}{(\epsilon + IS_k)^2}, \quad k = 1,2,3. \tag{21}$$

Here, IS_k is the smooth indicator on the small stencil and is suggested to be

$$IS_k = \sum_{l=1}^{2} \int_{x_{i-\frac{1}{2}}}^{x_{i+\frac{1}{2}}} \Delta x^{2l-1} \left(q_i^{(l)} \right)^2 dx, \tag{22}$$

where the function $q_i^{(l)}$ is the l-th order derivative of the interpolation polynomial on the small stencil of i. From a simple calculation, we can get

$$IS_1 = \frac{1}{4}(f_{i-2} - 4f_{i-1} + 3f_i)^2 + \frac{13}{12}(f_{i-2} - 2f_{i-1} + f_i)^2,$$

$$IS_2 = \frac{1}{4}(f_{i-1} - f_{i+1})^2 + \frac{13}{12}(f_{i-1} - 2f_i + f_{i+1})^2, \tag{23}$$

$$IS_3 = \frac{1}{4}(3f_i - 4f_{i+1} + f_{i+2})^2 + \frac{13}{12}(f_i - 2f_{i+1} + f_{i+2})^2.$$

For negative flux f^-, we only need to change the subscripts in Equation (20) and (23) from $i + 2$ to $i - 2$, from $i + 1$ to $i - 1$, from $i + \frac{1}{2}$ to $i - \frac{1}{2}$, and from $i - 2$ to $i + 2$, from $i - 1$ to $i + 1$.

For DNS of turbulence, WENO5 is too dissipative to capture small vortex structure. Based on the method in [7, 19], it is easy to get the seventh-order WENO scheme (WENO7). The detailed formula for WENO7, please refer to [12,19].

HYBRID WENO SCHEMES

In this section, firstly, we introduce the hybrid central-WENO scheme. The central scheme generally has not enough dissipation to suppress the oscillation near the shock. However, the WENO scheme has sharp resolution near the discontinuity but with too much dissipation. Hence, it is natural to combine two different schemes. And the numerical flux of a hybrid central-WENO scheme can be written as

$$\hat{f}_{i+\frac{1}{2}}^{hybrid} = \Omega \cdot \hat{f}_{i+\frac{1}{2}}^{central} + (1 - \Omega) \cdot \hat{f}_{i+\frac{1}{2}}^{WENO}. \tag{24}$$

We can take Equation (4) or (7) as the central flux. If you want to get more accurate approximation, it is also feasible to use eighth-order central scheme. When $\Omega = 0$, the numerical flux in Equation (24) becomes a standard WENO flux, but when $\Omega = 0$ the scheme is a standard central scheme. Using the same method as in Equation (24), we can also design a hybrid compact-WENO scheme. For example, take

$$f_i^{\prime WENO} = \frac{\hat{f}_{i+\frac{1}{2}}^{WENO} - \hat{f}_{i-\frac{1}{2}}^{WENO}}{\Delta x}. \tag{25}$$

And from Equations (12)-(15), we can calculate the derivative of the flux at the grid point. Then we can get a hybrid compact-WENO scheme as

$$f_i^{\prime hybrid} = \Omega \cdot f_i^{\prime compact} + (1 - \Omega) \cdot f_i^{\prime WENO}. \tag{26}$$

In paper [11], authors gave a different method to deduce a hybrid compact-WENO scheme. In this method, using Equation (16) and (18), the upwind flux at half-point is calculated by

$$\frac{1}{2}\Omega H_{i-\frac{3}{2}}^{\prime+} + H_{i-\frac{1}{2}}^{\prime+} + \frac{1}{4}\Omega H_{i+\frac{1}{2}}^{\prime+} = \Omega\left[\frac{1}{\Delta x}\left(\frac{1}{240}H_{i-\frac{7}{2}}^{+} - \frac{1}{12}H_{i-\frac{5}{2}}^{+} - \frac{11}{12}H_{i-\frac{3}{2}}^{+} + \frac{1}{3}H_{i-\frac{1}{2}}^{+} + \frac{31}{48}H_{i+\frac{1}{2}}^{+} + \frac{1}{60}H_{i+\frac{3}{2}}^{+}\right)\right] + (1 - \Omega)\hat{f}_{i-\frac{1}{2}}^{WENO}. \tag{27}$$

In Equation (24), (26) and (27), the weight of Ω should be automatically tuned between the smooth flow and the shock wave. In the paper [11], the authors used multigrid method to detect the shock, but the algorithm of multigrid is more difficult. Multi-resolution analysis is used in paper [14]. These methods are not direct. In the paper [15], Kim and Kwon gave a detailed study. But with the switching function transforming from the WENO scheme to central/compact scheme, the oscillation will appear and the stability of the scheme will become deteriorate. Based on our research, for the scheme (27), the Ω is set to be 0.72 for 1D problem. The including of the WENO in (27) can enlarge the numerical dissipation and stabilize the compact scheme.

TIME DISCRETIZATION

Up to now we have only discussed spatial discretization. After the discretization of spatial derivative, the original governing equation will become a system of ordinary differential equations. In this section we introduce the time discretization.

The ordinary differential equations can be simplified as

$$\frac{\partial U}{\partial t} = L(U), \tag{28}$$

resulting from a line method of the discretization of the governing equations (1). There are usually two types of methods to calculate the solutions of (28), and that is implicit or explicit discretization. A simple implicit second-order three level time discretization scheme is

$$\frac{3U_i^{n+1} - 4U_i^n + U_i^{n-1}}{2\Delta t} = L(U_i^{n+1}). \tag{29}$$

Since the scheme is implicit and the right-hand term is generally non-linear, the iterative method should be used to get the convergent solution. Another type of discretization method is explicit scheme. In explicit methods, Runge-Kutta method (RK) is very popular in the simulation of the compressible flow. The popular RK method generally has third or fourth order accuracy. For example, a three-step third-order RK method can be formulated as [12]

$$U^{(1)} = U^n + \frac{8}{15}\Delta t L(U^n),$$
$$U^{(2)} = U^n + \Delta t \left(\frac{1}{4}L(U^n) + \frac{5}{12}L(U^{(1)})\right), \tag{30}$$
$$U^{n+1} = U^n + \Delta t \left(\frac{1}{4}L(U^n) + \frac{3}{4}L(U^{(2)})\right),$$

and a four-step fourth-order RK method is

$$U^{(1)} = U^n + \frac{1}{2}\Delta t L(U^n),$$
$$U^{(2)} = U^n + \frac{1}{2}\Delta t L(U^{(1)}),$$
$$U^{(3)} = U^n + \Delta t L(U^{(2)}), \tag{31}$$
$$U^{n+1} = U^n + \frac{1}{6}\Delta t \left(L(U^n) + 2L(U^{(1)}) + 2L(U^{(2)}) + L(U^{(3)})\right).$$

For high-order scheme, the TVD (total variation diminishing) high order Runge-Kutta methods is another popular time discretization method. For shock wave, Shu's research show that the oscillations can be observed when the non-TVD Runge-Kutta method coupled with a second order spatial discretization [19]. Shu's paper gave the second-order TVD RK method

$$U^{(1)} = U^n + \Delta t L(U^n),$$
$$U^{n+1} = \frac{1}{2}U^n + \frac{1}{2}U^{(1)} + \frac{1}{2}\Delta t L\big(U^{(1)}\big), \tag{32}$$

and third-order TVD RK method

$$U^{(1)} = U^n + \Delta t L(U^n),$$
$$U^{(2)} = \frac{3}{4}U^n + \frac{1}{4}U^{(1)} + \frac{1}{4}\Delta t L(U^{(1)}),$$
$$U^{n+1} = \frac{1}{3}U^n + \frac{2}{3}U^{(1)} + \frac{2}{3}\Delta t L\big(U^{(2)}\big). \tag{33}$$

In our research, we use third-order TVD RK method (33) to discretize the half discretization equation (28).

NUMERICAL EXAMPLES

In this section, we will test the hybrid central/compact-WENO schemes. The first example is to simulate the interaction between shock wave and entropy waves [19]. The governing equation is the inviscid Euler equation as Equation (1) without viscous term. The calculation starts with the initial field as follows:

$$U_0 = \begin{cases} (3.857143, 2.629369, 10.33333), & x < -4 \\ (1 + 0.2\sin(5x), 0, 1), & x \geq -4 \end{cases}$$

In this example, the shock is moving into a density sine fluctuation field. The computational domain for this special problem is $[-5,5]$. We use the hybrid compact-WENO scheme (27). The numerical solution of density with $N = 401$ is shown in Fig. (**1**). The solution is approximated well to the exact solution calculated by setting $N = 1601$. In same figure, the result by WENO5 is also given, we can conclude the hybrid method can get more better result. The pressure is shown in Fig. (**2**).

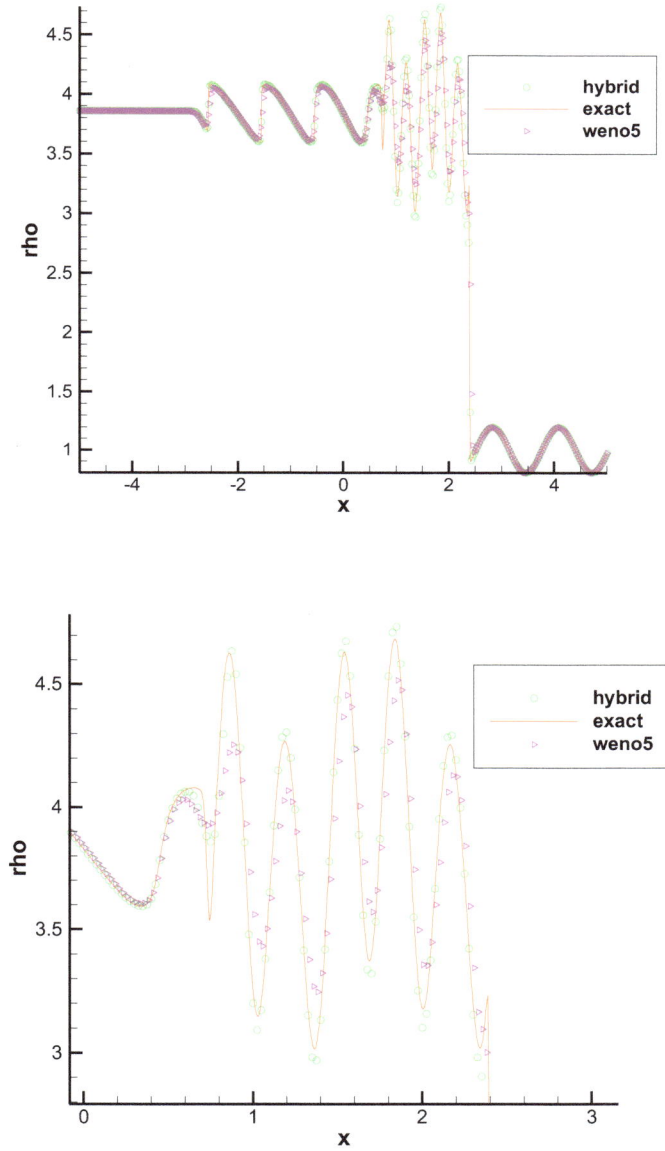

Fig. (1). The density solution of the Mach 3 shock-entropy wave interaction.

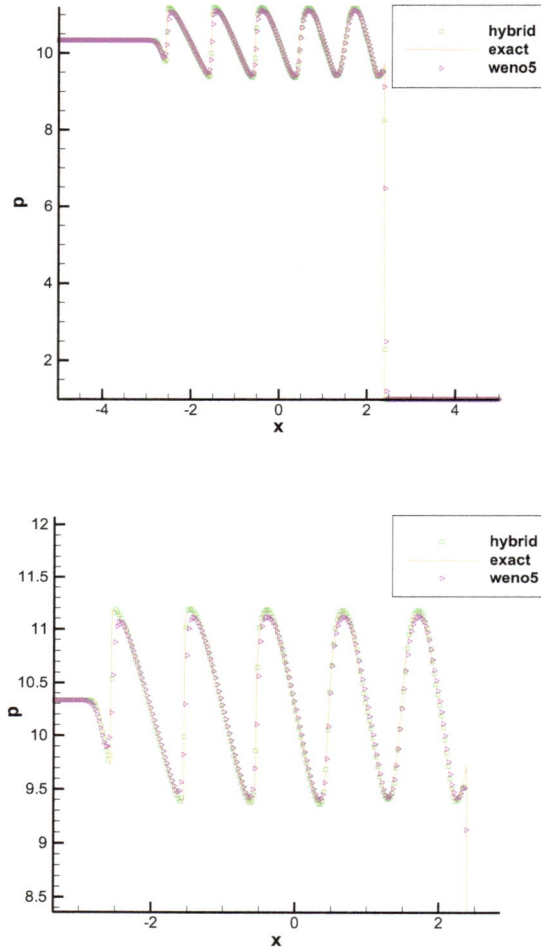

Fig. (2). The pressure solution of the Mach 3 shock-entropy wave interaction.

Second example is a shock tube problem. The set-up of the problem is a Riemann type initial data

$$U(x,0) = \begin{cases} U_L, & \text{if } x \leq 0 \\ U_R, & \text{if } x > 0 \end{cases}$$

and the Sod's problem

$$(\rho_L, u_L, p_L) = (1,0,1), \quad (\rho_R, u_R, p_R) = (0.25,0,0.1)$$

is solved by the hybrid compact-WENO scheme, the number of the grid points is 101. The density and pressure are shown in Figs. (**3** and **4**). The shock and contact discontinuity are sharply captured by the present method. The compare with WENO5 method is also reported in Figs. (**3** and **4**).

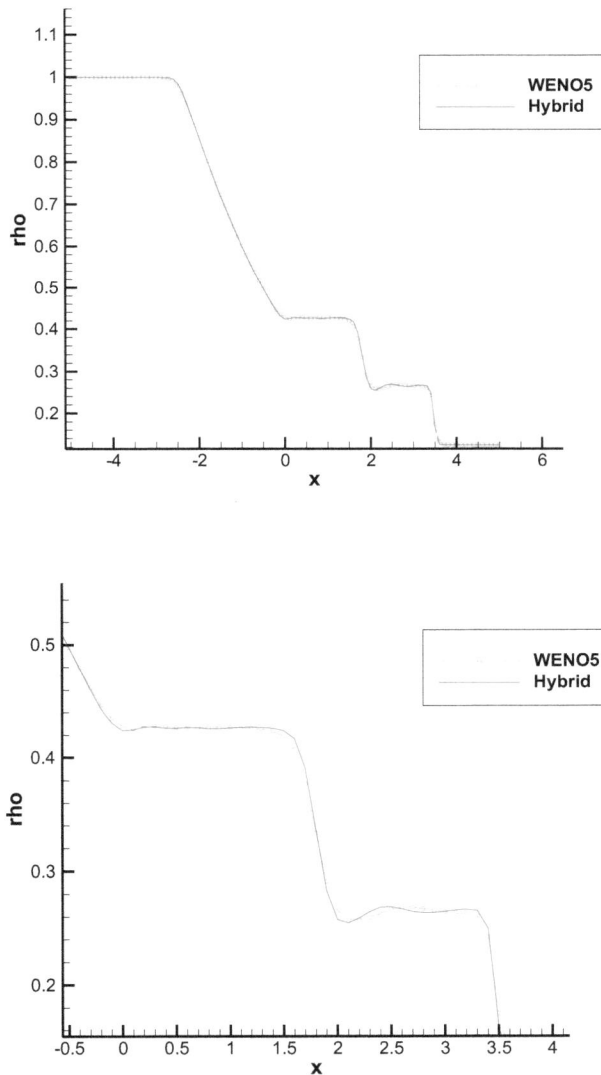

Fig. (3). The density of the shock-tube problem.

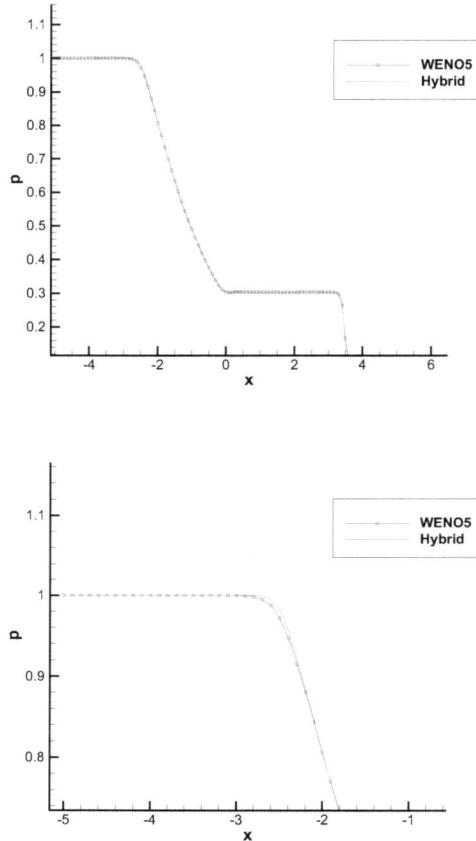

Fig. (4). The pressure of the shock-tube problem.

The third example, we will consider a two-dimensional Euler equation. In this case, an incident shock at attach angle of $35.241°$ with an inflow Mach number of 2 is simulated by the hybrid compact-WENO5 scheme. In order to prohibit the oscillation near the discontinuity, the upwind compact scheme coupled with WENO5 is used. The number of computational grids is 129×129. The final density and pressure contours are shown in Fig. (**5**). The shock wave is captured sharply as shown in the figure. For viscous compressible flow, the Reynolds number is 3×10^4 is considered. The second-order derivative term in the viscous Navier-Stokes equation is discretized by the high order central scheme. The final density, Mach number, pressure and temperature contours are reported in Figs. (**6** and **7**). It can be found the results obtained by the present method are reasonable.

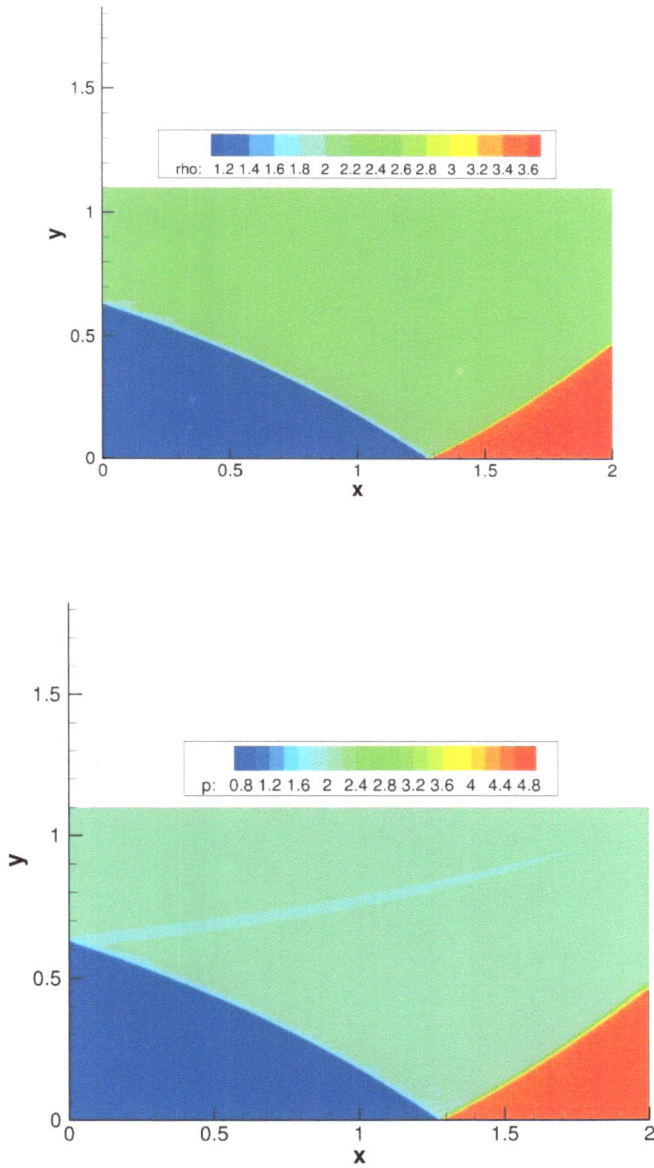

Fig. (5). The density and pressure contours.

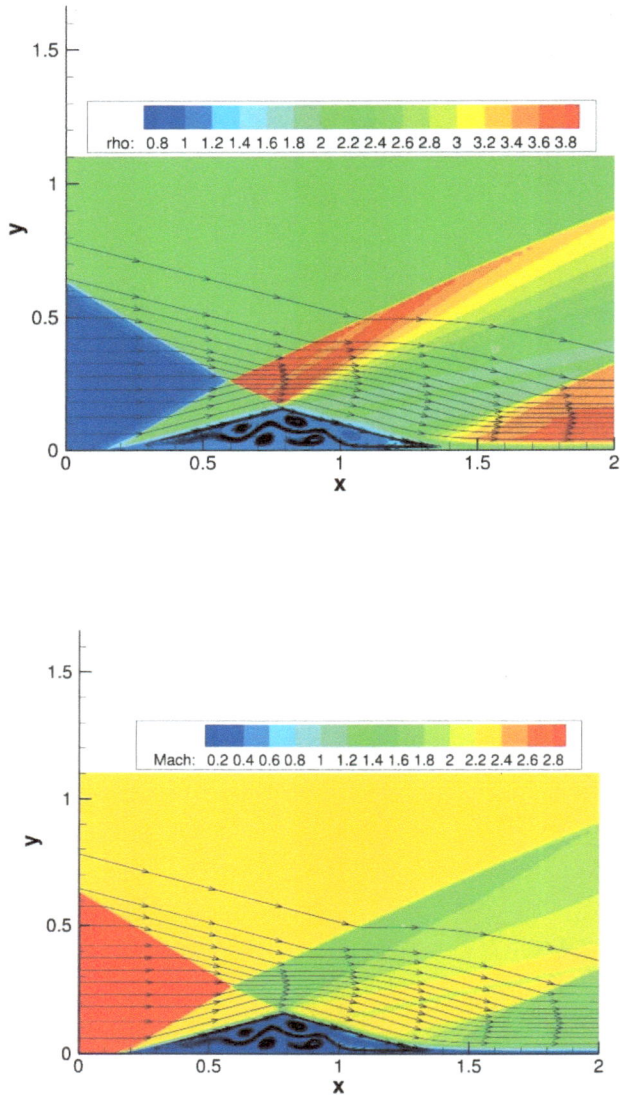

Fig. (6). The density and Mach number contours of viscous problem.

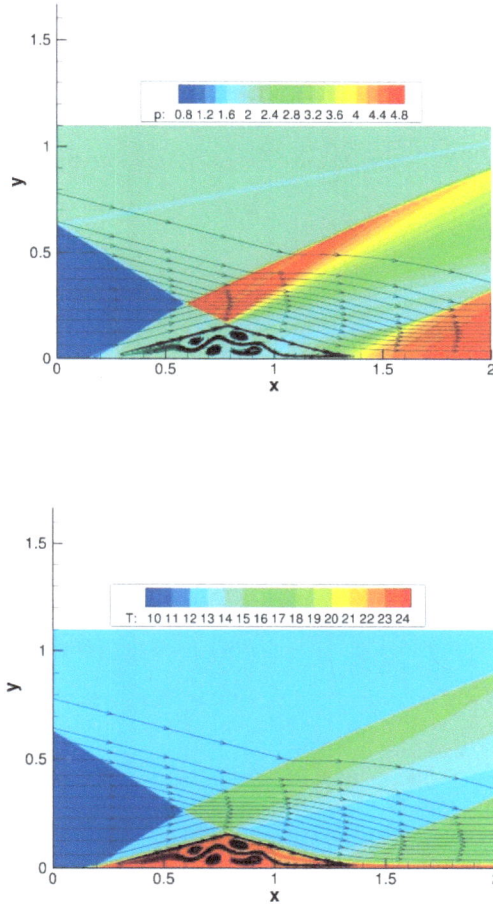

Fig. (7). The pressure and temperature contours of viscous problem.

CONCLUSION

In this chapter, the hybrid compact-WENO scheme is introduced, and some selected test cases are given to show the present method is feasible to deal with compressible flow. Although, we only consider some simple model problem, the extension to complex flow problem in three-dimension is hopeful and is being carried on.

CONSENT FOR PUBLICATION

Not applicable.

CONFLICT OF INTEREST

The authors confirm that this chapter contents have no conflict of interest.

ACKNOWLEDGEMENT

Declare none.

REFERENCES

[1]   J.S. Hesthaven, and T. Warburton, *Nodal Discontinuous Galerkin Methods*. Springer-Verlag: New York, 2008.
[http://dx.doi.org/10.1007/978-0-387-72067-8]

[2]   F.D. Witherden, A.M. Farrington, and P.E. Vincent, "PyFR: An open source framework for solving advection–diffusion type problems on streaming architectures using the flux reconstruction approach", *Comput. Phys. Commun.,* vol. 185, pp. 3028-3040, 2014.
[http://dx.doi.org/10.1016/j.cpc.2014.07.011]

[3]   Z.J. Wang, "Spectral (Finite) Volume Method for Conservation Laws on Unstructured Grids. Basic Formulation: Basic Formulation", *J. Comput. Phys.,* vol. 178, pp. 210-251, 2002.
[http://dx.doi.org/10.1006/jcph.2002.7041]

[4]   A. Harten, B. Engquist, S. Osher, and S.R. Chakravarthy, "Uniformly High Order Accurate Essentially Non-oscillatory Schemes, III", *J. Comput. Phys.,* vol. 71, pp. 231-303, 1987.
[http://dx.doi.org/10.1016/0021-9991(87)90031-3]

[5]   C-W. Shu, and S. Osher, "Efficient implementation of essentially non-oscillatory shock-capturing schemes", *J. Comput. Phys.,* vol. 77, pp. 439-471, 1988.
[http://dx.doi.org/10.1016/0021-9991(88)90177-5]

[6]   X-D. Liu, S. Osher, and T. Chan, "Weighted Essentially Non-oscillatory Schemes", *J. Comput. Phys.,* vol. 115, pp. 200-212, 1994.
[http://dx.doi.org/10.1006/jcph.1994.1187]

[7]   G-S. Jiang, and C-W. Shu, "Efficient Implementation of Weighted ENO Schemes", *J. Comput. Phys.,* vol. 126, pp. 202-228, 1996.
[http://dx.doi.org/10.1006/jcph.1996.0130]

[8]   H.X. Zhang, "Implicit, non-oscillatory containing no free parameters and dissipative (INND) scheme", *Appl. Math. Mech.,* vol. 12, pp. 107-112, 1991. [http://dx.doi.org/10.1007/BF02018075]

[9]   D.X. Fu, and Y.W. Ma, "A High Order Accurate Difference Scheme for Complex Flow Fields", *J. Comput. Phys.,* vol. 134, pp. 1-15, 1997.
[http://dx.doi.org/10.1006/jcph.1996.5492]

[10]   X. Deng, and H. Zhang, "Developing high-order weighted compact nonlinear schemes", *J. Comput. Phys.,* vol. 165, pp. 22-44, 2000.
[http://dx.doi.org/10.1006/jcph.2000.6594]

[11]   C. Liu, P. Lu, M. Oliveira, and P. Xie, "Modified upwinding compact scheme for shock and shock boundary layer interaction", *Commun. Comput. Phys.,* vol. 11, pp. 1022-1042, 2012.
[http://dx.doi.org/10.4208/cicp.250110.160211a]

[12]   D.X. Fu, Y.W. Ma, X.L. Li, and Q. Wang, *Direct Numerical Simulation of Compressible Turbulence*.

Science Press: Beijing, 2010.

[13]   N.A. Adams, and K. Shariff, "A high-resolution hybrid compact-ENO scheme for shock-turbulence interaction problems", *J. Comput. Phys.,* vol. 127, pp. 27-51, 1996.
[http://dx.doi.org/10.1006/jcph.1996.0156]

[14]   B. Costa, and W.S. Don, "High order Hybrid central-WENO finite difference scheme for conservation laws", *J. Comput. Appl. Math.,* vol. 204, pp. 209-218, 2007.
[http://dx.doi.org/10.1016/j.cam.2006.01.039]

[15]   D. Kim, and J.H. Kwon, "A high-order accurate hybrid scheme using a central flux scheme and a WENO scheme for compressible flow field analysis", *J. Comput. Phys.,* vol. 210, pp. 554-583, 2005.
[http://dx.doi.org/10.1016/j.jcp.2005.04.023]

[16]   N. Jha, and L.K. Bieniasz, "A Fifth (Six) Order Accurate, Three-Point Compact Finite Difference Scheme for the Numerical Solution of Sixth Order Boundary Value Problems on Geometric Meshes", *J. Sci. Comput.,* vol. 64, pp. 898-913, 2015.
[http://dx.doi.org/10.1007/s10915-014-9947-5]

[17]   J. Lin, Z. Xie, and Z. Tao, "Three-point explicit compact difference scheme with arbitrary order of accuracy and its application in CFD", *J. Appl. Math. Mech.,* vol. 28, pp. 943-953, 2007.
[http://dx.doi.org/10.1007/s10483-007-0711-x]

[18]   S.K. Lele, "Compact finite difference schemes with spectral-like resolution", *J. Comput. Phys.,* vol. 103, pp. 16-42, 1992.
[http://dx.doi.org/10.1016/0021-9991(92)90324-R]

[19]   C-W. Shu, Essentially non-oscillatory and weighted essentially non-oscillatory schemes for hyperbolic conservation laws.*Advanced Numerical Approximation of Nonlinear Hyperbolic Equations.* Springer: Berlin, Heidelberg, 1998, pp. 325-432.
[http://dx.doi.org/10.1007/BFb0096355]

SUBJECT INDEX

A

Adaptive mesh refinement 211
Advantages of Chebyshev nodes 70
Amplitude 86, 115, 146, 147
   and frequency 115, 146
   and growth rate 147
   functions 86
Angular momentum 239
Angular velocity 8, 28, 29, 42
  fluid-rotational 8
  pseudo average 42
  pseudo time-average 29
  spatial average 42
  spatial mean 28, 29
  time-average 29
Antisymmetric shear vector 18
Axisymmetric water jet 280

B

Base flow 63, 77, 81, 82, 85, 97, 100, 104,
   106, 107
  cylindrical 85
  quasi-rotational 104
  velocity profiles 104
Basic reference frame (BRF) 3, 156
Blasius 2, 76, 155, 179, 202
  boundary layer 2, 155
  solution 76, 179
  velocity 202
Boundary 57, 60, 61, 64, 146, 179, 205, 207,
   211, 229
  immersed 211
  outflow 75, 179
  ribbed 229
  up-stream 207
Boundary conditions 64, 75, 78, 80, 84, 90,
   94, 95, 191, 215, 216, 226, 245
  discretized 80
  incoming flow 216

periodic 191
  thermal wall 245
Boundary layer 60, 76, 116, 190, 191, 195,
   245, 265, 266, 278
  flat plate 76
  hypersonic 245
  laminar separation 116
  parameters 60
   supersonic 245
  transitional 116
  velocity profile 190
   zero-pressure gradient 266
Boundary-layer flows 60, 71
  flow transition 71
Boundary layer interaction 54, 293
  turbulence 293
Boundary layer transition 5, 25, 48, 51, 62, 63
  late natural 5

C

Calculation Procedure for Liutex 9
Cartesian tensor form 294
Cauchy-Stokes decomposition 1, 4, 9, 11, 12,
   18
Cavitation 233, 240
  bubbles 233
Channel 56, 155, 211, 226, 227, 228, 229,
   233, 234, 244
  cooling 227
  flow 56, 155, 244
  impeller 234
  inlet 228
  internal cooling 211
  ribbed 226, 227, 228, 229
  vortex 233
Characteristics 229, 243, 244, 245, 250, 259,
   266
  statistical 243, 244, 245, 250, 259
  streaky 229
  structural 266
Chebyshev 63, 64, 69, 79
  discretization 64, 79

F

www.ingramcontent.com/pod-product-compliance
Lightning Source LLC
Chambersburg PA
CBHW050808220326
41598CB00006B/147